CW00661504

Manchester

MAPPING THE CITY

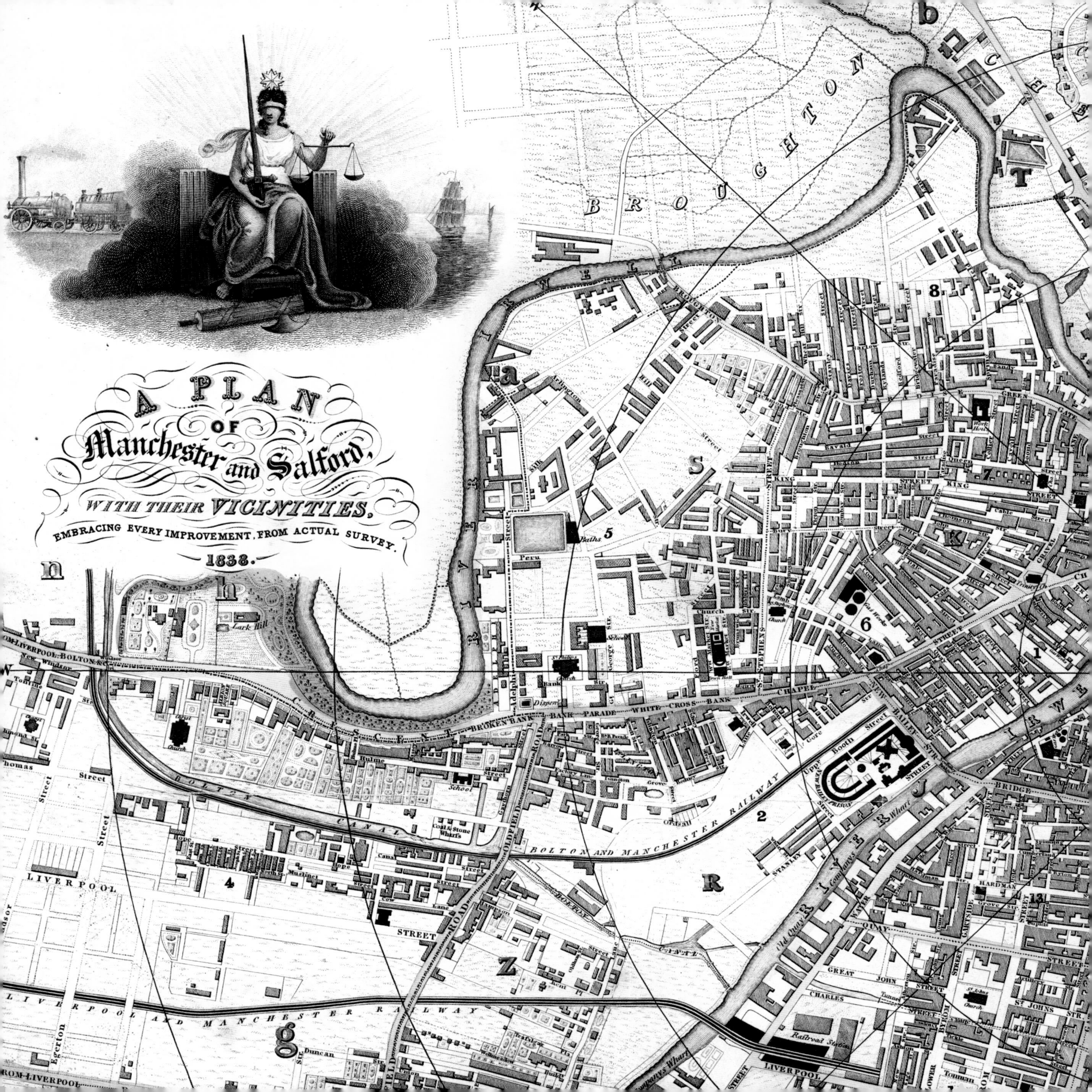
A PLAN OF Manchester and Salford, WITH THEIR VICINITIES,
EMBRACING EVERY IMPROVEMENT, FROM ACTUAL SURVEY.
1838.
BROUGHTON
LIVERPOOL AND MANCHESTER RAILWAY
BOLTON AND MANCHESTER RAILWAY
Lark Hill
Baths
NEW BAILEY PRISON
Railroad Station

Manchester

MAPPING THE CITY

Terry Wyke, Brian Robson and Martin Dodge

First published in
Great Britain in 2018 by
Birlinn Ltd
West Newington House
10 Newington Road
Edinburgh
EH9 1QS

www.birlinn.co.uk

ISBN: 978 1 78027 530 7

British Library Cataloguing-in-Publication Data
A catalogue record for this book is available on
request from the British Library

Designed and typeset by Mark Blackadder

PREVIOUS PAGE

J. Pigot, *A plan of Manchester and Salford . . .*
(1838) [AUTH]. The final title caption used by Pigot,
with the figure of Justice and a background
incorporating a railway engine and ship.

Printed and bound by PNB, Latvia

Contents

Introduction

'I am told there are people who do not care for maps' wrote Robert Louis Stevenson in an essay about *Treasure Island*. He went on to declare that they provided 'an inexhaustible fund of interest for any man with eyes to see or twopence-worth of imagination'. The purpose of this volume is to invite readers to use their eyes and imagination to look at a selection of the published and manuscript maps and plans of the rich and extensive cartography of Manchester, ranging from the eighteenth century to the present day.

The topography in which the city sits has been well captured by contemporary satellite maps, which show the partial amphitheatre formed by the Pennines to the north and east and the flatter land of the Cheshire Plain and south Lancashire to the south and west. Once-separate towns have gradually coalesced to form one of the largest of Britain's conurbations. The sequence of maps that we use here to explore Manchester's history helps to chart this growth and the changes that have transformed a relatively insignificant small town into the Cottonopolis of the nineteenth century and the conurbation of today.

Maps of the town and its wider region provide vital evidence of the changes that marked this growth. Its earliest published maps date only from the second quarter of the eighteenth century. Earlier plans of the town appear not to have survived, although there are still faint hopes that the lost survey of Manchester that John Dee engaged Christopher Saxton to produce in 1596 may turn up one day. The town's first real map, the eighteenth-century plan by Casson and Berry, included a small plan that purported to be of the town in 1650. But it was not until the 1790s that accurate large-scale maps were published in the shape of the two large-scale plans by Charles Laurent and William Green.

In the early nineteenth century, the speed of urban change meant that numerous maps were published, but were soon out of date. However, Manchester was richly served by the long sequence of accurate plans published by James Pigot in his trade directories. New larger-scale maps of the town were infrequent, an exception being Bancks's plan of 1832. However, in 1851 two landmark maps appeared. The first was by Joseph Adshead covering the township on the unprecedented scale of 80 inches to the mile (1:792). Hot on its heels came the Ordnance Survey plan of the borough on a scale of 60 inches to the mile (1:1056). They became important tools in planning and managing the town. Previously, major projects such as the widening of Market Street in the 1820s relied on plans commissioned from local surveyors, but subsequently, as the papers

---

OPPOSITE. F. Ramspott, *Manchester 3d Render Satellite View* . . . (2016) [AUTH]

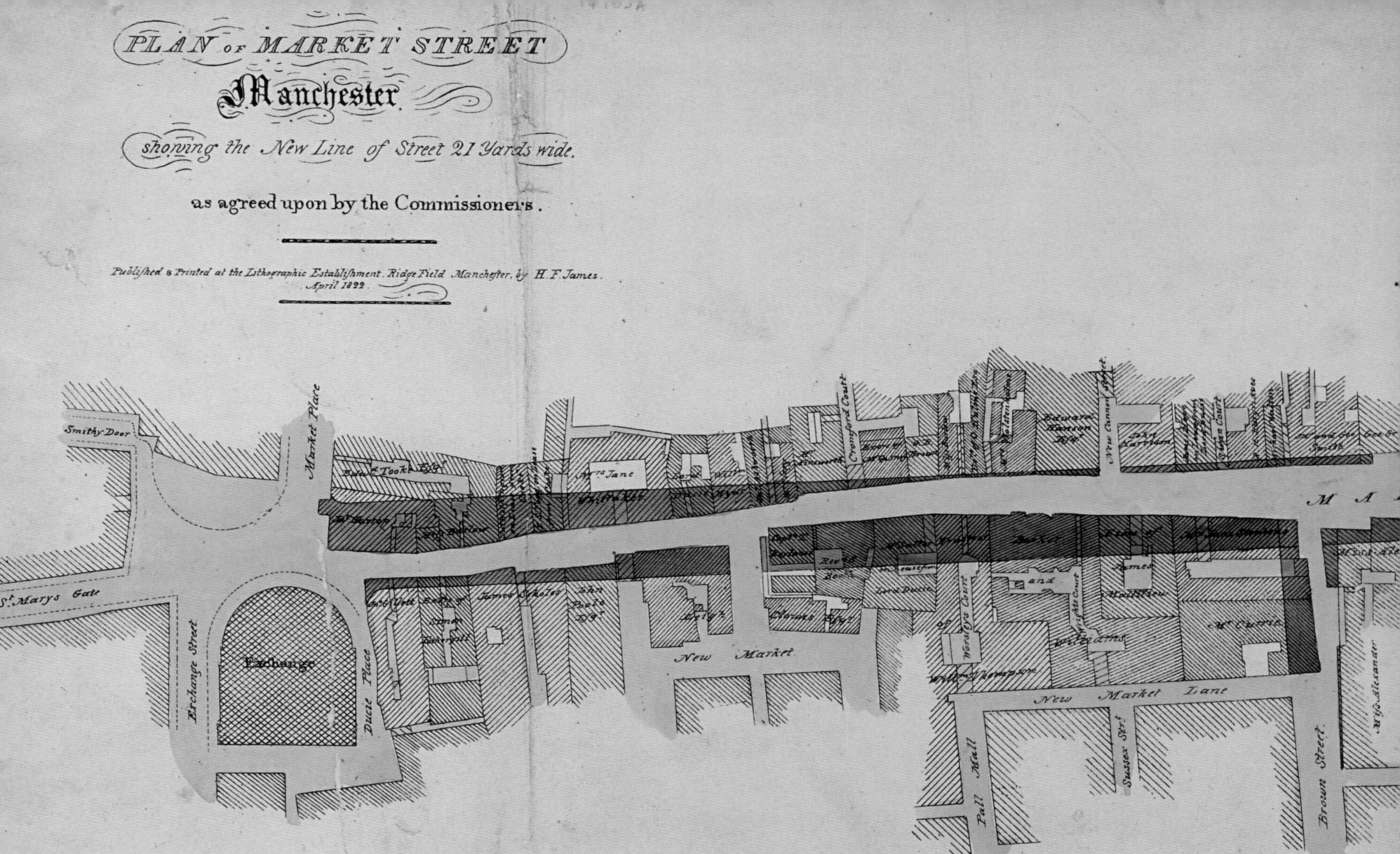

H.F. James, *Plan of Market Street . . . showing the new line of street 21 yards wide . . .* (1822) [MLA]
It is clear how narrow the original street was at its western end where it entered the town centre.

of council committees reveal, the use of Ordnance Survey sheets became routine. Maps also became important evidence in public debates, no more so than in the building of Manchester's much-admired town hall.

Throughout these years the bread-and-butter work of surveyors continued to be the making of property maps. In most cases their plans have not found their way into public archives, although there are exceptions, including, fortunately, some of the town's larger landowners and charities. The land market was a busy one, involving developments ranging from the running-up of cheap terraced housing for the working class in outcast Ancoats to the building of socially exclusive residential districts such as Victoria Park, plans that provide further insights into the much-discussed social divisions of the city.

The nineteenth century also began to see the publication of specialist maps. These included numerous maps charting the canal revolution – Manchester was the terminus of the country's first and last major canals (Bridgewater Canal and Manchester Ship Canal) – and the railway revolution that began with the opening of the Liverpool and Manchester

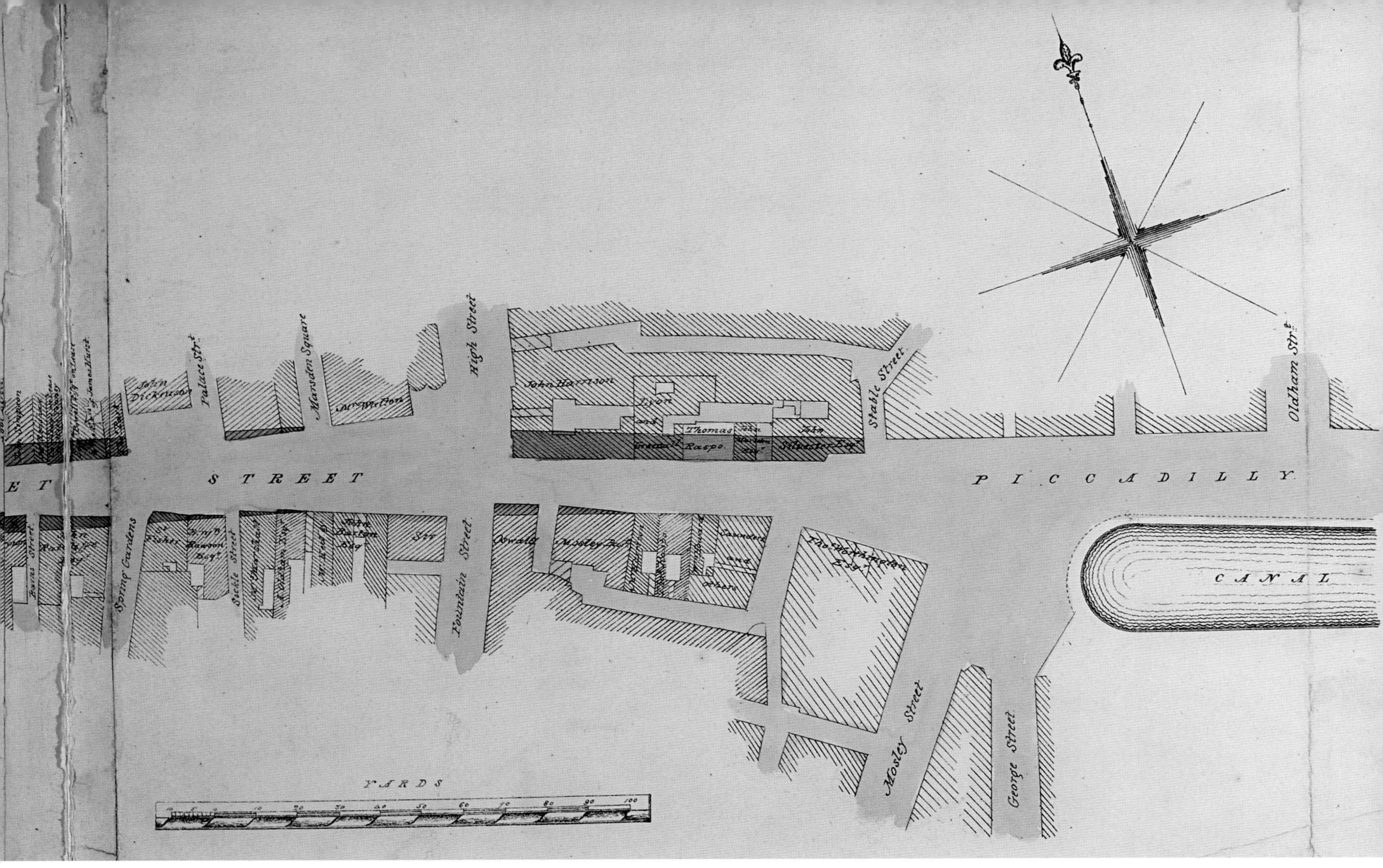

Railway in 1830. These years also saw the emergence of maps that contributed to the analysis of social problems. It was one of a small number of towns to map the first cholera epidemic, a small but essential step in studying the epidemiology of this lethal disease. Social reformers also recognised the usefulness of maps in highlighting social problems. Unsurprisingly for a city that was the headquarters of one of the foremost temperance pressure groups, the United Kingdom Alliance, there was a drink map identifying the location of licensed premises. By the Edwardian period, social reformers were more confident in using maps, exemplified by Thomas Marr's mapping of housing problems, although Manchester never had an equivalent of the poverty maps produced for London and York by Booth and Rowntree. Mapping the city's social issues continues to the present day, a recent example being a 2014 map aimed at providing rough sleepers with information about the services available to assist them.

Maps also became part of the ongoing leisure revolution. These featured in numerous guidebooks, coming of age in the cycling mania of the 1890s. Visitors to the famous Belle Vue

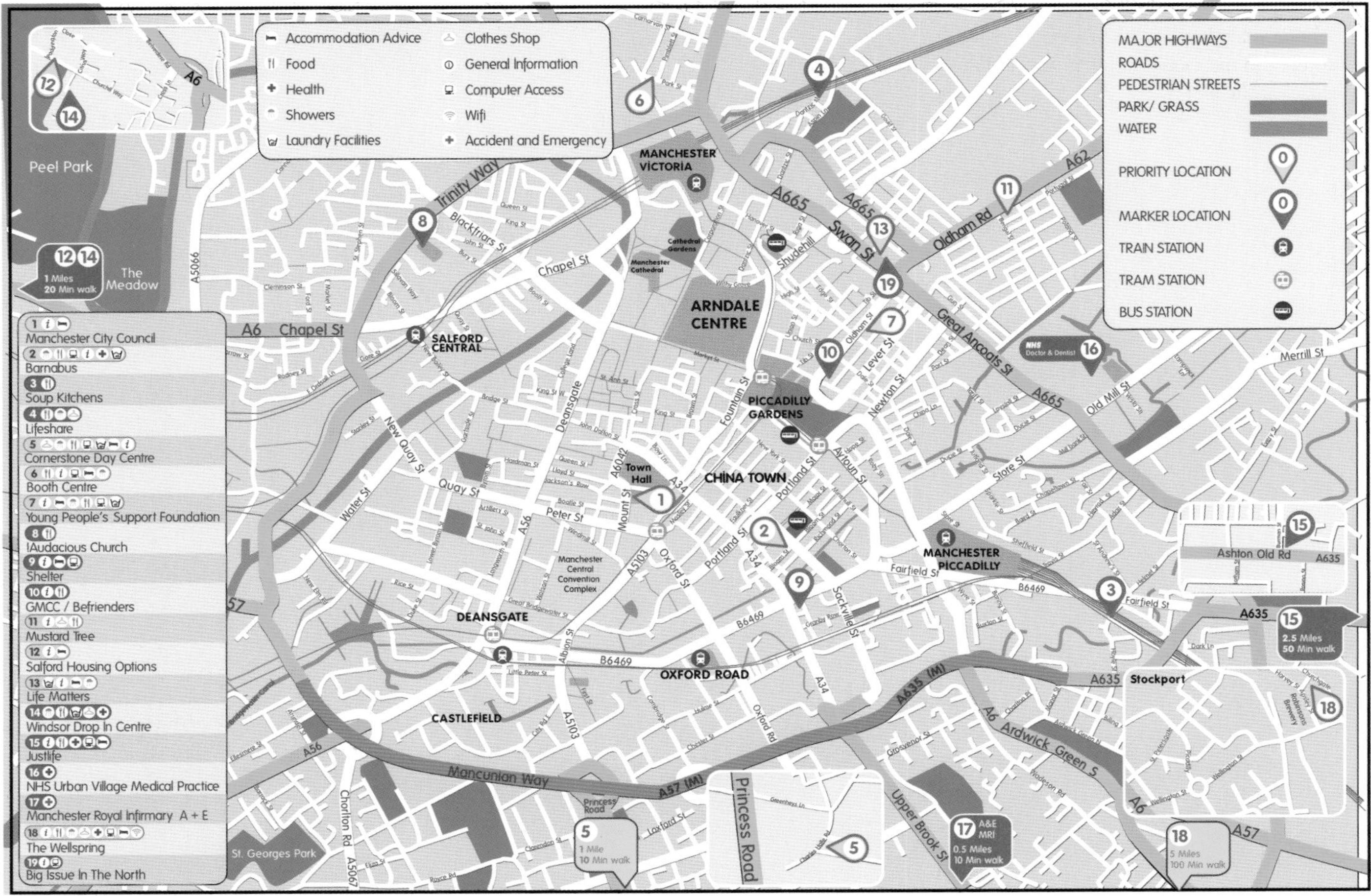

Rough Sleepers, Rough sleeper map (2014) [Jim Codd]

pleasure gardens were able to plan their day by consulting plans of the grounds.

Maps continued to be important in planning, and in the twentieth century some of the most memorable were those connected to the grand planning schemes which aimed to attack the city's 'irredeemable ugliness' or, even more ambitiously, to build a New Jerusalem. Rowland Nicholas's *City of Manchester Plan 1945* remains one of the cardinal documents of Manchester's post-war history. However, as in the street widening schemes of the previous century, actual change tended to be piecemeal rather than wholesale. A more concerted effort was evident by the 1990s as the city looked to revitalise its economic base and to regenerate large areas that had suffered the depredations of successive economic recessions and the impacts of de-industrialisation and globalisation.

Although Manchester's libraries have built up large collections of historical maps, interest in maps among historians of the city has been merely sporadic. Towards the end of the nineteenth century Charles Roeder, a member of Manchester's German community, tried to lift the collective amnesia shrouding earlier town maps, eventually persuading George Falkner to republish William Green's map of 1794. He knew

C. Roeder, *Roman Manchester* (1900) [MLA]
The relationship between the Roman fort in the south and the medieval centre around the cathedral in the north.

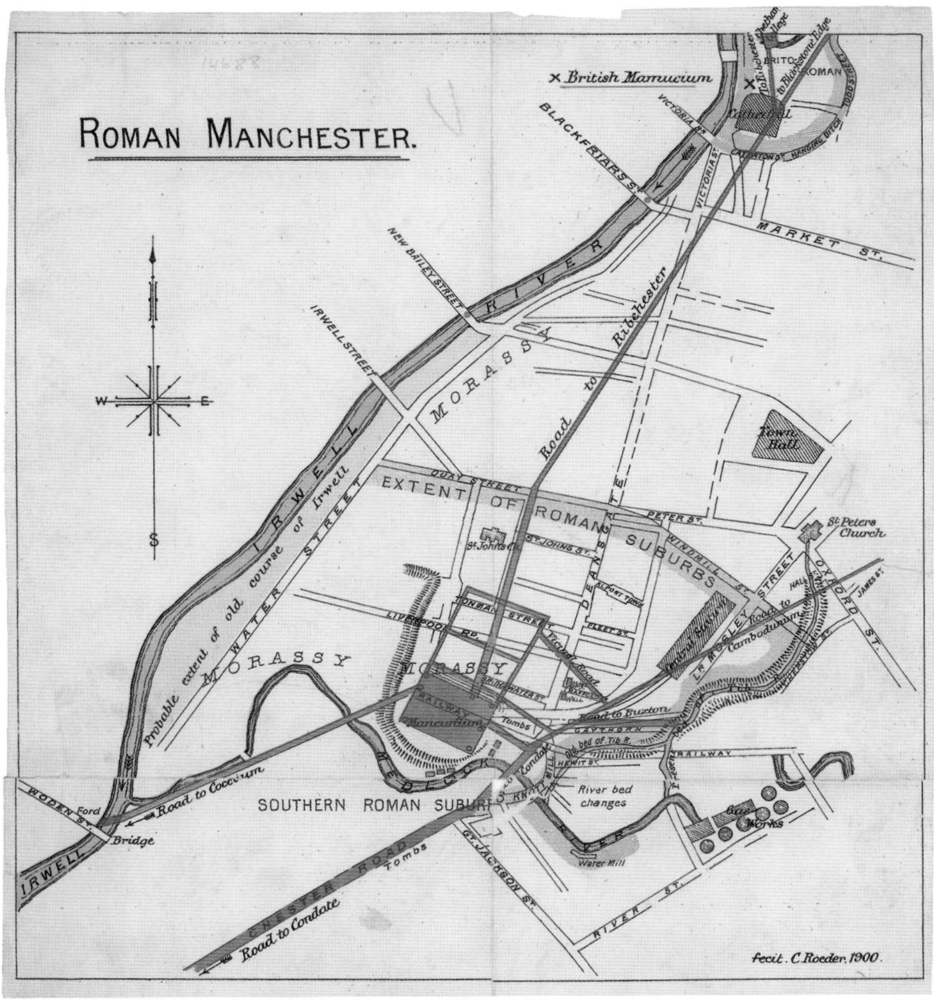

Green's map from having used it when investigating the Roman settlement of Mamucium.

Interest in historical maps continued to be sporadic for much of the twentieth century, which meant that the provenance of even some of the landmark maps of Manchester remained uncertain. Similarly, only skeletal biographies were available even for important surveyors such as Richard Thornton; and investigation of the map publishing work of printers such as Henry Blacklock and George Bradshaw (who became synonymous with railway guides) remains incomplete. More positive developments include a carto-bibliography complied by Jack Lee in 1957, but it was left to individuals like Neil Richardson to try to stimulate interest by republishing some of the historic large-scale plans of Manchester and Salford.

We hope that the wide range of maps and plans in this collection will encourage interest and generate closer research into the city's maps and map-makers, enabling them to be more generally integrated into the work of urban historians and others. To that end we have provided a brief introduction to each of the selected maps, our one-farthing addition to that vital 'twopence-worth of imagination' that is needed to appreciate all of the facets of the jewels in this cartographic treasure chest.

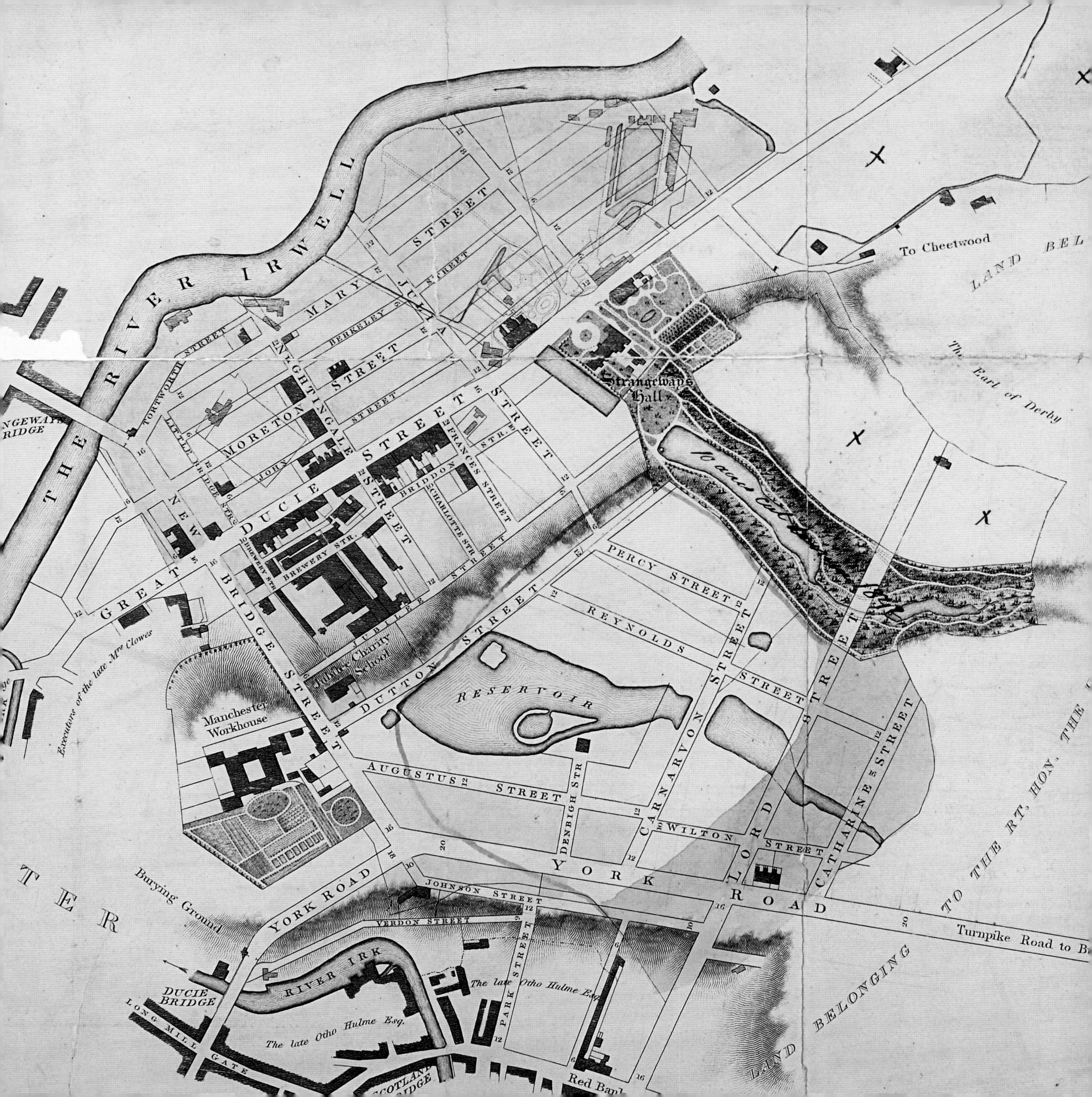

THE RIVER IRWELL
MARY STREET
JULIA STREET
BERKELEY STREET
TORTWORTH STREET
NIGHTINGALE STREET
MORETON STREET
LITTLE BRIDGE STR.
JOHN STR.
NEW BRIDGE STREET
DUCIE STREET
FRANCES STREET
BRIDDON STR.
CHARLOTTE STR.
BREWERY STR.
GREAT DUCIE STREET
JUBILEE STREET
Jubilee Charity School
DUTTON STREET
Strangeways Hall
To Cheetwood
The Earl of Derby
PERCY STREET
REYNOLDS STREET
RESERVOIR
CARNARVON STREET
LORD STREET
CATHARINE STREET
AUGUSTUS STREET
DENBIGH STR.
WILTON STREET
YORK ROAD
JOHNSON STREET
VERDON STREET
PARK STREET
Turnpike Road to B
Manchester Workhouse
Executors of the late Mrs Clowes
Burying Ground
RIVER IRK
DUCIE BRIDGE
LONG MILL GATE
The late Otho Hulme Esq.
Red Bank
LAND BELONGING TO THE RT. HON. THE

Acknowledgements and map sources

This cartographic account of Manchester's history would never have appeared were it not for Hugh Andrew's enthusiasm for maps and Birlinn's readiness to publish its series of city cartographies. In compiling the book, we have been greatly helped by many people. Local librarians and archivists have been especially generous in giving us their time and support – none more so than Michael Powell who has taken a close interest in the project from its outset and allowed us to use the rich collection of maps and plans at Chetham's Library. Special thanks are also due to Larysa Bolton, Helen Lowe, Dave Govier, Sarah Hobbs and Kevin Bolton at Manchester Central Library; Donna Sherman at Manchester University Library; the Heritage Imaging Team at the John Rylands Library; Chris Fleet at the National Library of Scotland; Duncan McCormick, Salford Local History Library; Emma Marigliano, Portico Library; George Turnbull at the Greater Manchester Museum of Transport; Mark Wylie, Manchester United Museum; Richard Dean at Canal Maps Archive; Karen Cliff, Trafford Local Studies; and Neil Spurr of Digital Archives. Many individuals have helped to clarify points and to source some of the more fugitive maps: Tom Bloxham and Fiona Small; Bob Bonner; Richard Brook; Derek Brumhead; Burnage Garden Village Management Committee; Gavin Elliott; Clare Hartwell; Nick Johnson; David Kaiserman; Michael Nevell; Mark Ovenden; Chris Perkins; Robert Poole; Lloyd Robinson; Ian Simpson and Rachel Haugh; and Andrew Taylor. Alan Kidd was hugely helpful in casting his expert historian's eye over the text. The scanning of most of the original maps was done by Stephen Yates, and much skilful graphic support was provided by Graham Bowden and Nick Scarle in the Cartographic Unit at Manchester University. Finally, but by no means least, we are especially grateful to Manchester Metropolitan University where encouragement and financial support have come from Sharon Handley, Pro-Vice-Chancellor, and Brian McCook and Melanie Tebbutt at the Department of History, Politics and Philosophy.

Map sources

The source of each map is shown, in square brackets, at the end of the map captions in each chapter. The following abbreviations are used:

AUTH Authors' collections
BL Copyright of the British Library
CHETS Chetham's Library
MLA Manchester Libraries and Archives
NLS Courtesy of the National Library of Scotland
UML University of Manchester Library

OPPOSITE. W. Johnson, *Plan of land in Strangeways* . . . (1823) [MLA] The Hall was later replaced by the Assizes and Strangeways Prison.

3
5
6

1728

Eighteenth-century panoramic views

Like other visitors to Manchester in the early eighteenth century, Daniel Defoe was impressed by its vitality. In his *A Tour Through the Whole Island of Great Britain* (1724–6) he was alive to the dynamism of its industries and trade. Long before cotton was spun in its steam-powered factories, the town was renowned for producing woollens and mixed fibre cloths, particularly fustians made of linen and cotton. 'Manchester cottons' as they were called were already traded widely throughout the country. While it is difficult to be confident about the size of 'the greatest mere village in England', Defoe was definitely employing the imaginative skills of the novelist in suggesting that its population was comparable to York and Norwich.

It was not until the second quarter of the eighteenth century that the first visual representations of Manchester were published, providing enticing evidence of its layout and buildings, evidence that pointed to a population of under 10,000, rather less than the 50,000 inhabitants that some claimed. In 1728 Samuel and Nathaniel Buck published the first panorama of the town – 'The South West Prospect of Manchester in the County Palatine of Lancaster' – in what became their monumental project to provide views of all the main towns in England and Wales. The drawing adopted the then familiar perspective in which the viewer was placed in a rural landscape at a distance from the town, from the Salford side of the River Irwell. Although Manchester was described in the accompanying caption as a town of 'handsome broad streets' with 'many noted buildings', only six buildings were named. Two of these were prominent on the skyline: the tower of the fifteenth-century Collegiate Church and the cupola of the recently built St Ann's. Holy Trinity in Salford, consecrated in 1637, was the only other named point on the skyline. Ongoing improvements in the town were a keynote, the caption drawing attention to the building of an Exchange and the investment to improve the navigability

---

OPPOSITE. S.& N. Buck, *The South West prospect of Manchester*...(1728) [Government Art Collection]

S. & N. Buck, *The South West prospect of Manchester* . . . (1728) [Government Art Collection]
The ferry chain across the Irwell led to the Spaw House, probably a spa bath.

of the Irwell. Manchester appears as a relatively small town, particularly when compared to the Bucks's drawings of other Lancashire towns, notably their impressive depiction of the busy port of Liverpool. Somewhat unexpectedly, none of the houses of the leading merchants and manufacturers were considered worthy of identification, even though this was an obvious means of securing subscriptions. Neither was much attention given to the architectural detail of the larger houses, the town seeming to comprise largely featureless houses. The view, however, did offer evidence of the layout of the town, especially the development to the south-west, although in the area between Deansgate and the Irwell there was still considerable open land or gardens, on which well-dressed couples were walking. This sense of the polite was further suggested by the inclusion of the Spaw House, a bathing establishment on the Salford side of the river which was reached by boat from the Manchester bank.

Six years later a second panorama of the town was published by the Manchester printer and bookseller, Robert Whitworth. He provided a far more detailed view than the Bucks and it may have been that some dissatisfaction with aspects of their view led him to produce a new prospect. That it was not a catchpenny copy was evident in Whitworth's decision to produce a view engraved on two rather than a single plate. It was dedicated to 'Hon. Ann Lady Dowager Bland, Lady of the Manor of Manchester', the daughter and heir of Sir Edward Mosley, a suitably courteous acknowledgement to one of the town's most wealthy and powerful families. Prominent in the view was St Ann's Church (consecrated in 1712) which she had provided, in part as an alternative place of worship for those unsympathetic to the Jacobite leanings of the Collegiate Church. However, she may not have seen the completed prospect as she died in July 1734.

Whitworth's prospect provided a sharper sense of the built-up area of the town and its urban frontier where the more affluent were building larger houses, with front and rear gardens. Some 25 sites and buildings are identified compared to only six in the Bucks's. The principal public buildings – the churches and Chetham's Hospital – were common to both prospects, though Whitworth was now able to include the Exchange with its neoclassical portico, another building financed by the Mosley family but which had still been under construction at the time of the Bucks's survey. Other commercial buildings are less easy to identify, as are notable public spaces such as the Market Place. The Irwell, of course, is prominent, but whereas the Bucks suggest that it was essentially for recreation and pleasure, Whitworth shows it more as an artery of trade, which helps to explain why the chain across the river by the Spaw House is absent.

R.Whitworth, *The South West prospect of Manchester and Salford* (1734) [CHETS]

Whitworth's prospect paid more attention to private residences, its scale allowing some architectural features to be identified. Suggestive of rising urban status, he identifies 11 houses, including some substantial properties such as Mr Brown's house, distinguished by its turrets. Cupolas were a defining element of three of the houses, suggesting an awareness of changes in architectural fashions. Street names are rarely provided, an exception being 'the 7 houses in Par[s]onage', an early example of the terrace which was to become an important housing type for the new urban middle classes. Whether the named houses were linked to their owners subscribing to the project is unknown but only one of them – Mr Marsden's house in Market Street – was to be included in the houses illustrated in Casson and Berry's plans published in the following decade.

As in the Bucks's prospect, there was still considerable open land close to the town. One such space was Dolefield, to the west of Deansgate, although since it was surrounded by buildings on three sides and stacked with piles of timber it seems clear that it was soon to be buried by the march of bricks and mortar. Indeed, in the following decades it was built upon, housing for a time the workhouse. In contrast to the activity on the left bank of the Irwell, Salford is depicted as a small settlement spreading out from the old bridge, a pattern that allows one to see why contemporaries regarded it as a suburb of Manchester.

That Manchester was in the throes of urban change was evident in a second version of Whitworth's prospect, which was added to the illustrations augmenting Casson and Berry's 1745 plan of Manchester. While one might have expected the original copper plates to have been used, changes were made for this smaller version. Of these, the inclusion of the hunting party on the Salford side of the river is the most eye-catching, but closer examination reveals that alterations were also made to the Manchester side, including new buildings.

In the absence of other visual representations of Manchester, the Bucks's and Whitworth's prospects have become key sources for interpreting the history of the town before the onset of full-blown industrialisation. However, their limitations should be recognised. Adopting the conventional Arcadian perspective of the urban inevitably determined their view of the town, while their very size limited the degree of detail, even if they had been willing to invest the time to record it. But no other comparable view of Manchester was published for over 60 years. Only in 1802 did a new view of the town appear – William Marshall Craig's 'View of Manchester from Mount Pleasant' – by which date other sources allow us to say that Manchester was then occupying a place in the urban hierarchy that would have surprised even Defoe.

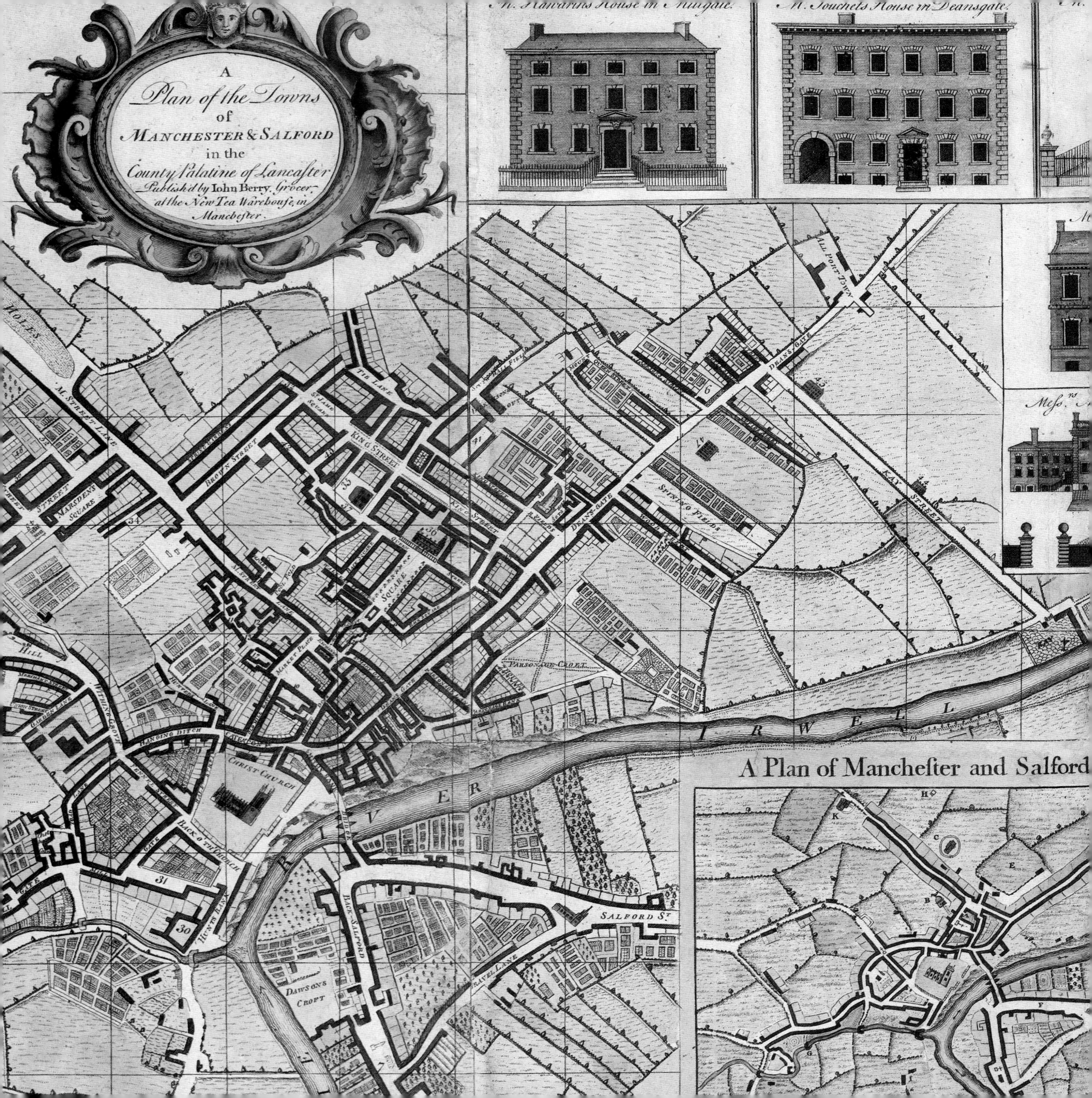

A
Plan of the Towns
of
MANCHESTER & SALFORD
in the
County Palatine of Lancaster
Publish'd by John Berry, Grocer,
at the New Tea Warehouse, in
Manchester.
M. Souchets House in Deansgate.
A Plan of Manchester and Salford
ALL PORT TOWN
DEANS GATE
KAY STREET
SPINNING FIELDS
KING STREET
BROWN STREET
TIB LANE
MARSDENS SQUARE
MARKET PLACE
PARSONAGE CROFT
HANGING DITCH
CHRIST CHURCH
BACK O'TH CHURCH
HUNTS BANK
I R W E L L
R I V E R
SALFORD St.
BACK SALFORD
DAWSONS CROFT

1746

The town's first authentic street map

> This Town and yt of Salford (wch is divided from it by ye River Irwell) seems to be but one in Extent, ye Streets several of them are large, open & well paved and within 30 years last past ye Town is become almost twice as large as it was before, so fast have its Inhabitants and their Riches increased . . . They are in particular known to be Industrious People; the reason of their being so numerous is ye flourishing Trade followed here for long time known by ye name of Manchester Trade, which not only makes the Town but ye Country round about for several miles populous, industrious & wealthy.

This was the boast of Russel Casson and John Berry who published the first useful street plan of the town in June 1741. As befitted a map dedicated to the 'gentlemen and tradesmen within ye towns', it proclaimed Manchester as a dynamic centre of manufacturing and trade, a place in which money was made and spent. The plan itself was framed by 12 views of local buildings, 6 on each side, in the then well-established form of urban cartography which can be traced back to at least James Millerd's plan of Bristol (1673). The orientation was to the east rather than the north, a decision which may have been determined by the line of the Irwell. It was drawn at a scale of 20 inches to the mile. The surveyor is unknown, but the plan was engraved in London by Benjamin Cole.

That there was a demand for such a plan was evident in the publication of four further editions by 1755. The second edition, published in 1745, made no changes to the plan itself but added significant new material, notably an amended version of Whitworth's prospect. There were also six new engravings of buildings which, with the omission of one, brought the total to seventeen. These additions resulted in obscuring some streets and a considerable section of the countryside that had been visible on the 1741 plan. The plan

---

OPPOSITE. R. Casson & J. Berry, *A plan of the towns of Manchester & Salford . . .* (1746) [MLA]

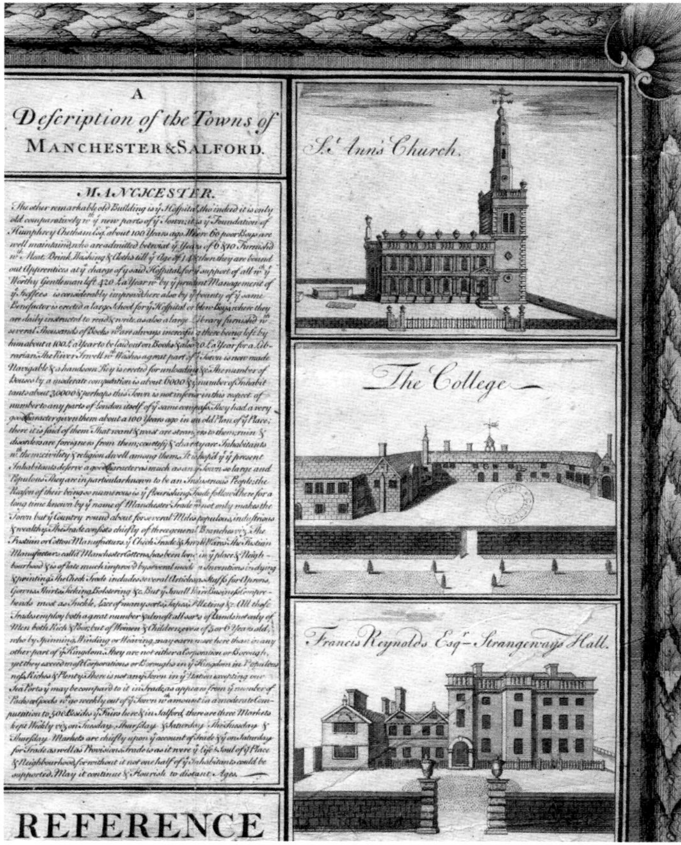

Views of St Ann's, Chetham's and Strangeways Hall.

must have sold well because in the following year, in spite of the uncertainties caused by the Jacobite rebellion, Berry advertised a new edition (price 1s 6d), available at his shop in the Market Place. It is this edition which is shown here. Again, the original plan remained unaltered, but new material was added in the form of a smaller unattributed map, *A Plan of Manchester and Salford taken about 1650,* as well as two new engravings – the houses of Miles Bower and Mr Johnson. A view of the interior of the Long Room, an important commercial meeting place which was 'furnished with ye best Maps, Plans, Terrestrial Globe &c', was removed, to bring the number of views to 18.

A fourth edition – 'A Compleat Map of the Towns of Manchester and Salford to the present year 1751' – was published solely by John Berry, whose address was now the New Warehouse, near the Market Cross. Although 'great additions' were claimed, the principal change was the inclusion of an engraving of the house of John Bradshaw, bringing a total of 19 views. Berry published a fifth version of the plan in 1755. Once more the central street plan remained unaltered, but two of the nineteen views were new: the engravings of Strangeways Hall and Mr Marriott's house in Brown Street being replaced by views of two of the town's newest buildings, the recently completed home of the Infirmary in Daube Holes/Lever Row (Piccadilly) and the yet-to-be consecrated St Mary's Church on Parsonage Croft, between Deansgate and the Irwell.

Important as the plan is in the history of Manchester, there is little direct evidence of its commissioning. The driving figure appears to have been Berry, whose recorded occupations – grocer, watchmaker, auctioneer and printer – exemplified the entrepreneurial spirit to which the town was laying claim. Berry clearly regarded the financial risks of producing and selling a map of the town as worthwhile, before eventually deciding to sell the plates.

The plan confirmed that Manchester was a substantial and expanding market town, a point underscored by the inclusion of the 1650 plan, though the estimate of 6,000 houses was as absurd as the claim of a population of 30,000. What was not fanciful was that Manchester was already, by the size of contemporary English provincial towns, a considerable settlement, evidence of its status and recent development being found in the plan's grid squares where one could locate the 118 principal streets, lanes and buildings identified in the reference key. But Berry was not alone in admiring the town. Writing in his history of the Jacobite Rebellion (published in 1746), James Ray, who knew Manchester at first hand (and had seen Casson and Berry's plan), depicted it as a thriving town, based on the high-quality cloths that it manufactured and traded at home and overseas. Manchester's trade had been boosted by the improvements made to the Irwell by the Mersey and Irwell Navigation Company, a point reinforced by the inclusion on

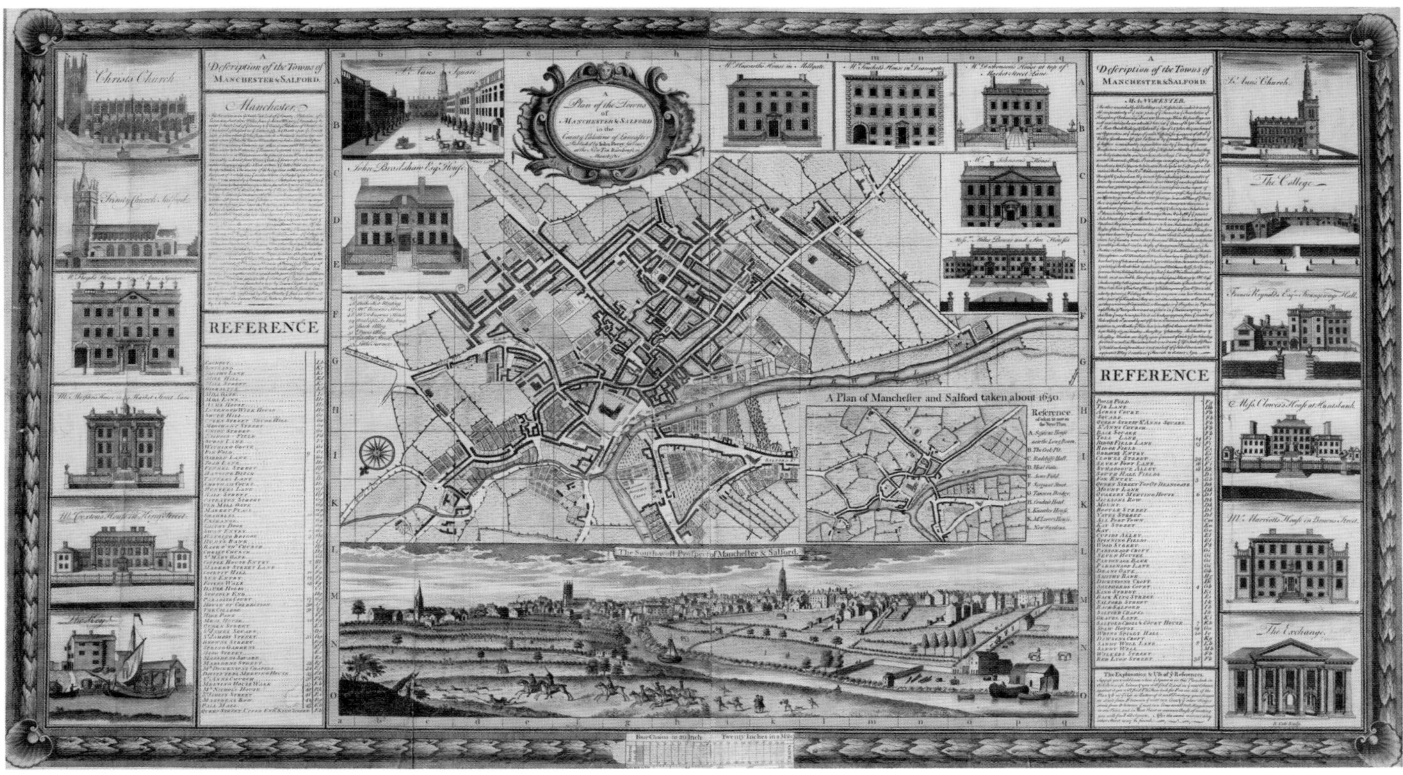

the plan of the much-reproduced view of the 'The Key', which shows a Mersey 'flat' docking at the quayside on which were two of Manchester's earliest purpose-built warehouses. Quay (Key) Street, running almost from the banks of the Irwell to Deansgate was itself a harbinger of modernity, its very straightness marking it out from the medieval muddle of older streets and lanes around the Collegiate Church and Market Place. Ray also observed of the town that its 'buildings are the most sumptuous of any hereabouts'. Some were depicted on the plan – large family residences, reflecting the architectural fashions of the time, marking their owners as cultured and affluent. New public buildings, exemplified by the classical Exchange, which was as much a social as a business centre, also reflected the shifts in tastes driving a provincial 'urban renaissance'. A further element in this new environment was the residential square – St Ann's and St James's – socially determined spaces that were in the process of being lined with town houses for a status-seeking middling sort whose wealth was derived from the burgeoning textile trade. The social geography of the town that was to be discussed so intensely in the following century was already in the making, as individuals chose to live beyond the built-up area of the 'old' town. John Dickenson's elegant house in Market Sted Lane (later Market Street) is now remembered, if at all, as being the residence of Charles Edward Stuart in the winter of 1745, but its more general significance was as one of the new homes of wealthy commercial families, located on what was then the outskirts of the town.

It now requires great imagination to recapture the impact that Casson and Berry's first map must have had on the inhabitants of Manchester, providing them with an entirely new representation of the landscape in which they lived. Above all, it represented a marker of Manchester's transition towards industrial modernity, a plan that acknowledged and approved the present rather the past.

Sugar Alley

Water Pitt

Green Platts

ORCHARD

Garden

Fewell Rook

A Rabbet or Poultry Pen

Back Door

Back Court

Draw Well

Barn

Court

Backside

Front

a Fewell or Rook

Dog Kennel

Easy House

Dunghill

called the Shude-[illegible]

[illegible] part of this House is an old fashioned firm timber Building [illegible]
[illegible] one half thereof new, all two Stories high and tiled. The long roof-house [illegible]
[illegible] that it also contains a Stable, a Barn, a Shippon, a Corn Chamber, an apartment [illegible]
[illegible] Room: All the other Out-housings are some thatch, or of brick, the Courts are [illegible]

[illegible] sides are some flagged, some paved, and some gravelld: the Garden, Orchard, [illegible]
Platts, and Walks, near the above, is in a firm, very convenient, neat, clean and
good condition.

1753

Shudehill and the Hulme Charity

One of the least discussed changes to the landscape in the years that saw Manchester transformed from a modest market town into an industrial metropolis was the disappearance of its trees. Writing in 1804, Joseph Aston recalled how in the previous decades many of the trees in Shudehill had been cut down, and as a consequence its rooks had been forced to find new places to roost. Shudehill can easily be identified on the eighteenth-century maps of Manchester, but their scale is such that they do not provide a more finely grained picture of the physical landscape. For that, manuscript estate plans, many drawn up as part of the legal process of buying and renting land, are more informative. Fortunately, one area of the town for which property plans have survived is for the area of Shudehill, Withy Grove and Fennel Street, providing more precise detail of its buildings, fields, roads and its trees, if not the birds that nested in them.

Shude Hill in the seventeenth century (the name was not yet conflated) was one part of the long road carrying traffic in and out of the centre on the eastern side of Manchester. This was an important entry point, a road along which travelled foodstuffs, goods and animals to the town's markets. Travelling in the 1750s from the top of Shude Hill at its junction with Miller's Lane and Green Lane to the Market Place took one from the open fields to the parish church in a distance of less than half a mile. The journey was downhill along a curving lane of variable width before reaching Withy Grove, Hyde's Cross and Fennel Street, though at this date the boundaries between streets were not as clearly defined as they were to become. It was in Shudehill in 1757 that the authorities decided to stop the crowds who were making their way into the town to protest over high food prices, the resulting deaths fixing the 'Shudehill Fight' firmly in both the official and radical histories of Manchester.

OPPOSITE. Hulme Charity, *James Hilton's messuage, near the Shude Hill* (1753) [CHETS]
Farming remained an integral part of the urban economy in the 1750s.

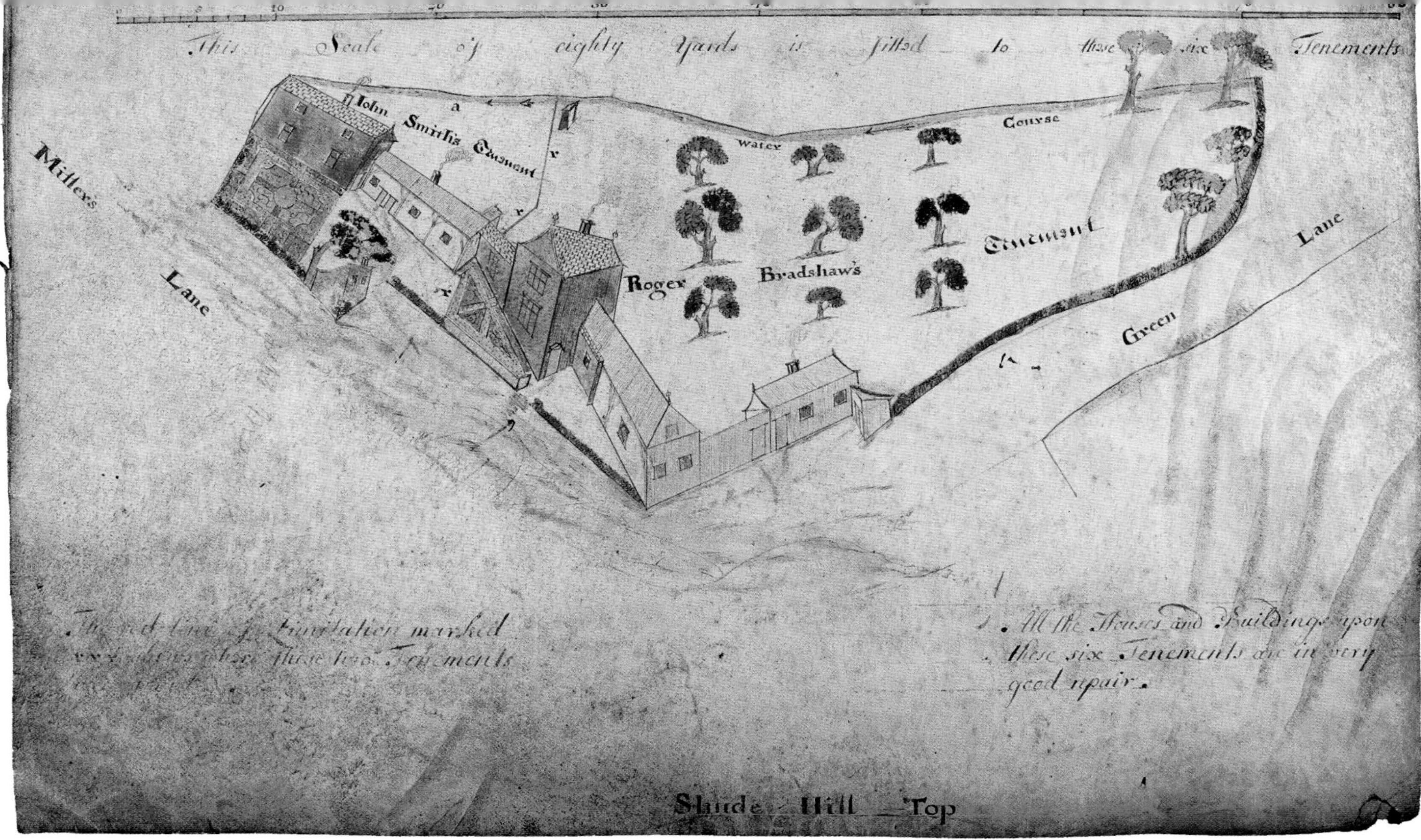

Hulme Charity, *Roger Bradshaw's tenement, Shude Hill* (*c*.1753) [CHETS]

William Hulme was one of the main landowners in this part of Manchester in the second half of the seventeenth century. He appears not to have been directly involved in the expanding textile industry, his wealth being derived from land he owned in Manchester and surrounding parishes, including Reddish and Prestwich. In the township of Manchester itself, his main landholdings were in Fennel Street, Withy Grove and Shudehill. Withingreave Hall (assumed to be the derivation of Withy Grove) was the largest of his houses, but it is doubtful whether he used it as a residence, preferring to live in Kearsley. Hulme died in 1691 leaving property to establish an education charity. It was to become, along with those founded by Hugh Oldham and Humphrey Chetham, among the best known of Manchester's endowed charities. Its main source of income came from renting land and property, an income that was to grow as the demand for building land increased. Such were its revenues that in the nineteenth century the charity's trustees were heavily criticised for not making a wider and more appropriate use of the funds, the pressure eventually forcing them to provide new educational buildings, including schools in Manchester and Oldham, both of which bore Hulme's name.

The earliest extant plans accompanying the leases and rent rolls of the Hulme Charity's lands in Manchester are from the early 1750s. Their importance as historical documents is heightened by the main buildings being depicted in three dimensions, providing a unique view of properties which can only be identified as footprints on the published maps of the town. Withingreave Hall was among the buildings depicted in this way. It is shown as a substantial L-shaped building with two gables, lattice windows and a slated roof, located on a large plot of land with a formal garden and an orchard. The outbuildings included a barn with outside stairs, and an assortment of smaller structures including an easy house (privy) and dog kennel. Trees are also shown on the Shudehill boundary side of the property. James Hilton, who was the tenant in 1753, used it as a farmhouse, also leasing from the charity a close of

8.5 acres, comprising four fields (Great Meadow, Little Meadow, Great Ley, Turner Meadow) located between Green Lane and Newton Lane (Oldham Street) at the top of Shudehill. In the absence of any evidence of textile manufacturing or finishing, it seems that Hilton was a farmer, a reminder that at this time agriculture remained an important part of the urban economy, even one that was being driven by the manufacturing and selling of textiles. Hilton had leased the property for 21 years, but on renewing the lease in 1751, the annual rent was increased from £28 to £35. Among Hilton's neighbours were Roger Bradshaw and John Smith, both occupying properties fronted by gardens, another feature of the townscape that was to become a rarity as industrialisation took hold.

Hulme Charity also owned land closer to the town centre in Hyde's Cross and Fennel Street. Here the drawings in the lease reveal that the main property on the land was one of Manchester's largest inns, the Boar's Head. It was a substantial four-storey building, facing Hyde's Cross, the architectural style dating it to the eighteenth century rather than earlier. Its row of nine sash windows also points to a large number of rooms. Contemporary accounts are few, but advertisements for the inn in the 1830s refer to 30 bedrooms 'usually made up'. The courtyard behind the building was entered through an archway at the front. Here were the service buildings including a pump house, brew house and kitchen. A dunghill is identified but, surprisingly, the stables, an essential part of any coaching business, are not specifically named. To the rear of the property there were also houses and shops, fronting on to Toad Lane.

These estate plans capture Manchester on the cusp of change. They confirm that, in what was already an important thoroughfare, by no means all of the land had been built on. The fields that Hilton rented at the top of Shudehill were all farmed. In the next 50 years, Shudehill and Withy Grove were to change. New commercial and retail buildings were erected, and many of the older black and white houses were demolished, as were the trees. The fate of Withingreave Hall has never been satisfactorily settled, though many have repeated the speculation that part of the hall became a public house – the Rovers Return – one of whose landlords asserted it to be

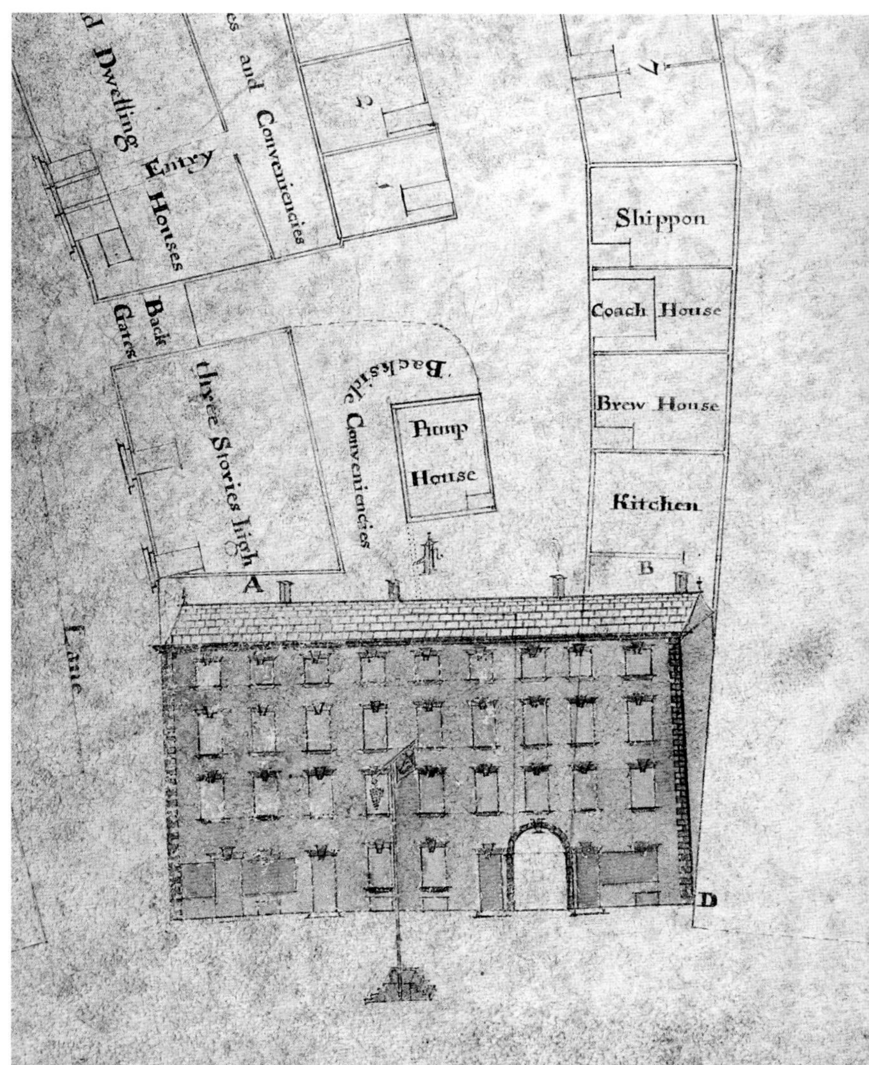

Hulme Charity, *Boars Head near Hyde Cross* (*c.*1753) [CHETS]
Urban growth saw inns become important as commercial centres and meeting places.

the oldest licensed pub in the city. Its known genealogy suggests otherwise. What is less disputable is that the Rovers Return on Shudehill was demolished in 1958 generating considerable local publicity, and that a short time afterwards its name was given by Salford-born Tony Warren to the public house in Granada Television's new soap opera, *Coronation Street*.

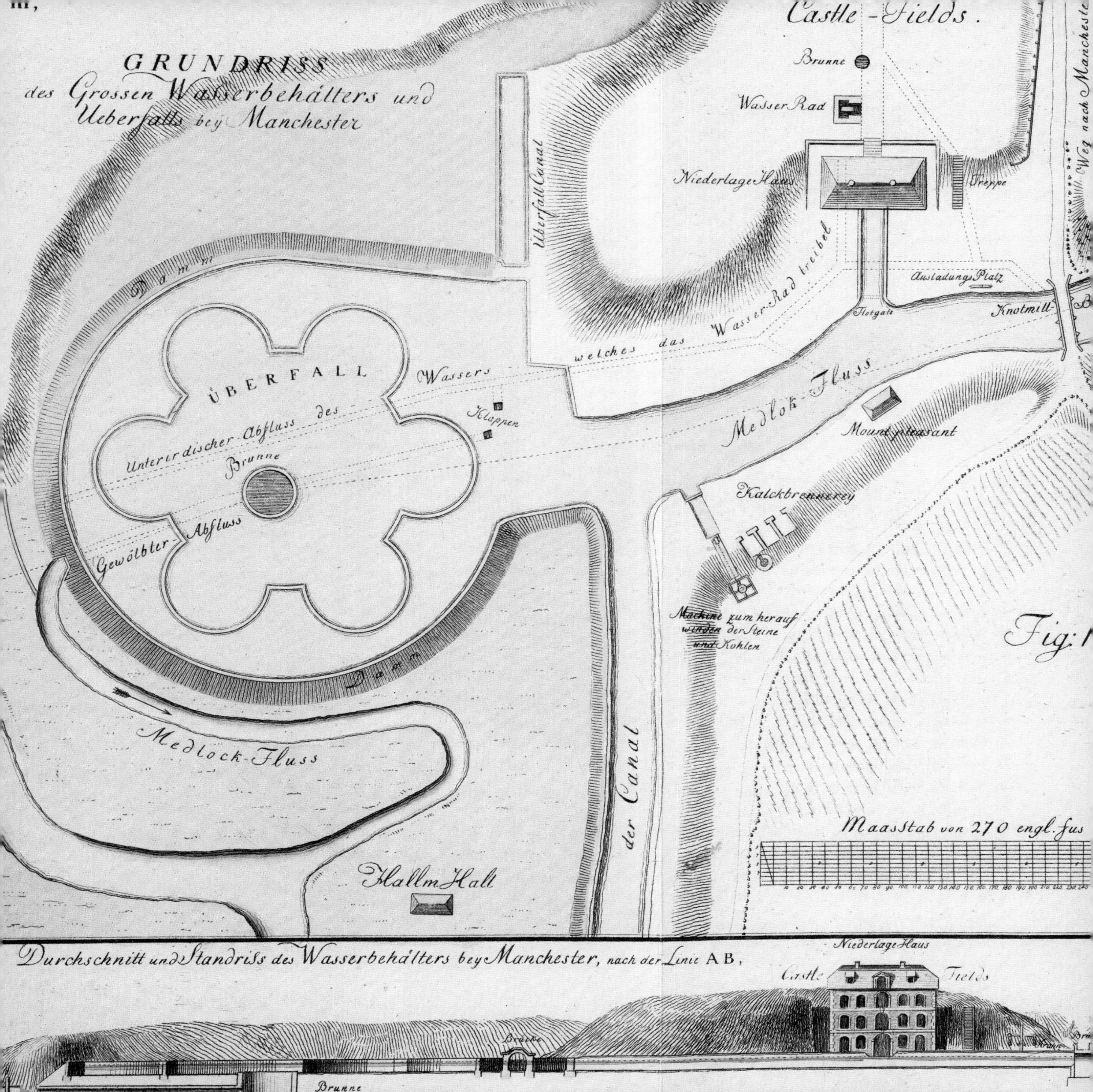
GRUNDRISS
des Grossen Wasserbehälters und
Ueberfalls bey Manchester
Castle-Fields.
Brunne
Wasser Rad
Niederlage Haus
Treppe
Weg nach Manchester
Überfall Canal
Damm
Ausladungs Platz
Flotgate
Knotmill
welches das Wasser-Rad treibet
ÜBERFALL
Wassers
Unterirdischer Abfluss des
Brunne
Klappen
Medlok-Fluss
Mount pleasant
Kalckbrennerey
Gewölbter Abfluss
Machine zum herauf winden der Steine und Kohlen
Fig: I
Damm
Medlock-Fluss
der Canal
Maasstab von 270 engl. fus
Hallm Hall
Durchschnitt und Standriss des Wasserbehälters bey Manchester, nach der Linie AB,
Niederlage Haus
Castle
Fields
Brücke
Brunne

1780

The first modern canal

The Bridgewater Canal was one of the great innovations of the eighteenth century, without which Manchester would not have been transformed into the world's first modern city. Its purpose was simple: to transport raw materials to Manchester, beginning with coal. Its success in reducing transport costs and widening markets was to be seen in the expanding list of raw materials (stone, timber), foodstuffs (grain, fruit, dairy products) and manufactured goods carried into and out of Manchester, without which the growth of the town would have been checked. As the world's first modern canal, it had a wider impact as a catalyst for a transport revolution, encouraging the building of other canals throughout Britain.

It was built in two stages. The first, between 1759 and 1765, linked the Duke of Bridgewater's colliery in Worsley to Manchester where the demand for coal was soaring. It was James Brindley, assisted by the duke's agent, John Gilbert, who famously met the complex engineering challenges of the route. These began far underground inside the Worsley colliery where the coal was loaded directly into tubs (standardised containers) in the barges (standardised length and beam). The water in the mine was used to fill the canal, thus turning what had been a persistent problem into an asset. The canal route followed the 82-foot contour line, obviating the need to build locks. However, the decision to abandon the original terminus in Salford in favour of taking the canal directly into Manchester necessitated building the Barton Aqueduct across the Irwell, an outstanding work of engineering that became one of the principal coordinates of the new industrial age. Unable to get as close to the centre of Manchester as he wished, Bridgewater located the canal terminus at Castlefield. It was a demanding site whose relief had suited the Romans as a defensive position but was far from ideal as a canal basin. Once again, Bridge-

---

OPPOSITE. J. Howgrewe (1780) Detail of Castlefield canal basin [UML]
The Medlock had to be diverted underground to create Castlefield basin.

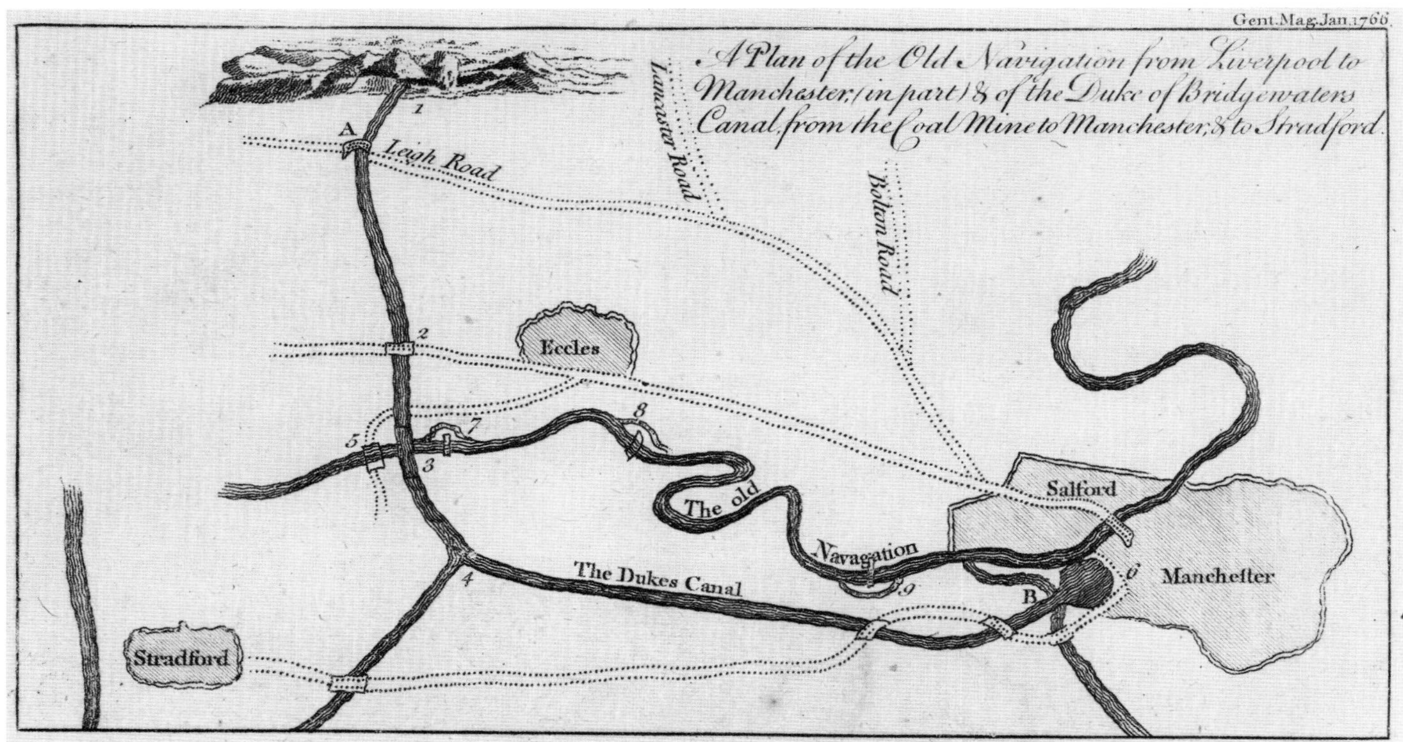

Gentleman's Magazine, *A plan of the Old Navigation . . . & of the Duke of Bridgewater's Canal* (1766) [UML]

water had to invest more money in the project, while Brindley was called upon to find solutions, notably controlling the River Medlock that crossed the land. The second and far longer stage of the canal was from Stretford to Runcorn where it met the Mersey, a route that put it in direct competition with the Mersey and Irwell Navigation. This section also had its engineering challenges, especially in constructing the locks at Runcorn. The canal was not finally opened along its whole length until 1776.

The economic impact of the Bridgewater was most obvious in the town's produce markets. By the time of the duke's death in 1803, consumers in Manchester had become used to purchasing not only cheaper coal but Irish butter and Welsh cheese that had been carried on the canal. A singular reminder of this trade is the survival of the Oxnoble inn at Castlefield, which took its name from a Norfolk potato, one of the foodstuffs unloaded at the basin. Other raw materials had travelled much further, notably the sugar and cotton which had been grown in the slave plantations of the West Indies. Manchester was to take a prominent lead in the protests and boycotts against Caribbean sugar that eventually resulted in the abolition of the slave trade, but it turned a blind eye to the origins of the bales of raw cotton being unloaded at Castlefield, cotton that fed the steam-driven factories of Cottonopolis.

The canal, especially the Worsley to Manchester section, also had a psychological impact. The Barton Aqueduct in particular astonished and fired the imagination of contemporaries. Manchester's first guidebook, written by James Ogden in 1783, began by proclaiming the entrepreneurial qualities of Francis Egerton, the duke whose money had financed the project, making it clear that the canal was the first place to visit in the town. One visitor, returning a second time to Castlefield, noted:

> . . . it is astonishing for a person who never sees anything of the kind to see the Business that's going on here. There's such Quantities of Slate, Timber, Stone & merchandise of all sorts. The warehouses are very Extensive, but they are pretty filled with one thing or other.

At the Worsley end of the canal, another visitor recalled it as being 'like a little Amsterdam filled with barges, timber yards, and limestone which is brought from Wales . . .' To most visitors the Barton Aqueduct epitomised the modern, and they left their responses in letters, notebooks, drawings and even verse. Some of this first generation of tourists were so curious that they might have been considered industrial spies. Johann Howgrewe, the military engineer and cartographer, wrote a detailed account for his German readers, illustrating it with maps and diagrams. He was especially interested in details of the engineering, the warehouses and the methods used to supply and control the water, the latter being easier to comprehend at Castlefield where Brindley had culverted the Medlock under the basin, than in the subterranean gloom of the Worsley colliery.

The Bridgewater was important in encouraging the building of canals elsewhere in the country. It was also the means of carrying the name of Manchester to a wider public. Numerous articles describing the canal were to be published. One in the *Gentlemen's Magazine* of 1766 was noteworthy as one of the earliest to include a map of the canal. In the following years maps were to be produced and printed of the projected and actual routes of all the main canals. By 1828 the Manchester printer and publisher, George Bradshaw, was issuing a series of maps of the country's 'canals, navigable rivers and railways', which confirmed Manchester as one of the hubs of the country's canal network. Canals running directly through or connected to Manchester included the Stockport (opened 1797), Ashton (1798), Peak Forest (1804) and Bolton and Bury (1808). The Rochdale Canal (1804), which had its Manchester terminus near Piccadilly, joined the Bridgewater at Castlefield, making it possible to carry goods by water from the North Sea to the Irish Sea. The Leeds and Liverpool Canal (1816) joined the Bridgewater at Leigh in 1820. Prospective investors must have been reassured to know that their routes connected them to Manchester.

Of these canals, the Rochdale was one of the most significant because it was the first trans-Pennine canal, providing the missing link in an east–west water route across the country. It was a spectacular feat of engineering, its 32-mile route taking it over the Pennines before finally terminating at Sowerby Bridge in Yorkshire, where it joined the Calder and Hebble Navigation. In Manchester, the Rochdale Canal Company established a basin between Great Ancoats Street and Dale Street, close to Piccadilly. The all-important link with the Bridgewater was completed by 1801, a distance of just over a mile but, given the rise in the land, it required nine locks.

The impact of the Rochdale Canal on the southern side of the town was immediate, opening up the adjacent land for development. Businesses located along the canal, recognising the economies that came from having a direct water link. Individual firms negotiated the construction of wharfs, while the canal was extended by arms, reaching as far as Newton Street and Portland Street. Thomas Leech, who had inherited land adjoining Dale Street, developed it into part of the larger basin, erecting warehouses and a cotton-mill. That the canal was a catalyst for development was especially evident in Ancoats where Adam and George Murray, and James McConnel and John Kennedy began to build what became two of the largest steam-powered cotton-mills before the canal was completed. These multistorey factories were to be among the first to employ over 1,000 operatives.

Although Bradshaw's canal maps included a small number of railway lines, few realised that the Canal Age was coming to an end. Canals shrivelled and withered as the railways took over the carriage of freight and passengers. But there were exceptions. The Bridgewater – the country's first modern canal

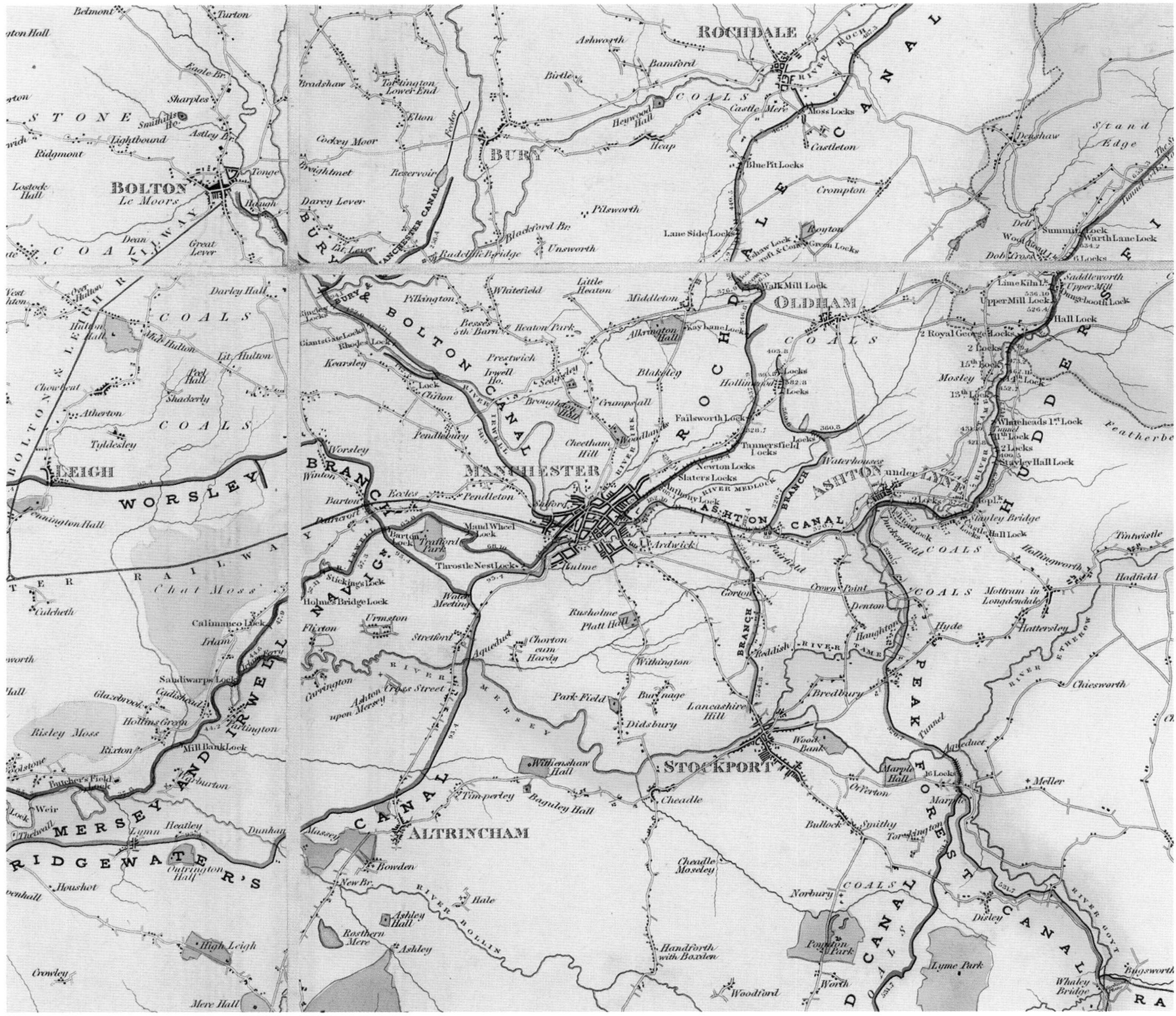

G. Bradshaw, *G. Bradshaw's map of canals . . . in the counties of Lancaster, York, Derby & Chester* (*c.*1831) [UML]

and the only one to be named after an individual – continued to be profitable into the twentieth century.

The Bridgewater Canal has continued to make history. Beginning in the 1980s Castlefield became an urban heritage park. Regeneration was suffused with the history of the place, recognising the enterprise and the achievements of Francis

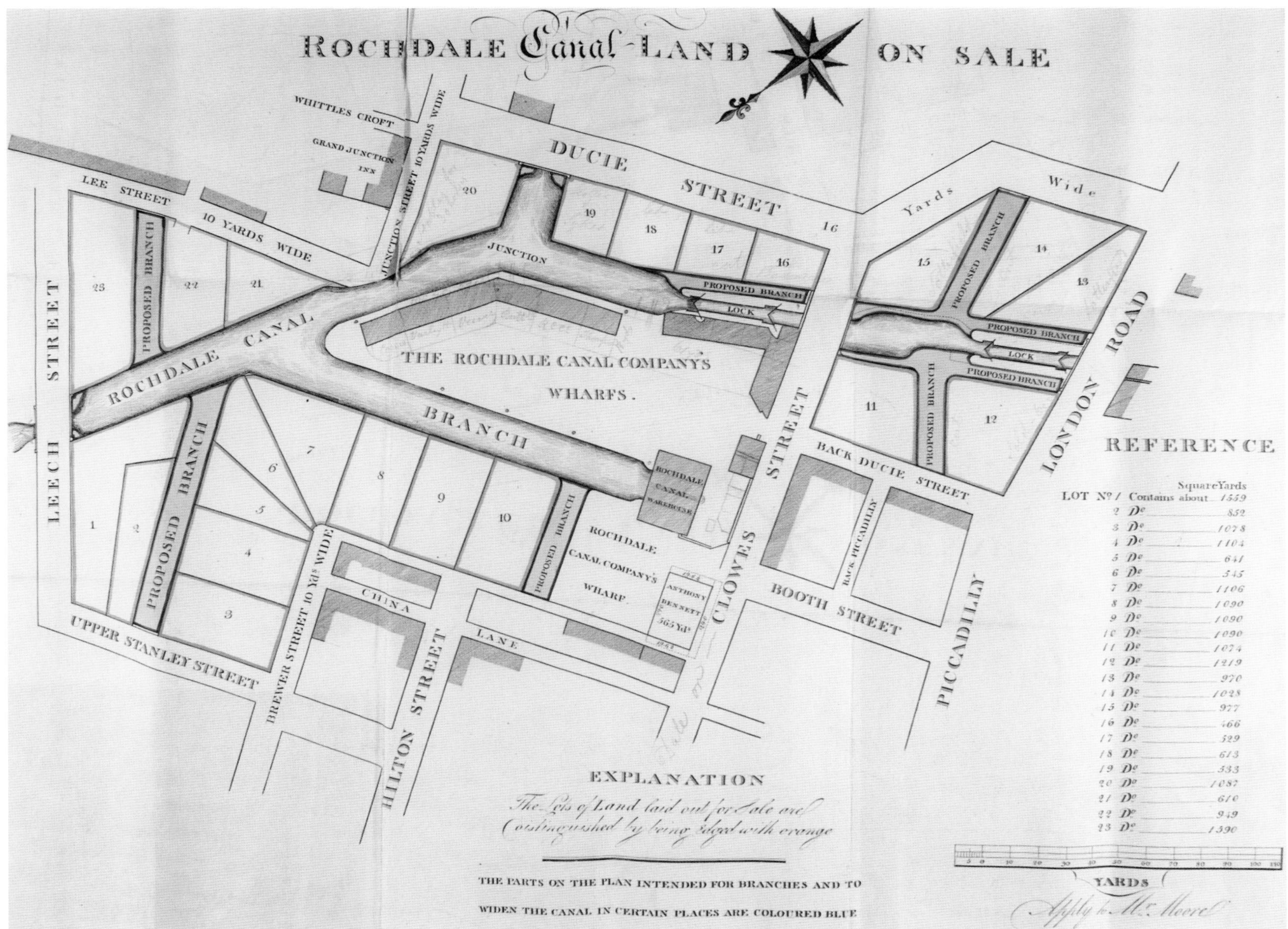

Anon., *Rochdale Canal land on sale* (*c*.1825) [MLA] The area shown is the Rochdale Canal Company's basin, Back Piccadilly.

Egerton, John Gilbert and James Brindley in transforming Manchester into the powerhouse of the industrial revolution. Extant buildings were awarded heritage status, money was invested in improving the landscape and new guidebooks were published identifying points of interest for a new generation of visitors.

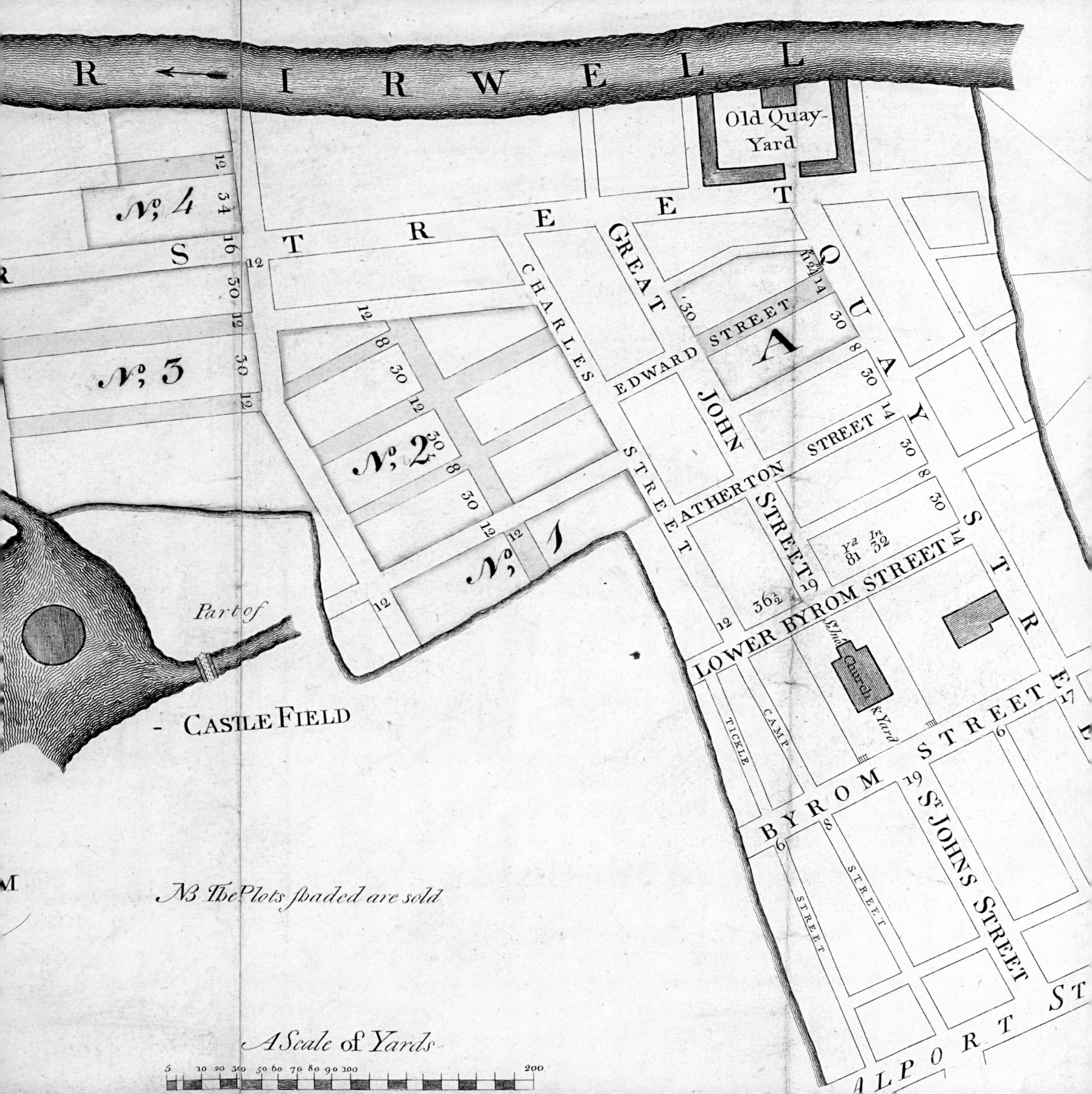

R I R W E L L
Old Quay-Yard
S T R E E T
Q U A Y
S T R E E T
GREAT
JOHN
STREET
CHARLES
STREET
EDWARD STREET
ATHERTON STREET
LOWER BYROM STREET
BYROM STREET
St Johns Church & Yard
TICKLE
CAMP
STREET
St JOHNS STREET
ALPORT ST
No. 1
No. 2
No. 3
No. 4
A
Part of
CASTLE FIELD
NB The Plots shaded are sold
A Scale of Yards
5 10 20 30 50 60 70 80 90 100 200

1788

The development of estate land in the growing town

Manchester's spectacular growth in the second half of the eighteenth century was part of the wider changes which transformed Britain from a rural into an urban industrial society. Growth was especially evident in the physical development of the town, in its eye-catching commercial and industrial premises and, of course, in the houses built to accommodate a population which by 1801 already exceeded 70,000.

The first stage in the building process was the release and development of land, a process that can be traced in the extant deeds and leases in the public archives. Some of the most easily identifiable schemes involved aristocratic landowners such as the Duke of Bridgewater and Lord Ducie, whose agents oversaw the surveying and the laying-out of land into proposed streets. As was the case in the land in Strangeways released by Lord Ducie, landowners in Manchester generally provided long-term leases in which restrictive covenants were included to ensure that subsequent developments did not jeopardise the general intentions of the scheme.

Older established Manchester families such as the Mosleys and Levers also began developing their lands as the pulse of industrialisation quickened. One of the larger housing schemes was that undertaken by the Byrom family, whose local status was evident in their chapel in the Collegiate Church. They owned land on the southern side of the town between the Irwell and Deansgate, and had supported the development of the Irwell in the early decades of the eighteenth century. Edward Byrom built a large family home on Quay Street, and the family had also protected their land by refusing to allow the Duke of Bridgewater to build his canal across it. Following Edward Byrom's death in 1773, the family fortune passed to his daughters, Ann (who married the barrister Henry Atherton) and Eleanora. Development of the estate continued, centred

---

OPPOSITE. T. Townley, *A plan of lands in Manchester the property of Henry Atherton Esq. & Miss Byrom* (1788) [MLA]
St John's church was the focal point in developing a high-class residential district.

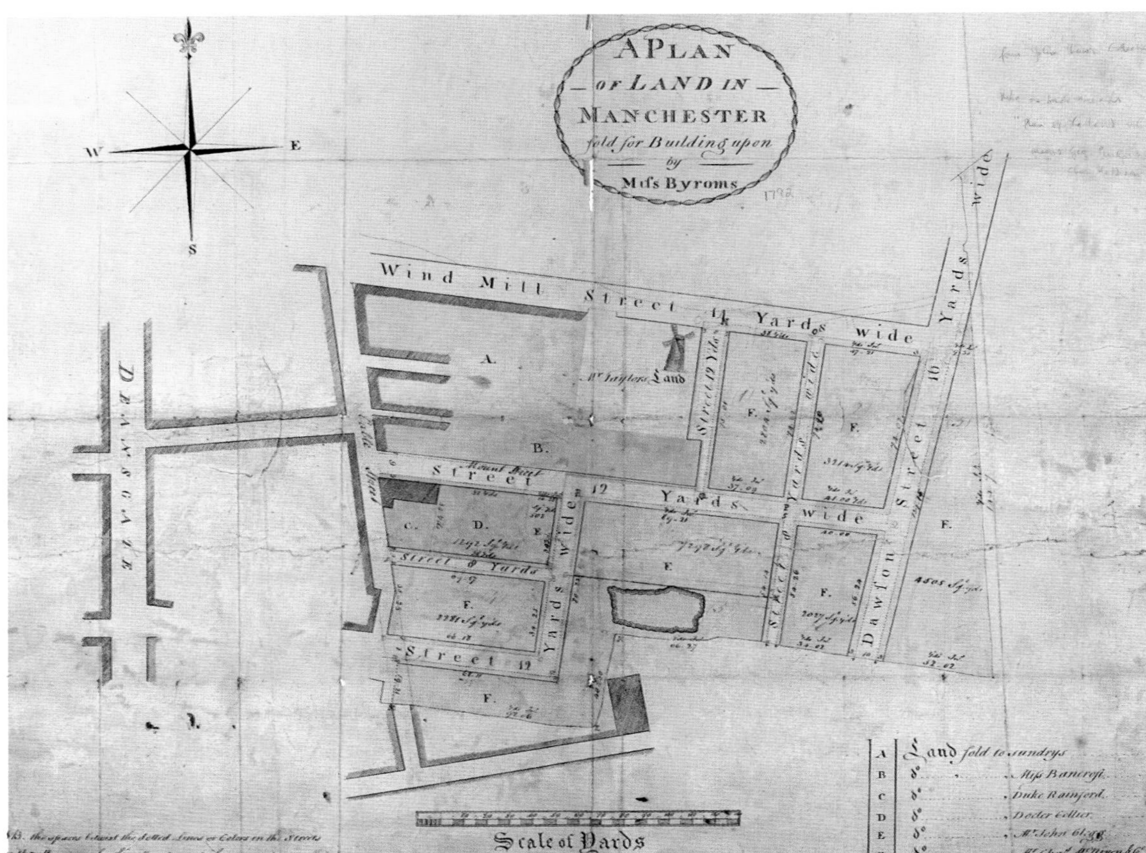

Anon., *A plan of land in Manchester . . . Miss Byroms* (1792) [MLA]

around the handsome church of St John's which Edward had built in the late 1760s. The intention was to establish an attractive residential district, whose principal streets were Great John Street and St John Street. However, this plan was to be overtaken by industrialisation, which resulted in buildings being erected along Deansgate and Water Street towards the increasingly busy Castlefield canal basin. By the time of Ann's death in 1826, not all of the land had been built on, even including parts of St John Street, developers having become increasingly aware that the town's better-off families now preferred their homes to be much further away from the noise and dirt of the centre. Another part of the family estate was a smaller block to the east of Deansgate where development was more piecemeal. Its northern boundary was Windmill Street which, unsurprisingly, took its name from the pre-industrial machine that was to become redundant in the age of steam power.

This failure to recognise the force and speed of urbanisation was also to be the *leitmotif* of the largest and most ambitious residential development in Georgian Manchester. This was in the area immediately south of the Manchester township, on a site in Chorlton-on-Medlock, where in 1792 William Cooper, George Duckworth, and Samuel and Peter Marsland purchased the Chorlton Hall estate with the intention, in Robert Owen's words, of 'building a new town upon it'. Grosvenor Square, modelled on London's aristocratic estates, was to be the centrepiece of the new development. But although Samuel Marsland built a family mansion in the square, other cotton manufacturers could not be persuaded to follow suit, and the subsequent years were to see the land given over largely to terraced houses of varying quality. All Saints church was built on what was planned to have been the residents' exclusive park. By the 1840s not only had the Marslands moved away – their house becoming the home of Manchester New College, the district's first institution of higher education – but other streets on the estate were being filled with cheaper houses, builders and occupiers ignoring covenants aimed at controlling development.

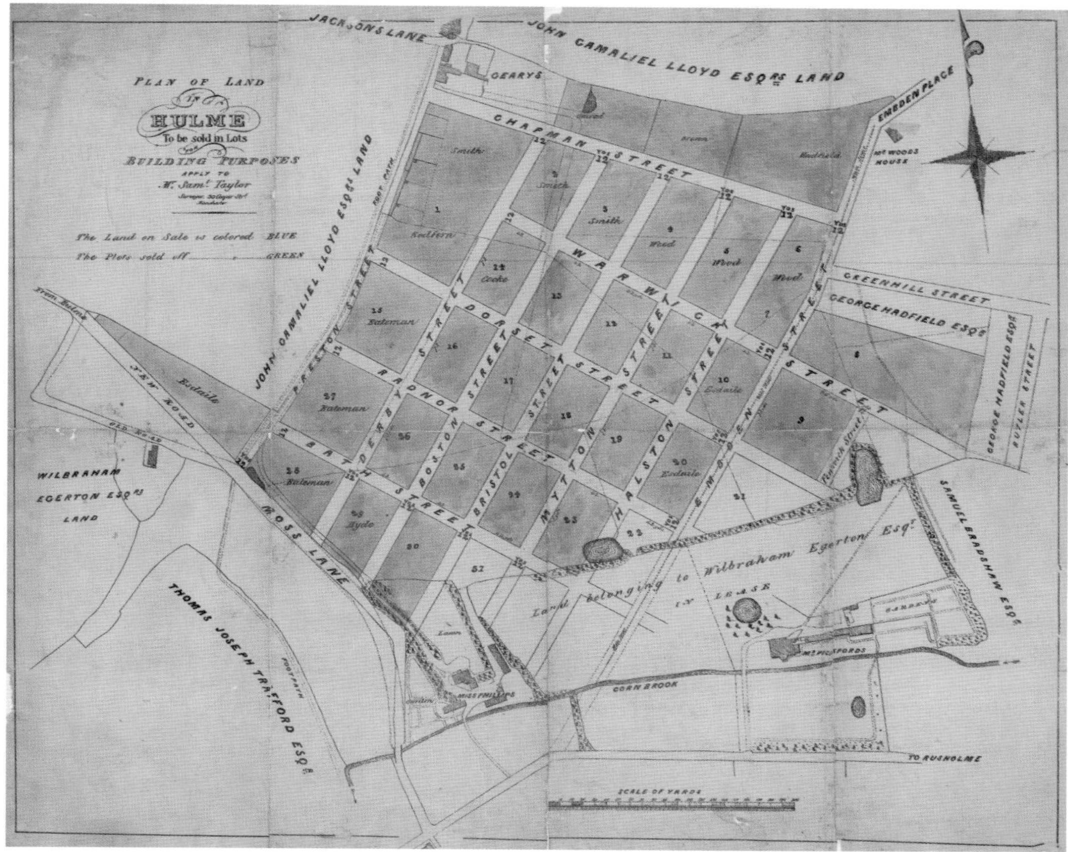

Samuel Taylor, *Plan of land in Hulme . . .* (*c.*1835) [MLA] The scramble to develop Hulme was well underway by the 1830s.

Land, of course, had become an important and profitable investment and some families had the good fortune to have inherited land in the boom town. Thomas Leech moved from Yorkshire to Manchester to develop two blocks of land that had been left to his family in the seventeenth century. This included Brownsfield, near Ancoats, rough pasture land which was on the line of the proposed Rochdale Canal. Leech built a cotton-mill and a wharf on his land, providing an income that allowed the family to live comfortably in what was then the distant village of Urmston. Other beneficiaries of rising land values included institutions such as the Booth Charities, which saw their revenues increase dramatically as land previously rented out for farming became ripe for industrial and residential development.

Manchester's building boom continued well into the nineteenth century. In the first half of the nineteenth century it was particularly evident in the township of Hulme, to the immediate south of Manchester. Within a generation, most of Hulme's 477 acres had been built over, the old field boundaries lost forever as its population increased from a mere 1,677 in 1801 to an astonishing 53,482 in 1851. Large and small landowners reaped the rewards of this boom, as did the attorneys and surveyors involved in the buying and selling of land. Each block of land had its own particular history, different factors determining both the timing and type of development. For example, the building plots offered for sale by the Manchester surveyor Samuel Taylor in 1834 were no doubt influenced by the recently cut new Stretford Road, a development that accelerated building in the central part of Hulme. James Esdaile, a Manchester hat manufacturer who owned small parcels of land bordering the new road, was one of the beneficiaries. Of all the participants in this long boom, surprisingly little is known of the actual builders, although, as in other northern towns, the building trades in Manchester appear to have been dominated by small-scale family businesses using traditional construction methods. Housebuilding was little affected by those revolutions in technology, organisation and scale seen in other sectors of the Manchester economy.

PICCADILLY
PORTLAND STREET
Portland Place
BLOOM STREET
CHORLTON STREET
SACKVILLE STREET
DAVID STREET
OXFORD STREET
Richmond Street
Major Street
Silver Street
Paradise Ct.
Aytoun Street
Auburn Street
Booth Street
Pitt Street
Gore Street
China Lane
Hope Street
Port Street
Hilton Street
Gt. Newton Street
Blossom Street
George Street
Jersey Street
Great Lever Street
Stevensons Square
Spear Street
Oldham Street
Dale Street
Levers Row
Infirmary Pond
Falkner Street
George Street
New York Street
Charlotte Street
Nicholas Street
Bond Street
Dickenson Street
River Tib
Wakefield Street
Great Marlborough Street
St. Peters Square
MOSLEY STREET
Independent Chapel
Cooper Street
Peter Street
Mount Street
South Street
Tib Street
Whittle Street
Oak Street
Turner Street
Thomas Street
Edge Street
Coop Street
Rope Yard
New Cross Street
SHUDE HILL
WITHY GROVE
HIGH STREET
CANNON STREET
Marsden Square
Fountain Street
York Street
Theatre Royal
Spring Gardens
King Street
Norfolk Street
Pall Mall
Cheapside
New Market
Cross Street
Alport Street
Byrom Street
Miller Street
Hanover Street
Long Mill Gate
Hanging Ditch
Fennel Street
Market Place
Exchange Street
St. Anns Square
Police
Deans Gate
Cumberland Street
Bridge Street
Quay Street
Garside Street
Water Street
Parsonage Bank
Walkers Croft
Apple Market
Old Bridge
Black Friars Bridge
RIVER
New Bridge
IRWELL
New York House
Chapel Street
Green Gate
Gravel Lane
Salford
Bowling Green
New Bayley Street
Stanley Street
Derby Street
Booth Street
Gore Street
Mason Street
Bolton Street
Brompton Street
PARK
CANAL

1793

Laurent and Green: copy or complement?

After the appearance in 1741 of the first Casson and Berry plan, there was a gap of over half a century before a new map of Manchester was published. Ironically, like the proverbial buses, two then appeared almost simultaneously, in 1793 and 1794. The absence of new maps is surprising since the Manchester and Salford townships were beginning to expand apace, spreading inexorably outwards with new streets and buildings springing up as they grew.

The two maps by Green and Laurent have long been surrounded by controversy. Charles Roeder, a German-born merchant in late Victorian and early Edwardian Manchester, saw Laurent's plan as a mere copy of Green's and his often-reiterated view has been widely accepted. More recent research has cast doubt on this to argue instead that there are distinct differences between the two plans and that each should be seen as a valuable contribution in its own right.

The two men could hardly be more different. William Green is best known as a Lake District artist, who drew widely admired views of its landscape. However, until his thirties he lived in Manchester and was a surveyor and teacher of drawing. He worked with William Yates on the large-scale map of Lancashire, and in 1787, when Yates's map was published, Green announced his intention to produce a map of Manchester and Salford. His surveying credentials were therefore clearly well established. Charles Laurent, on the other hand, is a shadowy character, about whom remarkably little is known. He was almost certainly French and styled himself as 'engineer and geographer', suggesting that he may have been

---

OPPOSITE. C. Laurent, *A topographical plan of Manchester and Salford* . . . (1793) [BL] South-east is at the top.

OVERLEAF LEFT. C. Laurent (1793) Detail of town centre, oriented with north shown roughly at the top.

OVERLEAF RIGHT. W. Green, *A plan of Manchester and Salford* . . . (1794) [UML] Detail of the town centre.

CHAPEL STREET
GREEN
Ditch
Cateaton Street
Smithy Bank
Old Millgate
Cannon Court
Hanging Bridge
Half Street
Collegiate Church
Church Yard
Apple Market
The College
H. of Correction
Hunts Bank
Old Bridge St.
Old Bridge
Lying in Hospital
Hardings Buildings
Maidens Row
Kings Head Yard
Cloth Hall
Dawsons Croft
Salford
Cross Court House
Berlley Square
Princess Street
Duke Street
Methodist Chap.
Old Shambles North Side
Kings Entry
R
I
V
E

Lord Ducie
St Mary Gate
Hunts Bank
Bridge
Bertles Square
Brewery
Exchange
Cross
Dawsons Croft
Green Gate
Duke Street
Norton St
Princess Street
Methodist Chapel
Cross Street
Brewery
Mess Barrow & Hardman
New Cloth Hall
Hunts Bank
Castle Inn
The College
Grammar School
Long Mill Gate
Burying Ground
Apple-Market
Collegiate
Church of Christ
Fennel
Kings Head Yard
Chapel Street
Stanearst
Lying Inn Hospital
Old Bridge
Old Bridge St
Cateaton St
Hanging
Door
Hill Gate
Hunters Lane
Water

a member of the elite *Corps des Ponts et Chaussées* where he would have acquired knowledge of surveying. However, apart from his Manchester map, there is no evidence of any other published map in which he was involved.

Both maps are fascinating artefacts, with masses of detail of the urban transformation that was being brought about by industrialisation. Green's plan is huge; published at the scale of 1:1800 on nine sheets and measuring 358 × 284cm; and oriented with north to the top. It is one of the largest of the urban plans that began to appear in the later eighteenth century as Britain followed the European revolution in scientific surveying and accurate vertical plans. It is embellished with an elaborate cartouche symbolising the industrial progress of the town: a female figure personifying Manchester, flanked by Britannia representing trade; and a kneeling female figure holding a cornucopia from which coins are scattered, signifying the wealth derived from industry; below are putti, one of whom holds a lyre and trumpet, suggesting the more artistic aspect of the town; on the left is a large mill with smoking chimney, and on the right a weaver's bobbin and shuttle and a beehive, which had come to symbolise Manchester's work ethic. Green chose to extend his survey beyond the two townships of Manchester and Salford, to include the adjacent townships of Ardwick and Chorlton Row (Chorlton-on-Medlock). This aimed to increase the number of subscribers, but also enabled him to show, with dotted lines, the proposed new streets on which landowners were capitalising on the building boom of the mushrooming town. The map's detail is remarkable: streets are all named; buildings are shown in outline; details of land use and gardens are indicated; landowners are identified on the outskirts of the town; factories of various kinds are specified; and the topography along the rivers is indicated with hachures. The plethora of proposed streets, especially in Chorlton Row and Ancoats, shows the continuing pressures for expansion and how much the town had grown in the half century since Casson and Berry's plan. It was no mean feat of surveying and, given that he did not work full time on the survey (since he had to continue to earn his living by teaching), it may not be surprising that Green took no less than eight years from start to publication.

Title and cartouche from Green

However, this long gestation was one element in his downfall as a commercial cartographer. The other was the decision to seek subscribers who would only have to pay when the map was finally published. Both allowed Laurent to steal the market. Laurent visited Manchester in late 1791 and advertised his intention to produce a new map of the town. He apparently left in early 1792, but visited again at some time in 1793. He initially offered to help Green to complete his survey, but Green turned this down, doubtless concerned – rightly – at the potential threat that it represented. Laurent published his map in 1793 and, to add insult to injury, his map plates were bought by the London printer, John Stockdale, who published the map at half the price of Green's and then used the map to illustrate Aikin's handsomely illustrated book *A Description of the Country . . . round Manchester* in 1795. Many of Green's subscribers must have been lost to him so that his map proved a commercial disaster. This doubtless influenced his decision to eventually move to the Lake District.

Roeder's view was that Laurent's map was a simple forgery, derived from having access to Green's work. Circumstantial evidence offers support to this: accurate surveying was a time-consuming process, which sits oddly with the short time that Laurent apparently spent in Manchester; and he could have copied Green's work when he met to offer his assistance, or he could have gained access to it via Green's lodger, a Frenchman, Anthony Felix Ciziz. These are persuasive views, but they could well be countered: in 1791, Green's surveying was incomplete and would have been in draft manuscript form, not easy to copy either for Green or Ciziz, and even by 1793 Green's surveying was incomplete; Laurent's claim to be both engineer and geographer suggests that he had experience of surveying; and there are examples of other surveyors able to produce large-scale plans of towns in remarkably short periods. A more convincing avenue to explore the accusation of plagiarism is therefore the tangible evidence of the similarities and differences between the content of the two maps.

Laurent's map was published at a scale of 1:3600, half that of Green's. It was engraved by the distinguished map-maker John Cary. Curiously, it is oriented with south-south-east to the top (so that visual comparisons between the two maps entail turning Laurent's map almost 180 degrees). This orientation could be interpreted as an attempt to conceal any suggestion of plagiarism. There are striking similarities with Green's map, for example the overall geometry of roads and rivers on the two is virtually identical and the great majority of road names are the same. Both maps are impressively up to date, identifying significant buildings only opened in the year immediately prior to their publication.

Nevertheless, there are some telling differences. There are 76 cases where streets are named by Laurent but not Green and only 29 where Green names a street not named by Laurent. Some of these differences are well illustrated in the Red Bank area in the north-east of the town where Laurent shows, and names, a number of proposed new streets that are only half indicated by Green. And while Green identifies many more industrial buildings, Laurent uses symbols to pinpoint large numbers of water pumps and corn mills, which are not shown by Green and can only have derived from on-the-ground exploration. The topography of the two also differs. Green shows a virtually flat landscape, except along the line of rivers, whereas Laurent shows a much hillier topography outside the main body of the town. This may reflect the engraving skills of John

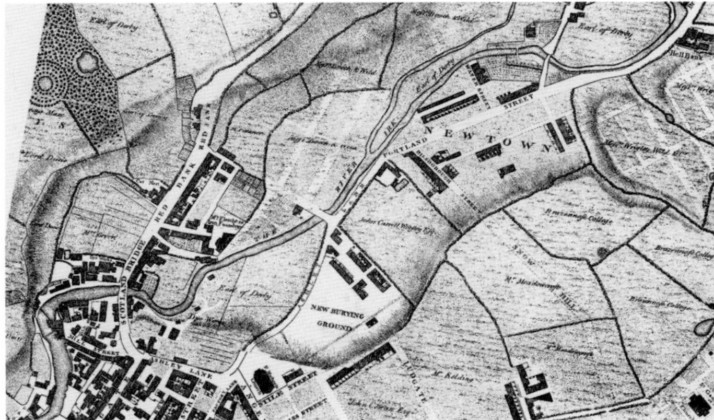

Red Bank area shown by Laurent (top) and by Green (bottom) [AUTH] The excerpt from Laurent has been rotated to put north at the top. The similarities and differences between the two plans are clear.

Cary, but Cary must have been given surveyed information from which to show such landscape detail.

These discrepancies throw doubt on Roeder's confident assertion of plagiarism. It may well be that Laurent was able to steal some of the big-picture cartography from Green's survey, but the differences suggest that he, or assistant surveyors, must have undertaken some original surveying on which his map drew. At the end of the day, the arguments about plagiarism could be considered simply diversionary. Both maps are splendid portraits of the rapidly growing town at a critical juncture in its growth. The differences between the two mean that both maps should be taken together to trace the details of the early years of the evolution of the world's first industrial complex.

STRANGE WAYS
GREAT NEWTON STREET
LEVERS ROW
PICCADILLY
PARLIAMENT STREET
BRIDGE STREET
KING STREET
CHAPEL STREET
QUAY
GARTSIDE
Castle Field
Coal Wharf
Castle Quay
DAWSON STR
SHUDE
TURNER STREET
CHURCH STREET
SWAN STREET
HANGING DITCH
CHORLTON ROW
SYDNEY STR
DOWNING STREET
Stephens Place

1809

Town directories and Pigot's Manchester plans

Nationally, the first trade directories began to appear at the end of the seventeenth century, but only covered London, which was then the only centre providing trade for the whole country. In the course of the eighteenth century, the pattern of trade changed. Industrial growth, better communications, the growth of population and its concentration in burgeoning towns helped to create the spread of home and foreign trade so that the market for trading widened, creating conditions in which there was a growing need for commercial directories which could put suppliers and customers in touch with each other.

Manchester's earliest trade directory was Elizabeth Raffald's 1772 directory which recorded the names and occupations of some 1,495 of the town's merchants and business inhabitants, about 6 per cent of the total overall population. Her data showed the importance of textile trades which comprised almost 30 per cent of the entries (the largest speciality being the fustian trade). Two further editions followed, in 1773 and 1781 (the latter covering both Salford and Manchester). Raffald was clearly enterprising and talented. Born in Doncaster, she was initially employed as a housekeeper and met her husband when both were employed at Arley Hall in Cheshire. They moved to Manchester in 1763 where she opened a confectionery shop, ran a cookery school, worked as an innkeeper and started an employment agency for servants. As well as compiling her directories, she published a highly successful cookery book, *The Experienced English Housekeeper*, in 1769, which appeared in numerous subsequent editions.

However, Raffald never used maps to accompany her listings of the town's inhabitants. Neither did other compilers of the few eighteenth-century directories of the town, such as

---

OPPOSITE. J. Pigot, *A plan of Manchester and Salford . . .* (1809) [MLA]
Townships are distinguished by colour.

Edmond Holme in 1788, and John Scholes in his directories of 1794 and 1797. Adding plans of the town to commercial directories was one of the singular contributions of James Pigot whose directories included a series of more than 20 plans of the twin towns of Manchester and Salford. They provide an almost biennial record of the growth and evolving shape of the town in the first half of the nineteenth century. The outward spread of its growth is well captured in the contrast between his 1809 and 1838 maps.

Pigot's maps of Manchester started with a fairly crude plan in 1804 drawn with north to the bottom, but they became increasingly detailed and handsome over time. The titles of each of his plans were set against the background of an illustrative cartouche for which he developed nine different designs. These included views of the town and of various buildings: a mill, the Exchange, the iron bridge over the Irwell, the new Blackfriars Bridge and the New Town Hall. The final design, shown here from his 1838 plan, shows a seated blindfolded Justice holding the sword and scales of justice and with a railway engine and ship in the background.

One of the distinctive features of his plans from 1824 onwards was his use of a fairly novel reference system with a spider's web of radii and circles to identify places on the map. The central point of this network was the Exchange, at the intersection of Market Street and Exchange Street. The virtue of this scheme is that the segments are smaller in more central areas where town density is generally higher and hence where more streets and significant buildings need to be recorded than in the more peripheral areas where density is lower. It also makes it easy to calculate distances from the centre of the town (although, of course, not from other origins). The drawback is that the reference system is more cumbersome than in the more conventional rectangular grid, which uses simple north/south and east/west alphanumeric identifiers.

His first plan in 1804 was produced for Dean & Co.'s Manchester Directory, but he began to produce his own local directories in competition with Dean from 1811, but then joined with Dean in 1815 to produce Pigot & Dean's Manchester and Salford Directory. He started his own national series of directories in 1814 and opened an office in London as well as his Manchester office, which was located in Fountain Street for over 40 years. He completed his directory coverage of the whole UK by 1823 and included smaller towns and villages as well as the major urban centres. In the 1830s, he brought his son into the business and – probably from 1839 – took into partnership Isaac Slater the engraver who had earlier been apprenticed to him and with whom he now traded as Pigot & Slater.

Pigot's plans of Manchester are rich in detail. Virtually all the existing streets are named, and intended new streets are shown by dotted lines often accompanied by their proposed names. Most of the highlighted public buildings are identified and include the wide variety of churches that multiplied in the first part of the century, as well as such prominent institutions as banks, places of entertainment, the relatively new phenomenon of post offices and much besides. Lee, who compiled a listing of Manchester maps, commented that Pigot's plans 'are invaluable records of the changing shape of our streets'.

An intriguing question is how Pigot compiled this impressive sequence of Manchester plans and kept them up to date and comprehensive. Neither he nor Slater claimed to be surveyors, so it seems unlikely that the plans drew on fresh surveys, which would have been an expensive addition to the cost. However, he employed agents who collected data for the directories about inhabitants and businesses and sought help from local notables with knowledge of their districts. He acknowledged this in one of his directories:

> . . . the proprietors, as upon previous occasions, have thankfully to acknowledge the prompt and valuable assistance afforded to them, from many talented gentlemen, in furnishing to their Agents information, upon various subjects . . .

It therefore seems probable that he drew on some of the existing formally surveyed plans of the town to provide a basic geometry and then used the information collected for his direc-

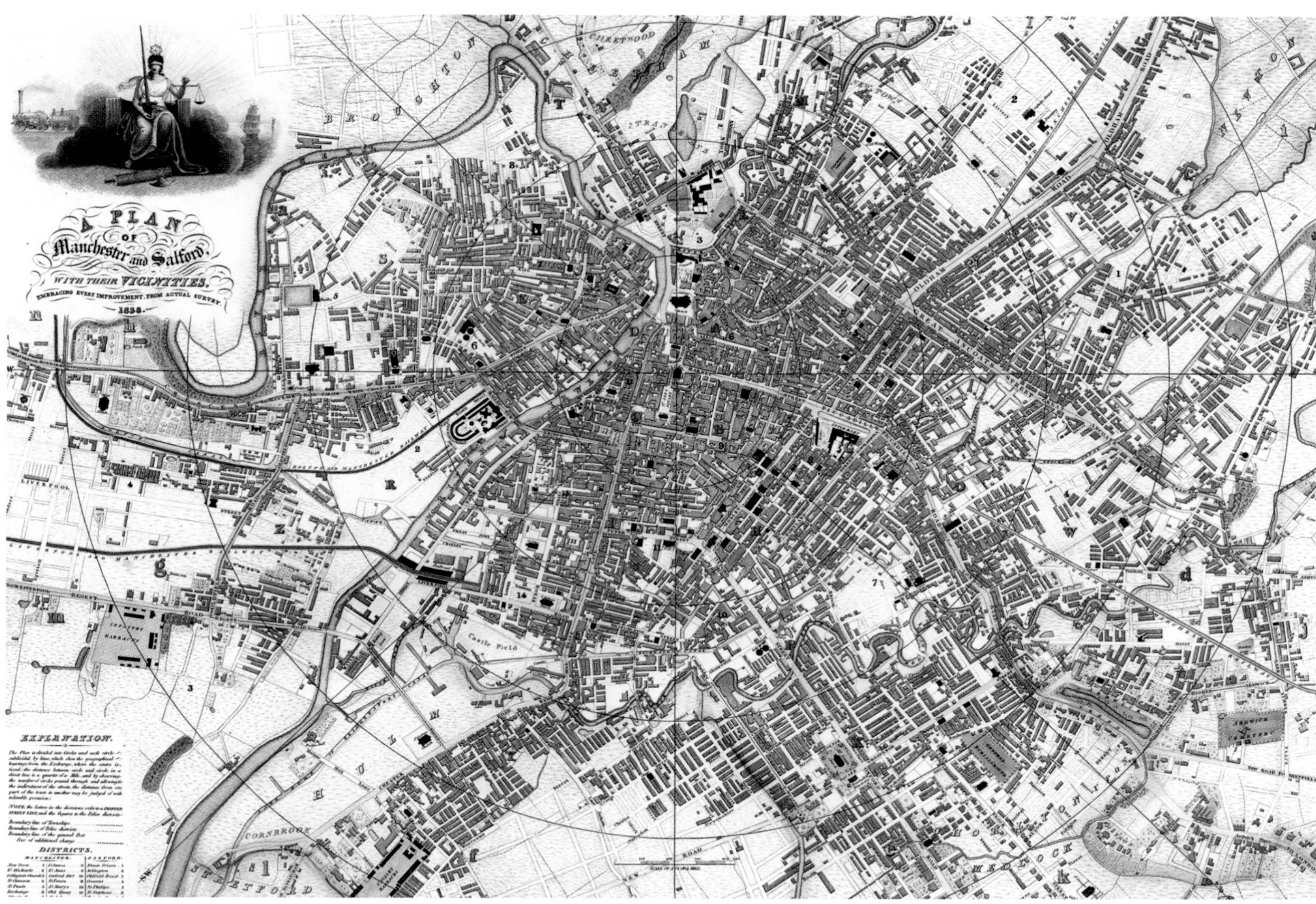

J. Pigot, *A plan of Manchester and Salford* . . . (1838) [AUTH] Pigot used circles and sectors to identify locations. The growth of the town since 1809 is clear.

tories as a way of ensuring that his plans were kept up to date.

Competition between the publishers of trade directories grew as the century progressed and Frederic Kelly increasingly came to command the market. Kelly had started a London directory in 1799 and subsequently expanded into southern England. Pigot was forced to retreat from his national coverage, and from the 1840s the firm's coverage was restricted to the northern counties of England together with Wales, Scotland and Ireland. After Pigot's death in 1843, the business was taken over by Slater who continued to publish local Manchester directories, initially using the plates produced by Pigot and later using new plates, although ones that were much less handsome than Pigot's, foregoing the pictorial cartouches and instead using a very plain title. The firm continued in the Slater family until 1892 when it was taken over by Kelly's Directories (although for some years the Slater name continued to be used). By mid-century, Kelly's had a national coverage and grew to be far and away the most popular of directories which continued to be produced until the outbreak of the Second World War.

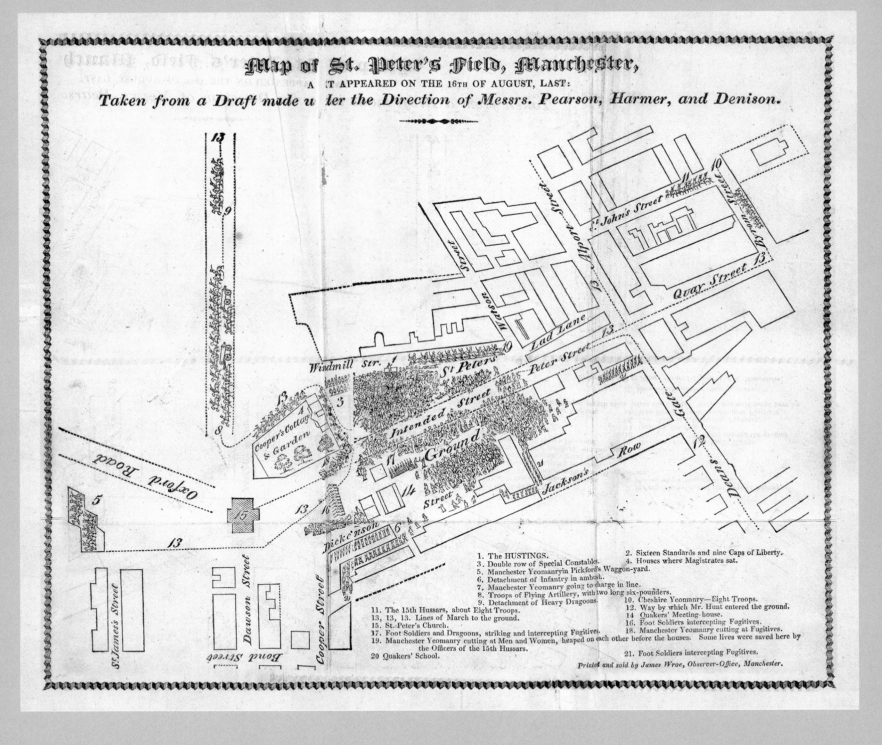

Map of St. Peter's Field, Manchester,

A IT APPEARED ON THE 16TH OF AUGUST, LAST:

Taken from a Draft mude u ler the Direction of Messrs. Pearson, Harmer, and Denison.

Windmill Str.
St Peters
Intended Street
Peter Street
Ground
Street
Lad Lane
Watson Street
Street
Alport Street
St John's Street
Byrom Street
Quay Street
Deans Gate
Jackson's Row
Dick'nson
Cooper's Cottage & Garden
Oxford Road
Cooper Street
Dawson Street
Bond Street
St James's Street

1. The HUSTINGS.
2. Sixteen Standards and nine Caps of Liberty.
3. Double row of Special Constables.
4. Houses where Magistrates sat.
5. Manchester Yeomanryin Pickford's Waggon-yard.
6. Detachment of Infantry in ambush.
7. Manchester Yeomanry going to charge in line.
8. Troops of Flying Artillery, with two long six-pounders.
9. Detachment of Heavy Dragoons.
10. Cheshire Yeomanry—Eight Troops.
11. The 15th Hussars, about Eight Troops.
12. Way by which Mr. Hunt entered the ground.
13, 13, 13. Lines of March to the ground.
14. Quakers' Meeting-house.
15. St. Peter's Church.
16. Foot Soldiers intercepting Fugitives.
17. Foot Soldiers and Dragoons, striking and intercepting Fugitives.
18. Manchester Yeomanry cutting at Fugitives.
19. Manchester Yeomanry cutting at Men and Women, heaped on each other before the houses. Some lives were saved here by the Officers of the 15th Hussars.
20. Quakers' School.
21. Foot Soldiers intercepting Fugitives.

Printed and sold by James Wroe, Observer-Office, Manchester.

1819

'The Peterloo Massacre'

No single event defined the politics of the nineteenth century more than the meeting held on Monday 16 August 1819 at St Peter's Field, an open piece of land which took its name from the nearby church. A crowd of men, women and children – estimated at some 60,000 – came to hear Henry 'Orator' Hunt speak on the urgent need for parliamentary and economic reform. The decision of the magistrates to send in the Manchester and Salford Yeomanry Cavalry to arrest Hunt and other speakers resulted in the death of at least 15 people and the injury of hundreds of others. The Peterloo Massacre, as it was quickly and cleverly dubbed by the radicals, confirmed that Manchester was experiencing as much a social as an industrial revolution, one that was giving birth to the politics of class.

Journalists provided the first accounts of the meeting. They were followed by other writers, cartoonists and poets, condemning or excusing the authorities depending on their political persuasions. There was to be no impartial account of Peterloo. All but lost in this flood of words and images were the maps of what the *Poor Man's Guardian* was later to refer to as 'the Plains of Peterloo'. Arguably, no political event in Britain had been mapped in such detail up to this time. But, as with the contemporary written accounts of Peterloo, none of the plans of St Peter's Field can be regarded as a neutral record.

The published plans became an integral part of the public arguments about what had happened on the day. The first plan to be published appeared in the radical *Manchester Observer* (23 October 1819) whose editor, James Wroe, provided the earliest detailed and impassioned account of the 'Manchester Massacre'. This plan appears to have been based on an existing one made on behalf of the radical lawyers, Charles Pearson, Henry Denison and James Harmer, for the inquests taking place into the Peterloo dead. It made no attempt to hide its political sympathies. It identified the principal buildings, the position of the hustings from which Hunt and others had

OPPOSITE. Observer Offices, *Map of St. Peter's Field, Manchester . . .* (1819) [Robert Poole]

briefly addressed the crowd, the location of the magistrates and the different troops, as well naming many of the surrounding streets. That it was a political map is evident in the key whose numbered locations included the 'Manchester Yeomanry cutting at Men and Women, heaped on each other before the houses' and 'Foot Soldiers and Dragoons, striking and intercepting fugitives'.

Plans were to be included in other Peterloo pamphlets, but establishing their provenance is problematic. A small general plan of the area featured in *An impartial narrative of the late melancholy occurrences in Manchester*, published in Liverpool by Henry Fisher, but while it included buildings such as the Friends Meeting House whose raised graveyard had provided people with an uninterrupted view of the field, prominence was also given to the Swedenborgian New Jerusalem Church in Peter Street, which barely featured in most narratives about the day. The maker of this plan is unknown, but given its rudimentary nature it seems unlikely that it was drawn by a surveyor.

It is worth noting that as these plans were being published, readers, especially those who did not know this part of Manchester, were able to see other views of the meeting and the surrounding buildings in drawings such as Thomas Thwaite's *A view of St Peters Plain Manchester on the memorable 16th August 1819, representing the forcible dispersion of the people by the Yeomanry Cavalry &c*. One of the earliest pieces of Peterloo memorabilia, John Slack's depiction of the meeting being 'dispersed by the civil and military power', may have been among the most widely circulated as it was printed on a handkerchief, sold to raise funds for the victims.

Lawyers, more particularly those representing the radicals, were quick to recognise the importance of establishing the geography of the area – the position of buildings, walls and gates, the lines of approach by the cavalry, the lines of escape for the crowd – when discussing the sequence of events in court. Harmer who, with Denison, played a decisive part in the inquest into the death of the cotton worker John Lees, made frequent use of a plan of St Peter's Field. It identified what were considered to be the main buildings, the positions of the yeomanry and cavalry, and features on the field such as the timber that was stacked close to the Friends Meeting House. Harmer used the map to establish where witnesses were standing in relation to the events they were describing. In one instance, a witness was describing events on the field from 200 yards away. This 'accurate plan' of St Peter's Field was reproduced in *The Whole Proceedings Before the Coroner's Inquest at Oldham on the Body of John Lees . . .* published by the London radical William Hone in 1820.

The authorities were to produce their own plans of St Peter's Field. By the time of the case brought by Thomas Redford against Hugh Hornby Birley and other members of the Manchester Yeomanry in 1822, plans had become an essential part of the legal courtroom arguments. Of the different plans used in the famous *Redford v Birley* trial only one – 'A plan of St Peter's Field in the town of Manchester with the avenues leading thereto' – was included in the published account. No mystery surrounds its provenance. It was drawn by no less a figure than James Wyld, the Geographer Royal, who produced a substantial number of fine maps for the army. Copies of his Peterloo plan were printed at the Lithographic Press, Quarter Master General's department in Horse Guards, Whitehall, a method of printing that Wyld is credited with introducing in England. The plan is dated 29 November 1819. In addition to its accuracy, one of its distinctive features is that it covered a far wider area than most other plans, presumably because it recognised the importance of showing the location of troops in different parts of the town on the day (for example in Portland Street and Byrom Street). Different versions of the plan exist, and it is unclear which ones were used in the courtroom. One extant copy includes a key which refers to the 'dense body of reformers' who stood in front of the hustings, and whose numbers and actions were to be closely discussed in various accounts of Peterloo. Wyld's general map of the field and its 'avenues' was to become the most widely reproduced and preserved of the Peterloo plans.

No memorial was allowed to be raised on the 'Plains of Peterloo', and by the 1850s new buildings, including the Free Trade Hall and the Theatre Royal, covered the area. St Peter's Field disappeared. Attempts to remember Peterloo on the site

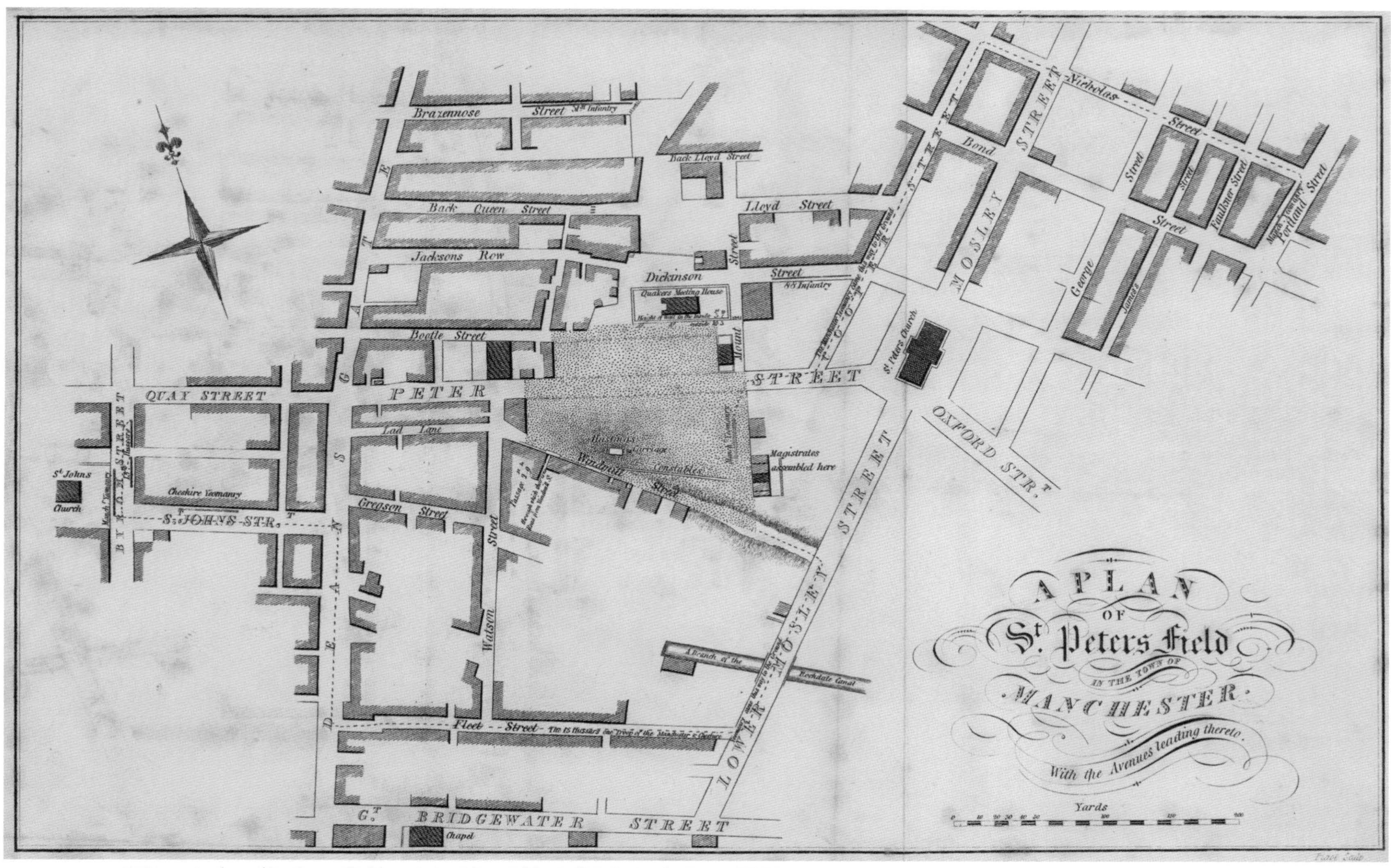

J. Pigot, *A plan of St. Peters Field . . .* (1819) [MLA] The 'official' plan by James Wyld.

were made but, as in the late 1960s when a proposal was put forward in the City Council that Peter Street be renamed Peterloo Street, they were largely unsuccessful. Peterloo was still capable of dividing Manchester. More recently, the ashes of 1819 have cooled, leading to a change in attitudes towards the democratic capital embedded in Peterloo: in 2010, the Peterloo Relief Fund Account book kept in the John Rylands Library was included by UNESCO in its UK Memory of the World Register, while in 2014 approval was finally given to erect a Peterloo memorial close to if not on the actual site of the massacre. Less discussed in the media was the rebranding of the area as Petersfield, an uncontested toponymic change arising out the wider policy of micro-branding the revitalisation of the city centre. This contraction and fusion of St Peter's Field is now a prominent feature on the free visitor maps available throughout the city.

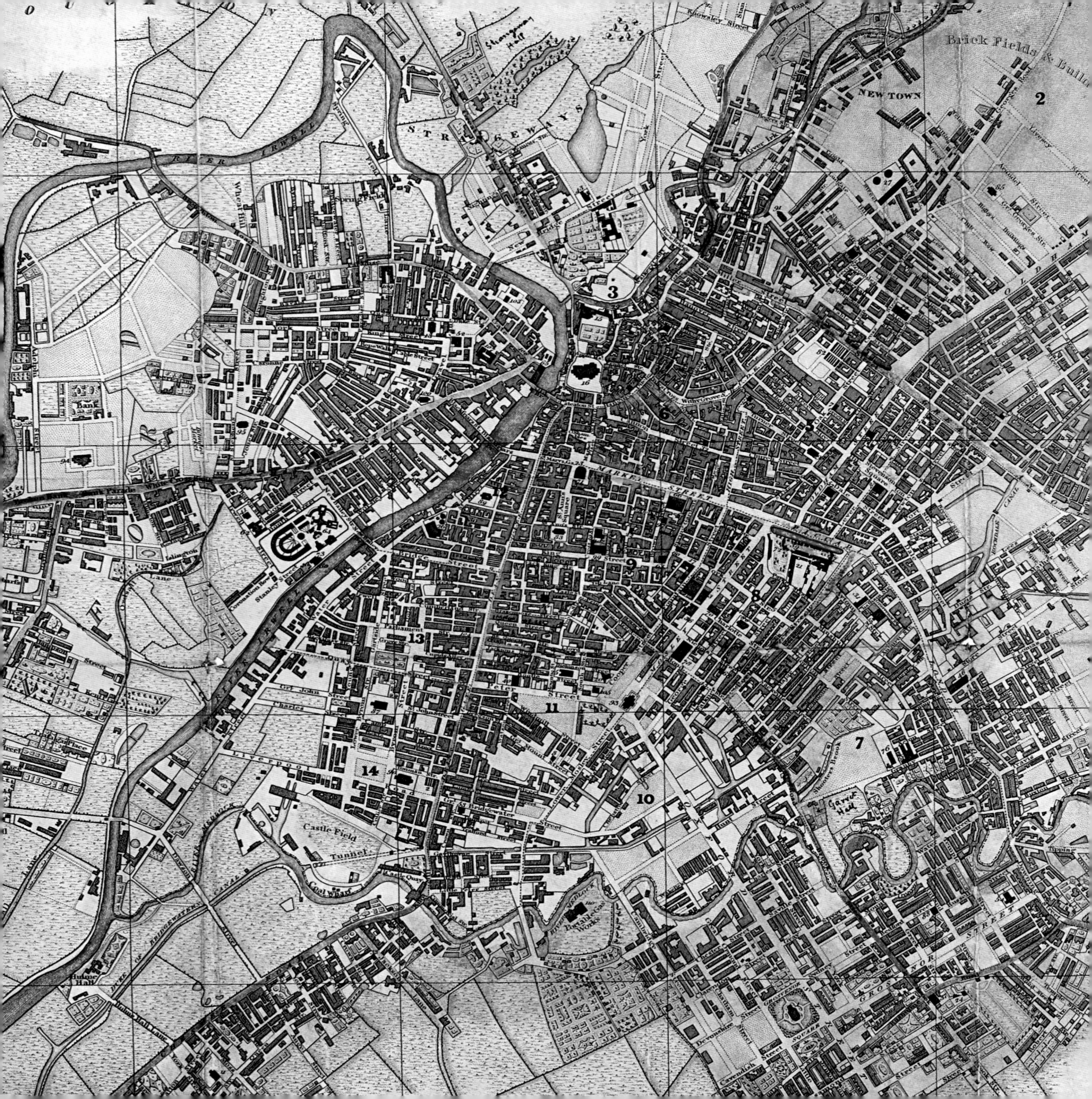

NEW TOWN
2
STRANGEWAYS
RIVER IRWELL
Knowsley Street
Spring Field
3
6
Market Street
Bridge Street
9
13
Parliament
Peter Street
Gt John Street
Charles
11
Liverpool
14
10
7
Castle Field
Tunnel
Coal Wharf
Islington
Stanley Street
Coronation Row
Garratt Hall
Hulme Hall
Hulme Hall Lane
Oldham
Cavendish Street

1824a

The mixed legacy of grand houses

William Swire's plan of 1824 was commissioned for Edward Baines's *History, Directory and Gazetteer of Lancashire*. Like Baines, Swire was based in Leeds, which probably accounts for the fact that it was he rather than a Manchester-based surveyor who Baines used not only for the Manchester plan but also a parallel plan of Liverpool and a smaller one of Rochdale. Swire later worked with W.F. Hutchings on a large-scale map of Cheshire in 1830. Despite the relatively small scale of his Manchester plan, it has a huge amount of detail, with a reference list of over 100 buildings as well as additional buildings named on the map itself. It is a handsome and finely engraved map with two views of Manchester and the frequently reproduced plan of the town in 1650.

One feature of the map's comprehensiveness is its inclusion of the ancient halls, which still survived in the town, as well as some of the areas in which 'stately edifices' were being built, such as Ardwick Green and The Crescent in Salford. Most of the old manor houses and their erstwhile aristocratic owners had disappeared by the early nineteenth century, remembered only through place names, but Swire's plan shows Hulme Hall, Strangeways, Ordsall Hill [*sic*] and Ancoats Hall still standing in the midst of the expanding industrial townscape. The subsequent fate of these halls and their estates varied widely. Some played important roles in configuring the town, others were subsequently demolished, leaving little physical trace.

Among those that disappeared is Hulme Hall next to the Irwell, a manor house which had existed since the twelfth century and whose owners had included the Prestwich and Mosley baronets. It was bought in 1764 by the 3rd Duke of Bridgewater to enable him to drive through his eponymous canal. The house fell increasingly into disrepair and was eventually demolished in 1840.

A second example is Strangeways Hall, for long the home of the de Strangeways family. It was bought in 1624 by a

---

OPPOSITE. W. Swire, *Manchester and its environs* . . . (1824) [MLA]

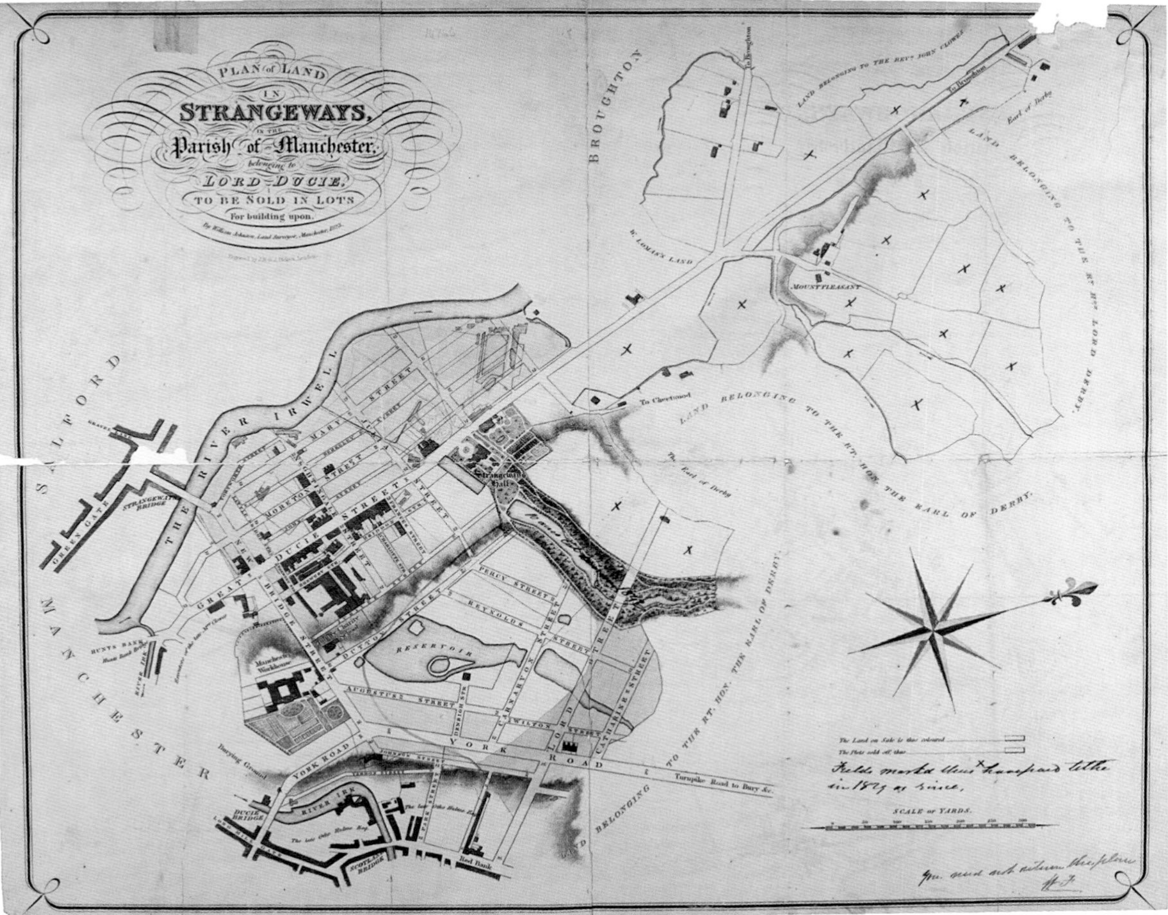

W. Johnson, *Plan of land in Strangeways* . . . (1823) [MLA] Names of some of the Ducie family are used for the streets platted for sale.

woollen manufacturer, John Hartley, and in 1713 left in trust to the Reynolds family, one of whom became 1st Earl of Ducie. One of its notable inhabitants, from 1808 to his death in 1811, was the radical politician Joseph Hanson who earned the moniker 'the weavers' friend' for his support of the handloom weavers hit by the wartime closure of Continental and American markets. Industrialisation encroached on the estate from the late eighteenth century, and the Reynolds leased parts of it to dye-works, printing works, brick-works and a brewery. They appointed an agent, William Johnson, to develop the land, and part of the estate was laid out with streets named after members of the Reynolds family – Frances, Julia, Catherine, Augustus and the earldom itself, Great Ducie. Initially, it attracted substantial houses – Friedrich Engels lived in a comfortable house on Great Ducie Street – but they were later demolished and replaced by industry and cheaper homes. The most dramatic change to the estate was the disposal of land around Walkers Croft and Hunts Bank, which was sold in 1838 to the Manchester and Leeds Railway Company for the construction of Victoria Station. The hall itself was demolished in 1858, and the city used the site to build the Assizes Courts and Strangeways Prison, both designed by Alfred Waterhouse. The neo-Gothic courts were eventually destroyed in the blitz of 1941; the prison still stands. What, effectively, is left of Strangeways is simply the name.

A third example is Ancoats Hall. Built in the early seventeenth century by Oswald Mosley, it had a striking site with gardens running down to the Medlock. By 1827 it was owned by George Murray, proprietor of the Murray Mills. He replaced the original hall with a large neo-Gothic house. As industry spread, the site clearly lost its attraction and the hall was bought by the Midland Railway to enable the construction of Ancoats railway station, which opened in 1870. From 1886, the hall was used to house the Manchester Art Museum founded by Thomas Horsfall, son of a Manchester-based cotton manufacturer. Horsfall's philanthropic work drew on his admiration for Ruskin's belief in the social benefit of art, of social engagement, and his strong commitment to social duty. This provided the basis for the establishment of the Manchester University Settlement, modelled on Toynbee Hall in London. It was the first such 'settlement' outside London and was initially based in Ancoats Hall, which it used as a student residence to bring students and the local community together. The Museum was taken over by the city, closed in 1953 and most of the collection was transferred to the city's Art Gallery. The Settlement moved to the Round House (Christ Church) in nearby Every Street. The hall was eventually demolished in the 1960s, and only vestigial traces remain on the ground.

Examples of the reuse of old halls include Ordsall Hall, the ancient Salford family home of the Radclyffes until 1662. By the late nineteenth century, it had become a working men's club and was later used as a training school for the clergy, which relocated to become the Manchester Theological College. The hall was bought in 1959 by Salford Council and opened as a museum in 1972. Other halls have survived as a result of Manchester buying land for public parks, as in the case of Platt Hall, Heaton Hall and Wythenshawe Hall.

A further survival – in Chorlton-cum-Hardy, well south of the area covered by Swire's plan – is Hough End Hall. It was built by Sir Nicholas Mosley in the sixteenth century when he became Lord of the Manor of Manchester. In its later life, it was used variously as a blacksmith's shop, farmhouse, restaurant and, most recently, a health education centre for the South Asian community. However, it is a rather poignant survival since it now sits forlornly overshadowed and hemmed in by

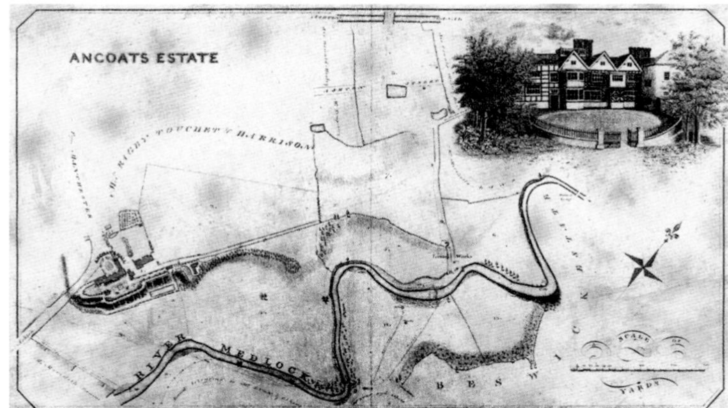

J. Hillkirk, *Ancoats Estate of Sir Oswald Mosley* (*c.*1804) [MLA] The road 'From Manchester' on the extreme left is (Great) Ancoats Street.

uninspiring modern office blocks, calling into question how planning permission was ever given for such development. In 1917, part of the estate had been taken over by the War Office and used for an aerodrome until 1924. It was during this period that the local Manchester architect John Swarbrick led a campaign to stop the hall from being demolished, and his campaign was instrumental in the founding of the Ancient Monuments Society in 1924.

Swire's plan also shows one of the nineteenth-century 'stately residences', Lark Hill Place, which overlooked the Irwell and bordered The Crescent in Salford. It was built in 1809 by Colonel James Ackers whose family's fortune came from fustian manufacture and land speculation. The house and its gardens were bought in 1846 by voluntary subscriptions, including support from Robert Peel, and handed to Salford Council, which used the Museums Act to establish it as the Royal Museum and Library, one of the earliest public libraries. A series of extensions were progressively added; for example, a new wing was built as a result of Edward Langworthy, a local textile manufacturer, leaving the princely sum of £10,000 in his will. By 1936, the original house had been replaced and the whole complex became the Salford Museum and Art Gallery and Local History Museum adjacent to the campus of Salford University.

A
Plan and Section
of an intended
RAILWAY OR TRAM-ROAD,
From
LIVERPOOL TO MANCHESTER
in the County Palatine of
Lancaster.
Surveyed by George Stephenson, Engineer.
20th day of Nov.r 1824.
Houghton
Leigh
Astley
Leigh Canal
Chorlton Fold
Monton
Pendleton
Bury & Bolton Canal
Cheetham
Newton
Salford
Manchester
Eccles
Parish
Eccles
Leigh
Parish
Bedford
Worsley Township
Betany Bay
Pennington Towns.p
Astley Township
Barton
Roscoe's Improvements
Pendleton Towns.p
Salford Township
Trafford Park
River Irwell
Culcheth Towns.p
Leigh Oak Hall
Culcheth Hall
Stocks
Culcheth
New Church
Chat Moss
Wilderspool
Bridgewaters Canal
Old Trafford
Chorlton
Part of Newton
Moss Side
Rusholme
Hulme Hall
Flixton
Irlam Hall
Irlam Ferry
Urmstone Hall
Urmstone
Stretford
Longford
Dog House
Chorlton
Croft
Glazebrook Moss
Risley Chapel
Hillock
Mill Houses
Hardy
Hough End
Withington
Heaton Norris
Ashton upon Mersey
Sale Moss
River Mersey
Barlow Hall
Cadishead
Glazebrook
Carrington Moss
Didsbury
Stockport
Fearnhead
Risley Moss
Woolstone
Rixton
Warburton
Rixton Hall
Martinscroft
Timperly
River Bollin
Wood Houses
Redish
Dunham
Altringham
Powder Mills
Thelwall
Duke of Bridgewaters Canal
Lymm
Dunham Park
Bollington
Bowdon
Croppenhall
Rosthern
Knutsford
Great Budworth

1824b

The coming of railways

It is possible that there were people in the crowds who attended the opening of the Liverpool and Manchester Railway on 15 September 1830 who had also watched the first barge cross the Barton Aqueduct on the Bridgewater Canal on 17 July 1761. To have done so was to have witnessed the beginning of the two great revolutions in transport whose impact was to be so profound, reaching far beyond Britain. Both of these transport innovations shared a common feature: they were driven by the trade that connected Liverpool and Manchester, and which was driving the growth of what were the two largest urban areas outside London.

The construction and opening of the Liverpool and Manchester Railway has become one of the landmarks of British history, so familiar that one can overlook that its *raison d'être* was simply to provide a more efficient and reliable means of communication than was available by either water or road transport between the two towns. The first survey of a railway route between Liverpool and Manchester was carried out by William James in 1822, a land agent and surveyor who recognised the potential of running steam carriages on lines. Visionary that he was, the scheme stalled when James was imprisoned for bankruptcy. Following the establishment of the Liverpool and Manchester Railway Company in 1824, George Stephenson was appointed as surveyor. A new survey was organised, though one which closely followed James's route. Stephenson's talent was as a practical engineer – he was to provide the best solution to the problem of designing and operating a steam-powered locomotive on a railway line – not as a surveyor. His surveying skills were basic and the survey of the intended line he produced had a number of flaws, which were highlighted by opponents of the scheme in the subsequent parliamentary proceedings. The project stalled. Another survey was undertaken. This was largely carried out by Charles Blacker Vignoles on behalf of the new surveyors of the line, George and John Rennie. This survey

---

OPPOSITE. G. Stephenson, *A plan and section of an intended railway or tram-road from Liverpool to Manchester . . .* (1824) [BL]

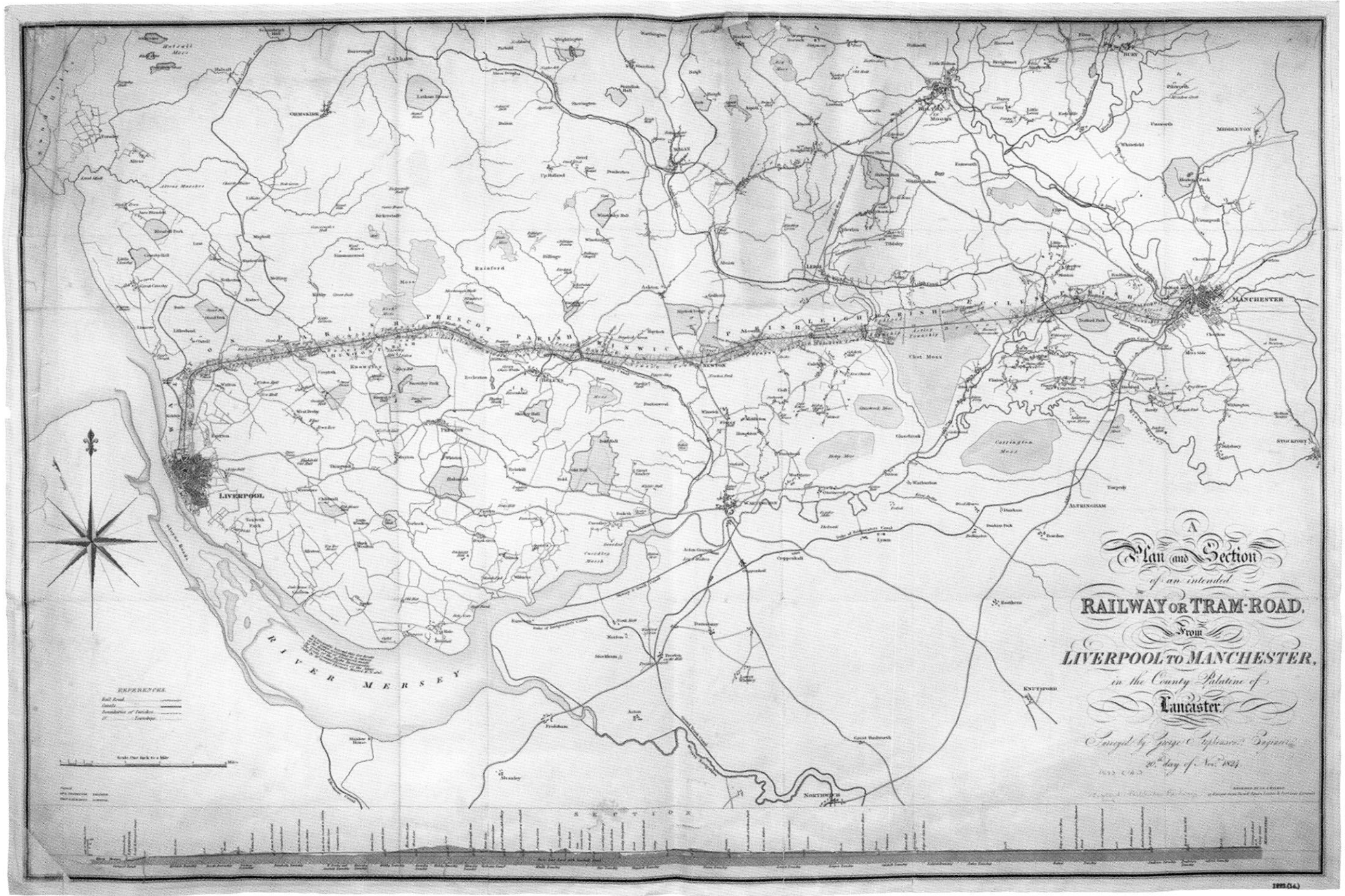

Detailed field boundaries are shown along much of the proposed railway line. Chat Moss clearly presented a major construction challenge.

(printed by James Wyld in 1826) met some of the objections to the route, the most vocal of which came from the canal interests. The bill was passed in 1826. Against expectations, the railway company reappointed Stephenson rather than allowing the Rennies to continue in control. It was Stephenson who was to lead the project to its completion.

Stephenson, like Brindley some 70 years before, had to provide solutions to the many problems encountered in constructing the line linking Manchester with Liverpool. Of these, the 'floating' of the track across Chat Moss is the best remembered, but there were other challenges that led to alterations being made to the agreed route. Selecting and acquiring a suitable site for the Manchester terminus had proved difficult. In 1829, agreement was reached with the Mersey and Irwell Navigation Company to build a stone bridge across the Irwell to carry the railway into Manchester. The new terminus was to be located on Liverpool Road, adjacent to the busy Castlefield canal basin. Because the bridge had to be tall enough to allow easy passage of the Navigation Company's boats, the railway line entered Manchester on a viaduct rather than at street level. The additional building work was to add to the costs of the project. It also cost lives as during the construction

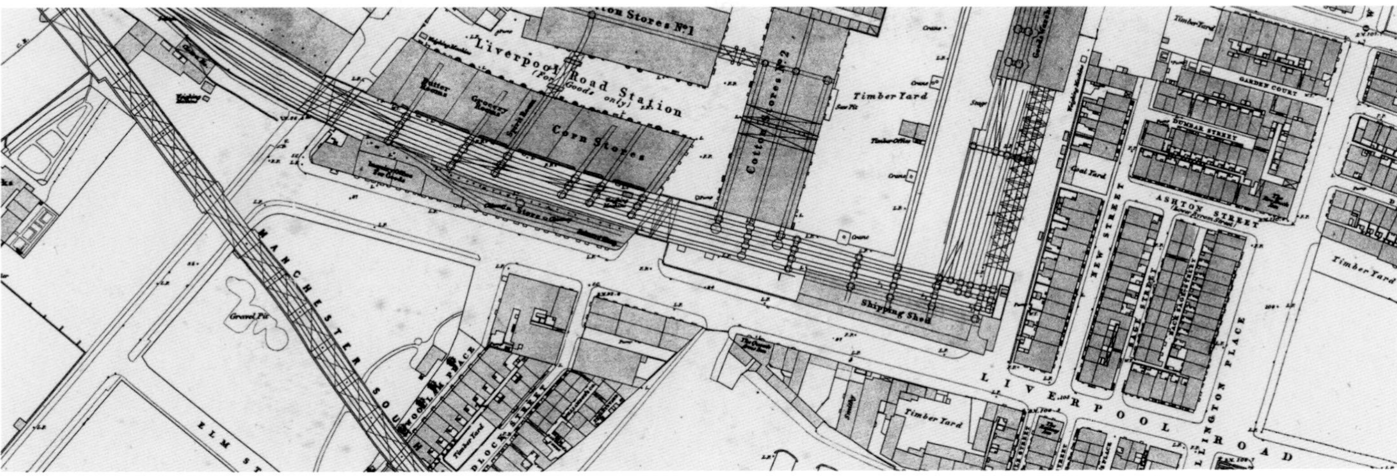

Ordnance Survey 5-foot [Liverpool Road Station] (*c.*1849) [Digital Archives] By this date Liverpool Road was only a goods station.

of the river bridge 12 workers were 'accidentally drowned', the first but not the last major loss of life in the building of the railways. The railway also required a bridge to be built in Water Street. This was noteworthy in that it made use of a newly designed type of cast-iron beam, the work of Eaton Hodgkinson and William Fairbairn, that was to become widely used in the construction of industrial and commercial buildings.

The railway station itself (now the oldest surviving one in the world) was a brick and stucco building facing Liverpool Road. It was architecturally unpretentious, giving no suggestion of how visually exciting this building type was to become in cities across the world. Indeed, the eventual inclusion of a sundial on the façade was a rather puzzling feature for a transport system that was already measuring journey times in minutes and which in a few years would be responsible for standardising the nation's clocks. The facilities inside the building were basic and differentiated according to social class. Opposite the passenger side of the station was a railway warehouse – the first in the world – a huge brick behemoth, which in design and construction was greatly influenced by the canal warehouses.

Passenger traffic had been considered secondary to freight in planning the station, but the line soon proved that this could also be an important source of revenue. Such was to be the speed of the railway revolution that other stations were soon opened in Manchester, stations that were to give greater attention to passengers. As early as 1844 these changes were to leave Liverpool Road operating only as a goods station. Additional land was purchased and over the coming decades more warehouses were built, which in their form and materials marked the rapid advances made in the design of railway warehouses. These warehouses can also be seen as representing the triumph of the railways over the canals. The early railway companies set out to compete with canals, but few expected that within a generation many of the canals would be taken over and even closed down by railway companies. This was not to be the fate of the Castlefield basin, but by the end of the nineteenth century it was to be crossed by a number of imperious railway viaducts which provided a dramatic visual metaphor of the result of the challenge thrown down by the Liverpool and Manchester Railway.

The Liverpool Road Goods station continued operating until the mid 1970s, throughout which time there was little alteration to the original station. This fortunate survival meant that the world's first passenger station and its warehouse were able to become the star attractions of the Greater Manchester Museum of Science and Industry when it opened on the site in 1983.

SALFORD,
C H E E T H A M
Majesty,
ENGRAVED BY J. & C. WALKER, LONDON,

1831

Spinning mills and the making of Cottonopolis

Manchester's wealth was initially generated in its cotton-mills and textile warehouses. Surprisingly, these buildings rarely feature on most of the published maps of the town. Indeed, it is impossible to point to an early nineteenth-century map that really captures the scale of cotton manufacturing and its significance to Manchester. One of few that at least acknowledges the existence of industrial sites is the *Bancks and Co. Plan of Manchester and Salford*, which is based on the detailed survey work of Richard Thornton and fine engraving of J. and A. Walker.

Purpose-built multistorey cotton-mills were established in Manchester from the 1780s onwards. Their success built on Manchester's long-established capacity as centre for the spinning of thread and weaving of cloth. While power from swift-flowing water had initially been significant to the region's textile industry, what really came to matter was the availability of coal to fuel the new steam engines in these mills. The transformation of Manchester from a thriving but small Georgian town into a major metropolis with global trading connections was closely bound up with technological innovations in steam power, cotton-processing machinery and the improved transport infrastructure brought about by canals. It undoubtedly also depended on the presence of talented and ruthless entrepreneurs like Richard Arkwright who radically reorganised the system of production to achieve significant economies of scale. Arkwright's pioneering cotton-mill in Manchester was built on land just outside the old medieval core at Shudehill in 1783 (shown on Bancks's map as 'Norris & Hodges's Cotton Mill'). At the time, it was the largest factory in the town, five storeys high. While it was pioneering in its use of steam power, it employed a Newcomen atmospheric steam engine, which could not generate the continuous rotary motion needed to drive the spinning frames, so was simply used as a pump to raise water to circulate through a closed reservoir system to turn a traditional waterwheel. The machines that were critical

OPPOSITE. R. Thornton, *Bancks & Co.'s plan of Manchester & Salford with their environs* . . . (1831) [UML]

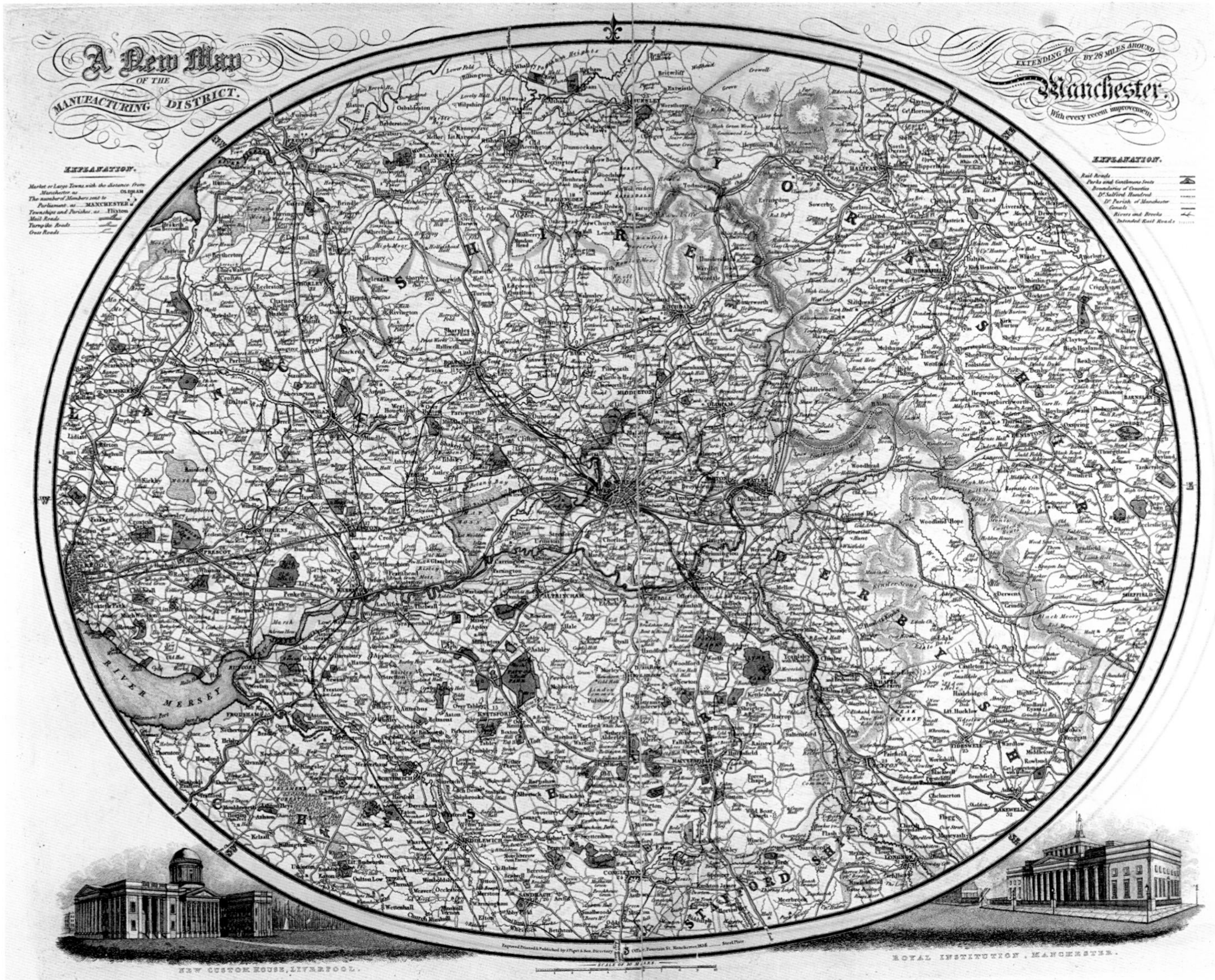

J. Pigot & Son, *A new map of the manufacturing district . . . around Manchester . . .* (1836) [MLA]
Despite its title, very little manufacturing is shown.

in powering the cotton-mills that, from the 1790s, characterised the industrial revolution in Manchester were the Boulton & Watt and other rotary action steam engines.

The iconic technology of the industrial revolution was the rotary steam engine – the essential component in the take-off of cotton production in Manchester away from the fast-flowing rivers of the Pennine hills. The numerous slender brick-built chimneys belching smoke were the most obvious architectural

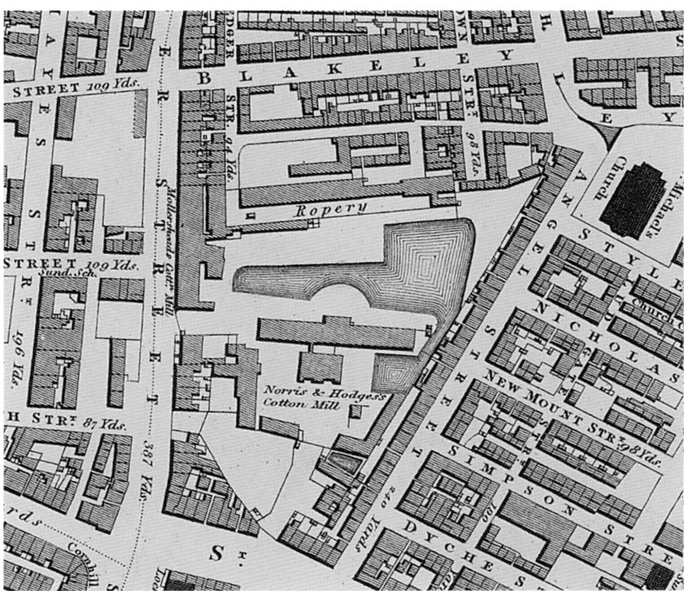

R. Thornton, (1831) [UML] Arkwright's factory in Angel Meadow.

marker of steam-powered textile mills. Two years after Bancks's map had been published, the French diplomat Alexis de Tocqueville wrote a vivid description of Manchester:

> A thick, black smoke covers the city. The sun appears like a disc without any rays. In this semi-daylight 300,000 people work ceaselessly. A thousand noises rise amidst this unending damp and dark labyrinth . . .

The most important area of cotton manufacture in Manchester at the time of Bancks's map was Ancoats. It was a planned industrial area with a grid of roads, and a dense concentration of factories, including several strikingly large spinning mill complexes, built side by side by competing firms. Ancoats in the 1830s must have been a forbidding industrial landscape, crowded with workers at shift changes and alive with the sounds of steam engines and the myriad spinning machines. The army of mill workers were housed in adjacent streets.

One of the leading firms in Ancoats was the partnership of the migrant Scots textile entrepreneurs James McConnel and John Kennedy. They built their first large spinning mill in 1797 on an undeveloped plot of land next to the route of the then proposed Rochdale Canal. It was over 50 metres long and seven storeys high and powered by a 16-horsepower Boulton and Watt rotary steam engine. A series of adjacent and interconnected mill buildings were constructed by the company in the following few years. These later mills were technically more advanced, tackling the risks of fire, using brick, slate roofing and stronger cast-iron pillars. On the extract of Bancks's map, the large McConnel and Co. mills can be seen along Union Street. At around this time, the company was said to have been the largest single employer in Manchester, overtaking their next-door neighbour and major competitor in cotton production, A. & G. Murray. Many of the other large mills were so-called 'room and power' concerns, housing numerous small producers. Single-owned mills like those of the Murrays were not the only form of business structure.

In the two decades after the map was published, Manchester became known as Cottonopolis. Raw cotton was Britain's largest import from 1825 to 1873, and vast amounts of value-added textile goods were shipped from Manchester to markets across the world.

While the multistorey cotton-mills in Manchester were highly mechanised in comparison to many other industries, they nonetheless required large numbers of workers to tend the machines. The nature of work in the cotton-mills was central to wider debates in the first half of the nineteenth century about the dramatic social changes caused by industrialisation and unprecedentedly rapid urban growth. The newness and scale of the mills in Manchester in particular, and their harsh working conditions, shocked socially minded commentators and politicians. Mill operatives were required to work long hours – 14-hour shifts were the norm, typically for six days a week. One of the worst features of the system was its exploitation of child labour. Campaigners strove for regulation of factories to improve conditions and legally limit working hours, particularly regarding the employment of children. However, they faced forceful counter-arguments as the rapidity of development and economic scale of the cotton

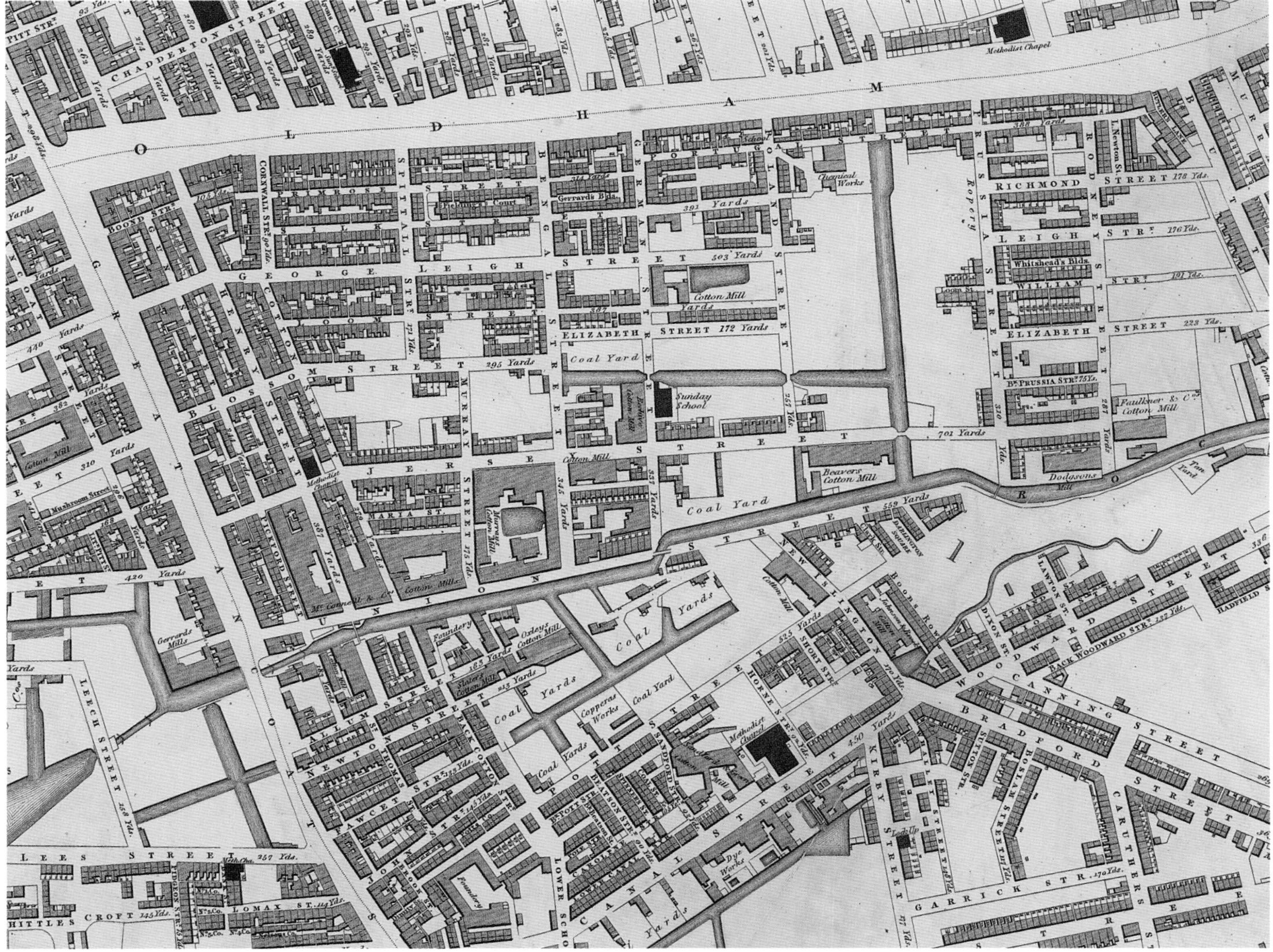

R. Thornton, (1831) [UML]. Ancoats had copperas works, dye-works, foundries, and chemical works, as well as its numerous cotton-mills and coal yards. There was a high proportion of back-to-back houses.

trade provided powerful testament against regulation that might thwart 'God-given' progress.

Despite resistance from factory owners, the Peel Act of 1802 . . . *for the preservation of the health and morals of apprentices and others, employed in cotton and other mills . . .* started the process of regulating working conditions. It was followed by a series of Factory Acts that limited the employment of children, reduced maximum working hours and made workplaces safer.

If the spinning mill with its smoking chimney symbolised the early years of the industrial revolution, the architectural structures that best characterised the commercial transformation of Manchester into Cottonopolis were the massive warehouses built in the city centre. They were not utilitarian

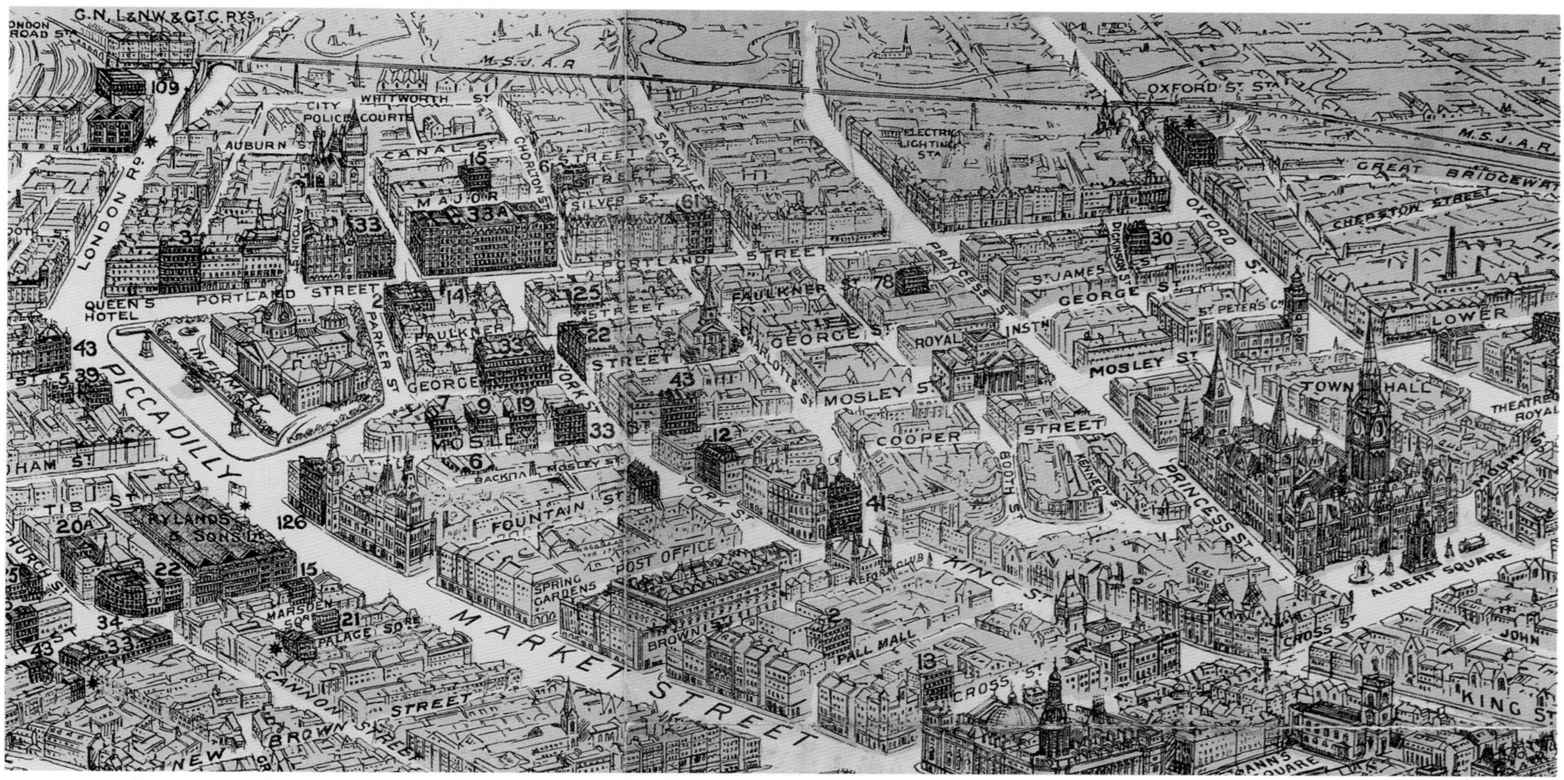

Warehouseman and Draper Magazine, *Bird's-eye view of textile Manchester* (1898) [AUTH]
The lack of open spaces reflects the speed of growth of the city.

storage sheds, but places to display and sell cotton products, as well assembling and making up orders, packaging them and arranging shipment. Architecturally, many of the largest commercial textile warehouses were like merchant 'palaces'. The grandest example was undoubtedly the S. & J. Watts & Co. warehouse on a prime site on Portland Street. Its flamboyant design – described by some as Venetian in style – was loved and loathed in equal measure when it was completed in 1856. Its eponymous owner, James Watts, came from a retailing background rather than cotton manufacturing, and a vast array of textile products were on sale inside his imposing warehouse.

An intriguingly detailed bird's-eye view of 'Textile Manchester' at the end of the nineteenth century shows the concentrations of large commercial warehouses in the city centre. Many were in prime locations on major thoroughfares, reflecting their status and economic significance. It is interesting to speculate whether firms paid for placement on the bird's-eye map, which was quite common in commercially funded cartography. For example, the Rylands & Sons warehouse on Market Street is specifically named. They were one of largest textile companies at the time, whereas the Watts Warehouse on Portland Street (no. 33a) is depicted but not named on the map itself. The commercial core on the bird's-eye is noticeably clear of factory chimneys and smog, which would have been a touch of artistic license, but the scene does capture the collar of the railways around the centre, with the ring of stations from Victoria and Exchange in the foreground to London Road, and Central Station aligned as a kind of transport horizon.

MANCHESTER.

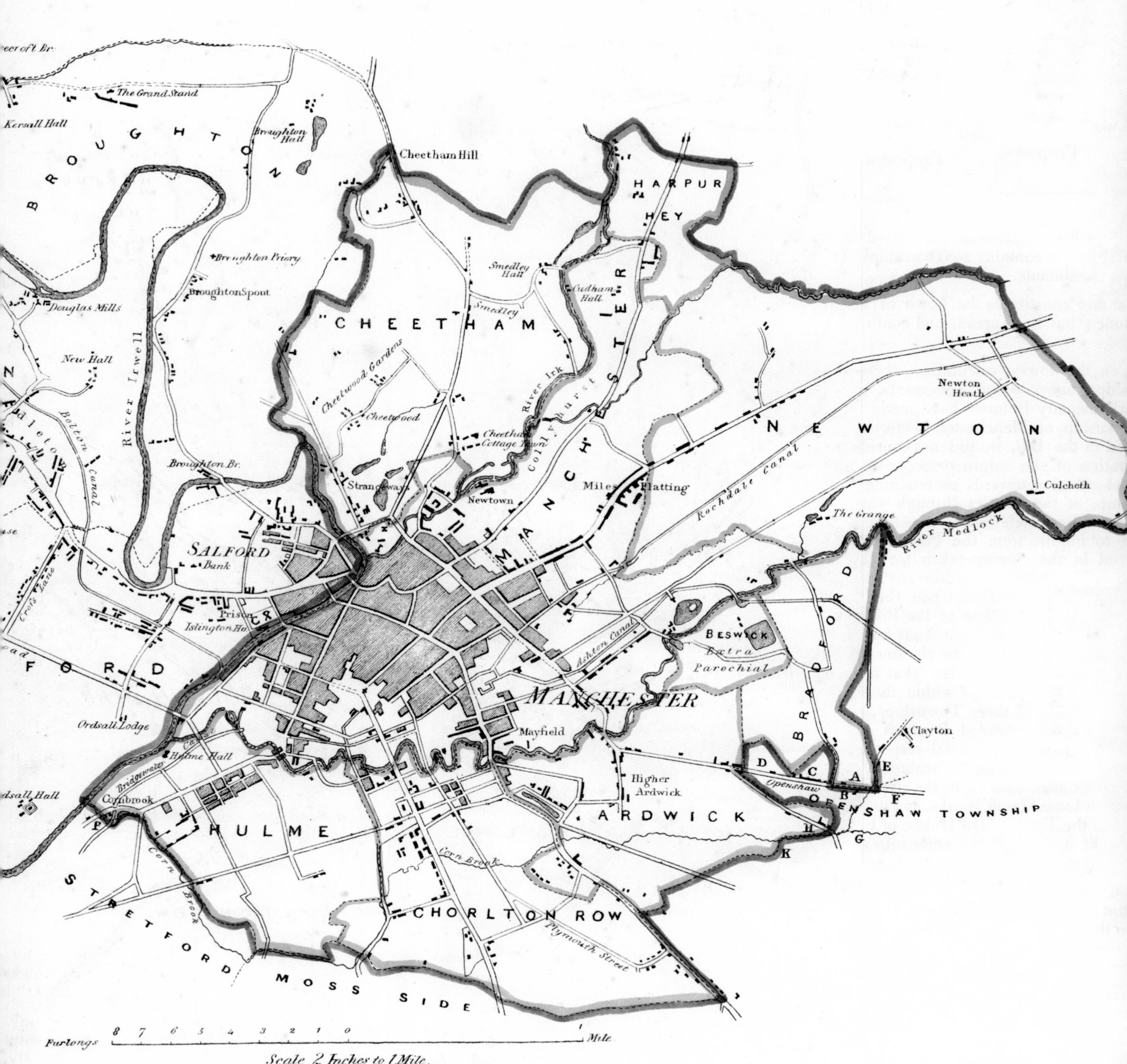
The Grand Stand
Kersall Hall
BROUGHTON
Broughton Hall
Cheetham Hill
HARPUR HEY
Broughton Priory
Smedley Hall
Cudham Hall
Broughton Spout
Douglas Mills
CHEETHAM
Smedley
New Hall
Cheetwood Gardens
Cheetwood
River Irwell
River Irk
Newton Heath
NEWTON
Cheetham Cottage Town
Collyhurst
Broughton Br.
Bolton Canal
Strangeways
Newtown
Miles Platting
Culcheth
Rochdale Canal
The Grange
River Medlock
SALFORD
Bank
Prison
Islington Ha.
Cross Lane
Ashton Canal
BESWICK
Extra Parochial
BRADFORD
MANCHESTER
Ordsall Lodge
Mayfield
Clayton
Hulme Hall
Bridgewater Canal
Higher Ardwick
Openshaw
Cornbrook
OPENSHAW TOWNSHIP
ARDWICK
HULME
Corn Brook
STRETFORD
CHORLTON ROW
Plymouth Street
MOSS SIDE
Farlongs 8 7 6 5 4 3 2 1 0 1 Mile
Scale 2 Inches to 1 Mile.

1832a

Lines on maps: Dawson and parliamentary boundaries

The Great Reform Act of 1832 generated an invaluable series of maps of towns across the UK, showing their existing and proposed parliamentary boundaries. Despite the political heat that it generated in Parliament and across the country, the outcome of the Act's aim to create a fairer representation of Members of Parliament and a wider electorate was a very limited revolution. It extended the right to vote only at the margin: the electorate remained wholly male and qualification to vote was based on the value of a resident's house. So, while the total electors rose from 435,000 to 653,000, this was nevertheless a tiny fraction of the then total England and Wales population of 14 million. However, the Act started a process that eventually led to universal suffrage in the twentieth century. The biggest change came in the geography of parliamentary representation, which went some way to acknowledge the impact of industrialisation on the demography of the country. It did this by classifying towns into categories: category A abolished parliamentary representation in 56 of the previous boroughs, including 'rotten' boroughs such as the notorious Old Sarum; category B reduced representation from two to one MP in 30 boroughs; category C created 41 newly enfranchised boroughs, 19 to have a single MP and 22 to have two MPs. The remaining 115 boroughs retained their existing representation. This process significantly changed the parliamentary geography: three-quarters of the old boroughs which were disenfranchised or which lost one of their MPs were in the South West and South East, whereas well over half of the newly created boroughs were in the three northern regions.

The Manchester region was a major beneficiary. Like other of the burgeoning towns in the north, Manchester became a new borough with two MPs, Salford became a new borough with a single MP. Elsewhere, in what is now Greater Manchester, Wigan, previously the area's only parliamentary borough, kept its two MPs; and six of the other main towns were newly

OPPOSITE. R. Dawson, *Manchester* (1832) [AUTH] The awkward boundary of Openshaw is on the lower right.

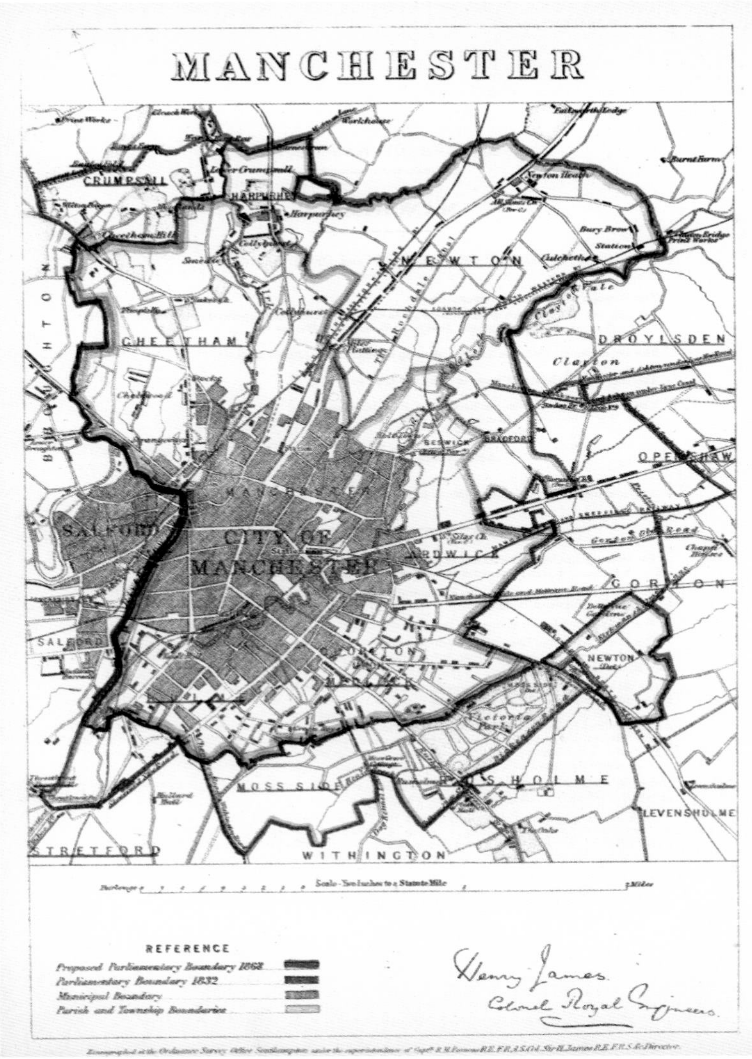

H. James, *Manchester* (1868) [AUTH] The 1868 extension of the parliamentary boundary.

created as parliamentary boroughs: Bolton, Oldham and Stockport each got two MPs and Ashton-under-Lyne, Bury and Rochdale each got one. From having only 2 MPs before reform, the area therefore had 14 subsequently.

Maps played an important part in determining winners and losers in this process. They were used to delineate the boundaries of the 'built-up areas' of existing boroughs and thereby to determine which should be abolished or suffer a reduction in representation. For the retained boroughs, their old and (where relevant) new borough boundaries were outlined, and for newly created boroughs their boundaries were delineated for the first time, often taking some account of existing administrative areas such as townships.

The maps are generally referred to as being by Robert Dawson who was appointed to the Boundary Commission with special responsibility for producing maps of the Commission's proposals. He signed his name – 'Rbt.K.Dawson, Lieut. R.E.' – on each of the reform maps. Dawson had been employed in the Ordnance Survey for over 40 years as a surveying draughtsman, following his father. For the great majority of his maps, he drew directly on existing published OS maps. However, for most northern towns there was then no existing OS coverage since, by 1831, the published OS 1-inch maps comprised 52 sheets covering only an area roughly south of a line from Cardigan to Ipswich together with much of Lincolnshire. In some cases, Dawson could draw on the original, as yet unpublished, OS survey drawings, which extended coverage further north. However, where OS surveys did not exist, he had to commission new surveys or draw on existing plans by private surveyors. It is not clear in every instance where his plans came from. Within the Greater Manchester area, the plans of Bolton, Bury, Rochdale and Wigan came from new surveys by Richard Thornton and James Smith, but the Manchester and Salford plans do not specify any origin.

The reports on Manchester and Salford were conducted by two Inspectors from the Boundary Commission and, as the map shows, they opted to use the existing townships as the jigsaw pieces to define the boundaries of the new parliamentary borough. Most of the built-up area lay within Manchester itself, but to this were added eight other townships; Cheetham, Newton, Harpurhey and the five townships south of the Medlock – Ardwick, Bradford, Beswick, Chorlton Row (Chorlton-on-Medlock) and Hulme. They noted that a good deal of new building had been taking place on the southern edge of the two townships of Hulme and Chorlton Row, in some cases stretching into Moss Side, and that the recent opening of a new road across Hulme to link with Chester Road was likely to encourage further future building within Hulme itself, 'in a direction not much built upon'. They agonised over the oddly-shaped portion of Openshaw, which ran between the townships of Bradford and Ardwick for just over half a mile. The configuration of the boundaries meant that someone living at point A in Bradford would be entitled to vote, whereas someone living directly opposite on the other side of the road at point B would not, nor would those living at points C and D, despite their being closer to the centre of Manchester than point A. However, given their focus on townships as building blocks, they used the existing township boundaries to define the borough and argued that there was no case to include either the whole of Openshaw or Moss Side. They also noted that the residents in Bradford had argued that their township should be included since most residents worked in collieries that sent coal to the cotton-mills and other factories in Manchester, a view with which the Inspectors agreed.

For Salford, the Inspectors used the same principle of focusing on townships as building blocks. They defined the new Salford parliamentary borough as consisting of the townships of Salford, Broughton and Pendleton. The sole issue they identified was the small (effectively, detached) portion of Pendlebury which lay to the north of the Irwell and which they concluded should be added to the three townships to form the new Salford borough.

Dawson continued his involvement with mapping, both with the 1835 Municipal Corporations Act and through his appointment in 1836 as an assistant commissioner on the Tithe Commutation Commission. Following the 1835 Act, Manchester applied to become a municipal corporation and this status was conferred on the town in 1838. Dawson died before the next major reform of parliamentary boundaries, in 1868, and his role was performed by Colonel Henry James, another Royal Engineer from the Ordnance Survey. Manchester was extended with parts of Openshaw and Gorton to the east, Moss Side and parts of Rusholme (covering Victoria Park) and an

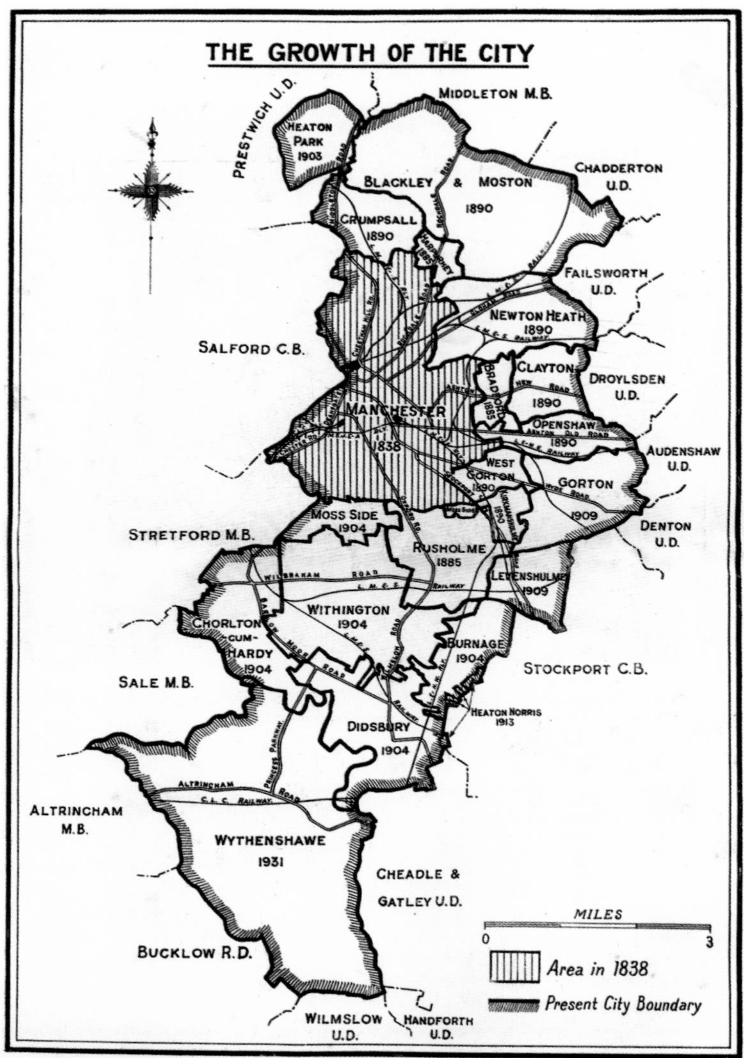

Manchester Guardian, *The growth of the city* (1938) [AUTH] The 1931 incorporation of Wythenshawe was Manchester's last incursion into Cheshire.

extension into Stretford to the south, and with a small part of Crumpsall to the north. This process of extending the town's municipal boundaries continued as its population grew, although significantly – with the sole exception of Wythenshawe in 1931 – never into north Cheshire which consistently resisted any attempted encroachment.

Plan
and
SECTION
OF THE
Manchester Race Course
Shewing the Situation of the Stables
and other
ACCOMMODATIONS
Including a Plan of the
KERSAL MOOR
BOOTHS
Sand Hills
F Ins
168. 7¾
White Hill
F Ins
177. 2½
Foot Road
Grand Stand
Turf Tavern
Bowling Green
Cottages
Bayley's Farm
Booths & Stands
Booths & Stands
Winning & Starting Post
Distance Post
¼ Mile Post
Foot Road
F Ins
177. 2½
SHOW FIELD
Show Field Brow
F Ins
172. 9¾
Stalls & Booths
Clough
Road to Prestwich
Print Works
Foot Road to Prestwich
To Bury
Toll Bar

1832b

Moving the starting line: 'Manchester' race courses

Horse racing in Manchester was something of a misnomer since all of the major racecourses that bore the town's name were in Salford. The sport's origin in the twin cities can be traced back to the seventeenth century, but it was not until the eighteenth century that regular meetings began to be organised on Kersal Moor in Salford. In spite of attempts to prohibit racing, motivated by the belief that it encouraged idleness and dissolute behaviour among the lower orders, racing survived and thrived. By the early nineteenth century, the three-day Whitsun meeting at Kersal Moor had become a much-anticipated part of the local holiday calendar, attracting people of all classes throughout the region. They came to watch and bet on the races, and to enjoy the attractions and diversions that turned 'Karsey Moor' into a boisterous fair. Controlling the crowds on an open course – 'to keep the Foot People out of it between the Distance and the Starting Chair'– proved difficult, at least until the course began to be roped off. Dogs were another problem, until their owners were warned that the animals would be destroyed if found on the course on race days. By the 1830s, as Twyford and Wilson's plan indicates, improvements had been made to the actual course as well as providing facilities including a grand-stand. Even so, policing the enormous crowds remained a challenge.

Partly in response to the popular carnival of the Kersal Moor races, the Earl of Wilton decided to establish a more select race course on his estate in Heaton Park. Thomas Egerton Wilton was a renowned sportsman and was to be one of the 'gentlemen jockeys' who rode at Heaton Park. Strict measures were taken to ensure that the meeting was socially exclusive, admission being by ticket and then only to those who arrived in carriages or on horseback. Puffed in the press as 'beyond dispute the best and most important private races in the kingdom' and even

---

OPPOSITE. Twyford & Wilson, *Plan and section of the Manchester race course . . .* (1832) [Salford Local History Library] The course lay in Kersal Moor in Salford.

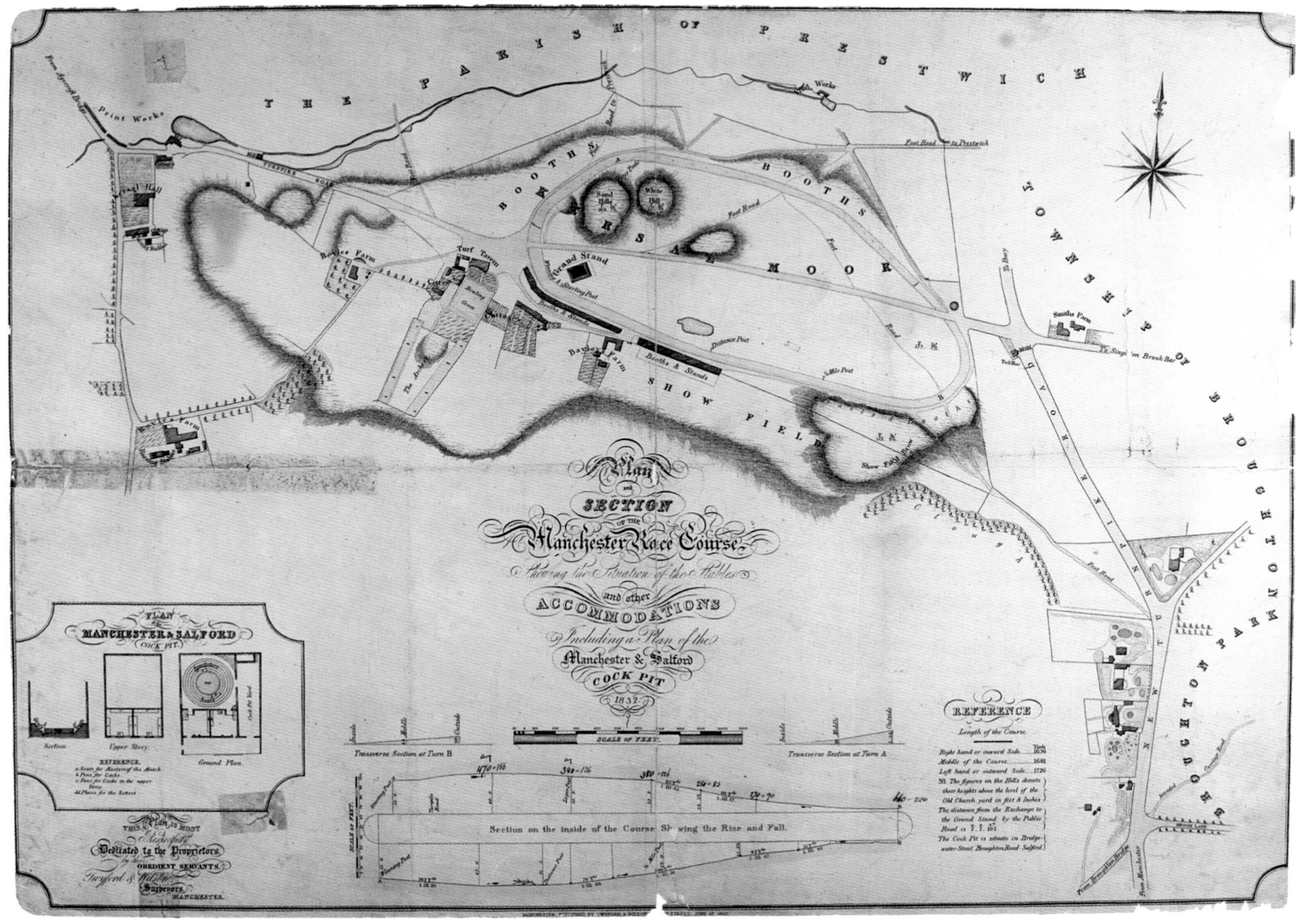

flattered as the 'Goodwood of the North', the course – a mile and a quarter in circumference – was rather basic in its facilities. In spite of inviting the aristocracy and gentry who had attended the St Leger meeting at Doncaster in the previous week, the Heaton Park meeting was unable to attract the breadth of support that was to make the Duke of Richmond's private course at Goodwood an important date in the calendar of the leisured class. Races were held at Heaton Park from 1827 to 1838, after which they were transferred to Aintree under the patronage of the Earl of Sefton and Lord Stanley.

Racing was also to come to an end at Kersal Moor. In 1846, the new joint owner of the land – Lieutenant Colonel William Legh Clowes – declined to renew the lease because of his moral objections to the sport. The racecourse was quickly re-established a short distance to the south, on land in the loop of the River Irwell, close to the Bury New Road, Higher Broughton. This land was owned by the Fitzgerald family and it took its name from the castellated house built by the family. Considerable work was required to open the Castle Irwell course in time for the Whitsun meeting in 1847. What some considered a picturesque location had its disadvantages, notably its proximity to the river, which made the course vulnerable to flooding. It had three grandstands, with the second-class stand capable of holding almost 5,000 spectators. Admission, however, was closely controlled, spectators having to pay for entry to the stands.

Horse racing continued to be popular. Local breeders and owners such as the Manchester cotton spinner Thomas Houldsworth were popular figures, better known for their racing than their political colours. Manchester claimed Kettledrum, the Derby winner of 1861 owned by Colonel Charles Towneley of Burnley, as its own. It brought good fortune to Edward Hulton who took 'Kettledrum' as his tipster name in the *Sporting Chronicle*, his racing sheet becoming the foundation of the newspaper empire based at Withy Grove, whose presses always found space and time to print the popular racing papers demanded by punters. Betting, of course, was an integral part of the sport. It had its critics but, as in earlier times, anti-gambling pressure groups failed to take account of the excitement and hope that betting gave to the lives of working people. Legislation aimed at curbing betting – the Betting Act of 1853 and the Street Betting Act of 1906 – had little impact on either the occasional or inveterate gambler.

Castle Irwell, however, proved to be a temporary home for the racecourse. When the estate passed into the hands of J.F. Purcell Fitzgerald, a temperance supporter, he refused to renew the lease. The Manchester Racecourse Company needed to find a new location. This was a 100-acre site between Regent Road and the River Irwell in Salford, a site they decided to purchase rather than risk renting. The new course was opened in time for the Whitsun races in 1868. It proved to be popular, with the main meetings – at Easter and in the autumn – attracting large crowds. Prize money was increased and new races introduced: the first November Handicap being run in 1876 and the first Lancashire Plate in 1888. However, in the early twentieth century the racecourse was compelled to move once again, this time not for moral pressures but because it stood in the way of plans to expand the Manchester Ship Canal docks. After tortuous negotiations, the Ship Canal Company

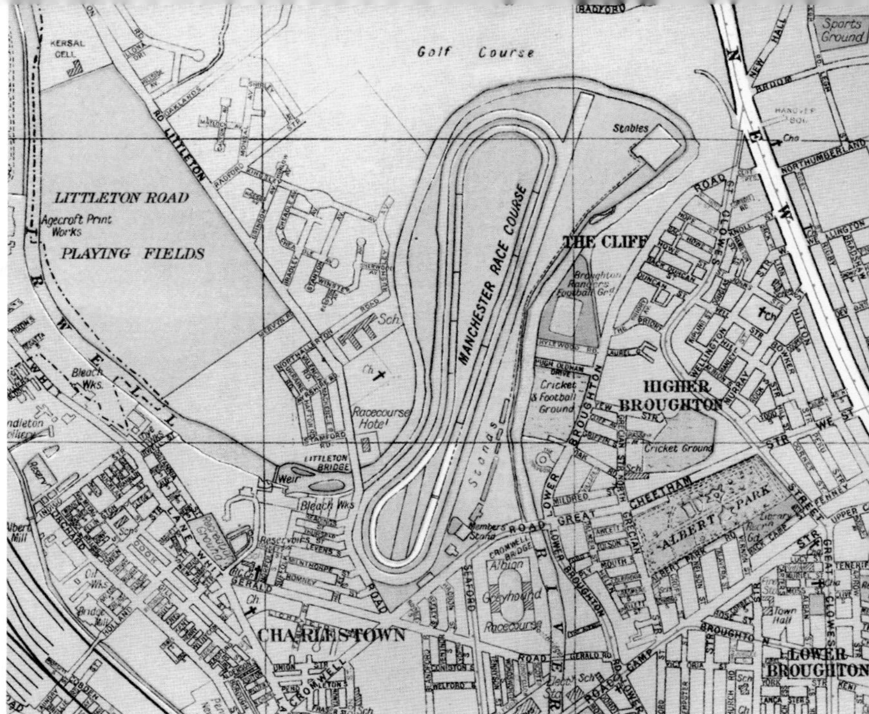

A. Gross, *Manchester* (*c.*1923) [AUTH] The Castle Irwell course was south of Kersal Moor.

purchased the racecourse site for £262,500. The Manchester Racecourse Company, under the chairmanship of John Edward Davies, decided to return to Castle Irwell, which they were now in a position to purchase. They demolished the Fitzgerald house and built a new course, stands and stables. The first race meeting was at Easter, 1902. The course was well served by Salford Corporation trams, although as in earlier times there were still those who preferred to walk. When in the early 1930s the November Handicap, regarded as the last major race of the flat racing season, was linked to the Irish Hospitals Sweepstake, it became one of the biggest gambling races of the year.

Manchester remained a popular course, and even innovative – in July 1951 it held the first evening race meeting in the country. By the 1960s, urban courses such as Birmingham and Manchester were attracting the interest of developers, and in November 1963 racing came to an end in Manchester when shareholders accepted an offer to sell to a London property company for £500,000. Manchester has remained without a racecourse from that time. Plans announced in 2004 to reopen a Manchester racecourse, once again in Salford, were successfully opposed, but on this occasion largely for reasons of traffic congestion and noise rather than moral concerns.

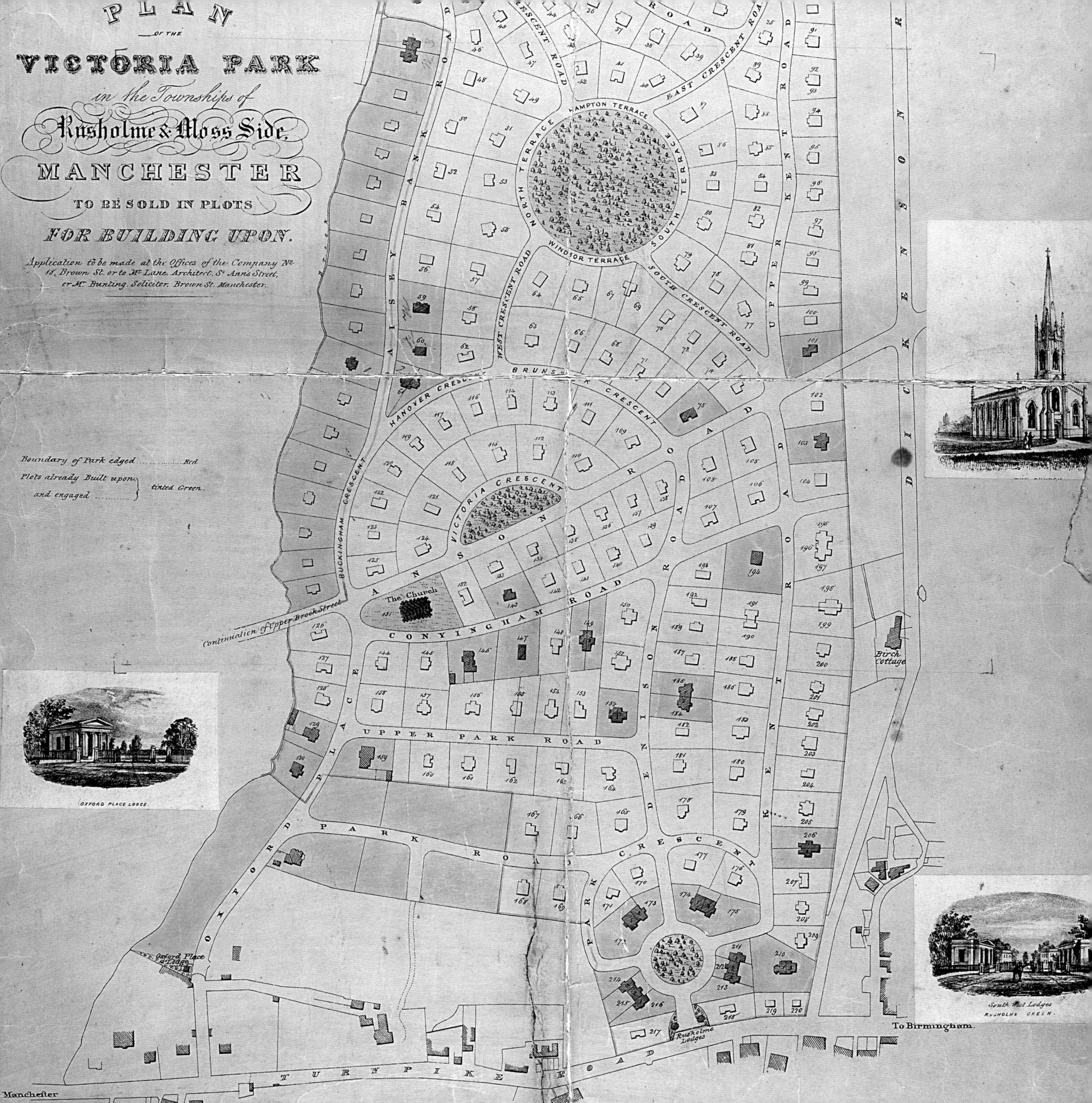
PLAN
OF THE
VICTORIA PARK
in the Townships of
Rusholme & Moss Side,
MANCHESTER
TO BE SOLD IN PLOTS
FOR BUILDING UPON.
Application to be made at the Offices of the Company No. 18, Brown St. or to Mr. Lane Architect, St. Anns Street, or Mr. Bunting, Solicitor, Brown St. Manchester.
Boundary of Park edged .... Red
Plots already Built upon and engaged .... tinted Green.
HAMPTON TERRACE
NORTH TERRACE
SOUTH TERRACE
WINDSOR TERRACE
EAST CRESCENT ROAD
WEST CRESCENT ROAD
SOUTH CRESCENT ROAD
CRESCENT ROAD
UPPER KENT ROAD
DICKENSON ROAD
AISEY BANK ROAD
BRUNSWICK CRESCENT
HANOVER CRESCENT
VICTORIA CRESCENT
BUCKINGHAM CRESCENT
ANSON ROAD
CONYINGHAM ROAD
The Church
Continuation of Upper Brook Street
UPPER PARK ROAD
PARK ROAD
PARK CRESCENT
DENISON ROAD
KENT ROAD
OXFORD PLACE
Oxford Place Lodge
Birch Cottage
Rusholme Lodges
TURNPIKE ROAD
To Birmingham
Manchester
OXFORD PLACE LODGE
South West Lodges
RUSHOLME GREEN

1837

Victoria Park: a gated enclave

As Manchester became ever more a place of warehouses and industry, its residential structure changed. Earlier, many merchant, manufacturing and professional families had created fashionable addresses in areas 'in town' such as Mosley Street. Over time, a growing number began to move away from the increasing smoke and bustle of the burgeoning town. Social segregation increased as the incentive to move grew ever more compelling. The Crescent in Salford, the higher ground of Broughton and the 'rural' environment of Ardwick Green were some of the areas in which the 'carriage class' began to settle. So, when Victoria Park was conceived in the 1830s, there was nothing new about the concept of 'suburban' living in Manchester. What was different was the Park's scale and that it was one of the earliest examples of a gated community planned to ensure privacy for those rich enough to afford it.

The Park was located on a large site bought in 1836 by a small group of individuals including the Manchester architect Richard Lane. Lane's plan in the 1837 Prospectus sketched out curving tree-lined avenues and generous plots on which large houses could be built. The Prospectus was clear in its appeal to those wishing to move away from town. It would meet the needs of the '. . . very considerable number of the inhabitants [who] must abandon their present abodes, in consequence of the rapid conversion of dwelling houses into warehouses . . .'

The venture was officially opened in July 1837 with a flourish that included a procession of the town's great and good followed by a banquet. However, the initial financing proved shaky. Not only did its use of a tontine scheme fail to meet the high cost of the land, but the cotton industry was slipping into one of its recurring depressions which lasted from 1838 to 1843. As early as August 1837, the Victoria Park Company met to end the tontine scheme and dispose of its holdings as best it could.

OPPOSITE. R. Lane, *Plan of the Victoria Park . . .* (1837) [MLA] The extract covers the western section. The views of buildings were pasted on as later additions.

Spiers suggested that until 1843 the Park consistently had more unoccupied than occupied houses and that occupied houses only grew from five to fourteen. However, by 1845 the landowners established a Trust aimed at preserving the area as a private park. Tolls were charged for vehicles passing through the entrance gates, initially as an attempt to prevent them from using the estate to bypass the tolls on the turnpike Wilmslow Road.

Under the Trust, the Park began to develop anew. By 1850, 65 houses had been built and by 1899 there were 128. Merchants and manufacturers dominated throughout the period, and German households grew from 15 per cent to over one-third by 1860. Professional households grew significantly after 1870 to reach 17 per cent by 1885. Charles Hallé, founder of the Hallé Orchestra, and Ford Madox Brown, who painted the murals in the Town Hall, both lived in Addison Terrace. Richard Cobden lived at Crescent Gate between 1845 and 1848. The Pankhursts lived in Buckingham Crescent in the 1890s. Summerville in Daisy Bank Road was briefly the home of Sir Harry Smith who became Governor of Cape Colony and was commemorated along with his wife by the naming of the South African towns of Harrismith and Ladysmith.

However, 1900 was the Park's zenith. From the turn of the century many smaller houses began to appear. In part, this was the inevitable outcome of the expansion of the city, but also a result of the machinations of Sir William Anson, Warden of All Souls in Oxford, who had inherited land in the Park and in Birch estate to the south. He sold some land to builders and in 1898 embarked on a protracted battle with the Trustees both over access to the Park via the Anson Road gate, and against the building tie that restricted houses to a minimum value. The sanctity of regulated access via the gates became increasingly hard to sustain, and numerous smaller houses were built north of Dickenson Road and in a gridiron pattern in the area of Langdale Road and Kensington Avenue.

The Park's seclusion was also undermined by battles over access for trams. The city's first proposal to build a tramway via Anson Road in 1902 was defeated, but in 1920 a second proposal was agreed in return for the city leaving the gates at Anson Road standing and paying the Park £100 per annum. The tram effectively split the Park in two. There was an even more convoluted tussle over access for cars, again prompted by Anson who demanded that residents in his Birch estate should be allowed to drive cars along Anson Road. The legal battle focused on whether cars should be considered as 'vehicles' or whether the original stipulation only applied to horse-drawn carriages. By 1905 the issue was settled in favour of cars, which were allowed access if they paid tolls.

It was inevitable that the Park began a rapid demise. In her biography of Manchester, Rachel Ryan bemoaned the outward march of the town which she saw as a series of stages through which the city gobbled up the once grand houses:

> When I was a child there was something wholly gloomy, for instance, about Rusholme . . . The rich had not deserted it entirely. A few families still lived in the largest houses in Victoria Park. But the third phase of the cycle was visibly approaching. Several big houses were empty and could find no purchaser; meanwhile their gardens grew rank and uncared for . . . I first felt the melancholy of departed splendour when I peered, not at ruined castles or still banners hanging in a deserted chapel, but through slatternly evergreens at the empty mansions of Victoria Park, Manchester.

From the turn of the century, more and more houses were acquired by institutions. The Xaverian College moved into Firswood in 1903. In 1905 Summerville became a Unitarian College. Numerous properties became student housing. Some were newly built, as with Dalton Hall, which in 1882 became the country's first purpose-built student hall. The sprawling complex of Hulme Hall, whose first block was built in 1907, replaced existing houses on Oxford Place. Other university halls simply used existing mansions, as with St Anselm's, which had been the home of Sir Arthur Schuster, the mathematician and physicist. Ashburne House was donated to the University as a women's residence. Langdale Hall – built in 1846 for Edward Langworthy, who owned a major cotton firm in

Ordnance Survey 25-inch (1893) [UML] Victoria Park. The contrast between the density of housing in the Park and the surrounding areas is dramatic.

Salford – became a student hall in 1910. Denison House had the unusual distinction in 1911 of providing a temporary home for over 50 women, including some of the country's most prominent suffragettes, who hid there to avoid filling in the census. By 1930 Anson Road had three nursing homes, three halls of residence, two University annexes, a hotel, a school and a home for Christian women.

The Park was eventually absorbed into the city and the Trust formally abolished in 1954. Since 1972, a conservation area has demarcated part of the original extent of the Park, excluding the areas of smaller houses in the east and south, and the restaurants, takeaways and kebab houses along the so-called 'curry mile' on Wilmslow Road. A number of the buildings are formally listed (including the Grade I Arts & Crafts Christian Science church, designed by the architect Edgar Wood). Inevitably, however, it has become increasingly difficult to visualise the mansions set in secluded green splendour that the Park was once able to boast.

CATHEDRAL
CHURCH
COLLEGE
INFIRMARY
SMITHFIELD MARKET
EXCHANGE
ARCADE
OLDHAM STREET
GREAT ANCOATS STREET
PICCADILLY
LONDON ROAD
MARKET STREET
WITHY GROVE
HANGING DITCH
CATEATON ST.
CORPORATION STREET
SHUDE HILL
CHURCH STREET
HIGH STREET
KING STREET
JOHN DALTON STREET
PETER STREET
Free Trade Hall
MOSLEY STREET
LOWER MOSLEY STREET
OXFORD STREET
PORTLAND STREET
GREAT BRIDGEWATER STREET
BRIDGEWATER STREET
ROCHDALE CANAL
ASHTON CANAL
STOCKPORT
Birmingham
Travis
Tipping
MEDLOCK
SWAN STREET
LONG MILLGATE
TODD STREET
CHARTER STREET
ST. MARY'S GATE
DEANSGATE
COOPER STREET
ST. PETERS SQUARE
YORK STREET
Parker St.
George St.
Sheffield
Ashton
Granby
Eagle Quay
Russell
A
B
C
D
F
G
H
I
O
P
W
3
6
7
8
9
10
12

1849

The scourge of cholera

Maps were to become an important tool in the investigation and analysis of disease in the nineteenth century. In Britain, it was the first cholera epidemic of 1831–2 for which doctors began to use maps to record the location of cases as they struggled to understand the causes of the disease.

By the summer of 1831, Britain was preparing itself for what seemed to be the inevitable arrival of cholera, the disease having advanced inexorably westwards across Europe. As in other towns, Manchester's response was guided by the recommendations of the central Board of Health. A Special Board of Health, comprising doctors, clergy, leading citizens and local government officials, was established in the town. It faced an enormous task. Manchester was in the grip of unprecedented industrial growth, its population continuing to expand at an alarming rate. The preliminary returns of the 1831 census indicated the population of Manchester, Salford and the adjoining suburbs numbered some 238,000 compared to 98,000 in 1801, making it the largest urban area outside London. The Special Board set about its preparations by removing waste and filth that had accumulated over the years. Hospitals were established to isolate and treat the diseased, while plans were put in hand for the swift and safe burial of cholera victims. In carrying out this work, the town was divided into districts, members of the Board being appointed to investigate and report on living conditions in each area. Maps were used in this work, but the existing published maps were often of limited use because of their scale and the speed of the town's growth, which meant that they did not record the many new streets that had been built. Fortunately, Bancks's large-scale 1831 map of Manchester and Salford, published in January 1832, provided the Special Board with detailed up-to-date cartography.

Doctors were to be one of the few middle-class groups who

OPPOSITE. J. Leigh and N. Gardner, *Cholera districts in Manchester* (1849) [CHETS]
The incidence was highest in the poorest areas of the city.

came into direct contact with cholera. For some of those at the sharp end of the crisis, it provided them with the opportunity to study the disease more closely and advance their own theories about its causes and transmission. In Manchester, these included the physician, James Phillips Kay, who served as honorary secretary of the Special Board. His investigations into the town before and during the epidemic led him to write *The Moral and Physical Condition of the Working Classes Employed in the Cotton Manufacture in Manchester*, of which two editions were published in 1832. It made innovative use of the statistical data collected by the Special Board to describe and analyse the wider social problems of the town and has rightly been described as 'one of the cardinal documents of Victorian history'.

Another Manchester doctor who was to make use of the findings of the Special Board was Henry Gaulter, honorary physician at the recently established Chorlton-on-Medlock Dispensary. His interest in the disease led him to travel to Sunderland where the first cholera cases in England were reported in October 1831. He and other doctors were now able to confirm the terrifying nature of the disease that had been reported by their colleagues on the Continent. As cholera spread, they became all too familiar with the different stages of the disease, which began with violent vomiting, watery diarrhoea and dehydration, followed rapidly by collapse and then, in many cases, death. Detailed as the medical descriptions of the disease became, there was no agreed explanation of its cause and no agreed treatment. To the frightened poor, what the medical profession called *cholera morbus* was simply the blue vomit.

The first cases of cholera in Manchester were confirmed in May 1832, and by the end of the epidemic over 600 people had died. Gaulter's interest in the disease led him to publish a study of the Manchester epidemic in order to establish its cause. He carried out a close analysis of the first 200 cases of the disease based on the information recorded in the register of cases kept by the Special Board, which he supplemented by interviewing families of the dead. He also plotted the location of where the 200 individuals had died on a hand-drawn map, which indicated that the disease was particularly concentrated in the poorer working-class districts of the town, identifying densely populated pockets of the old town such as Allens Court, as well as some of the more recently built areas of the town. These included 'Little Ireland', close to the Medlock, where living conditions had been so appalling that the Special Board had carried out a detailed investigation before cholera arrived. Gaulter concluded from the evidence of the Manchester epidemic that the prevailing miasmatic theory was the most convincing explanation of how the disease started and was transmitted.

Gaulter's pioneering research was accompanied by an 'illustrative chart' that was published in *The Origin and Progress of the Malignant Cholera in Manchester* in 1833. It was one of numerous such studies into the disease and received little notice. He was not the only doctor to map the disease in a local community in the 1832 epidemic, but his methodology was an early attempt to establish and understand the spatial and social epidemiology of the disease. Gaulter did not live to see further developments in the mapping of epidemic diseases as he died in 1833, not from cholera but consumption.

When cholera returned to Britain in 1848, doctors were more familiar with the idea of mapping disease. In 1848, Thomas Shapter's retrospective study of the first cholera epidemic in Exeter also included the mapping of cases. In the following year, Manchester suffered its second and more lethal cholera epidemic. It was investigated by John Leigh and Ner Gardiner whose published account was also accompanied by a map, though in this case individual cases were not marked on it, rather those areas of the town where the disease had been most severe were identified in colour marked on one of Slater's directory plans. These and other studies of cholera are now largely forgotten. It was to be John Snow's mapping of Broad Street in Soho in London during the cholera epidemic of 1853–4 that provided evidence to support the then contentious hypothesis that cholera was a waterborne disease that was to be recognised as the landmark study in medical cartography in the nineteenth century.

OPPOSITE. T. Physick, *Chart . . . of the first 200 cases of the cholera in Manchester* (*c.*1833) [Royal College of Physicians of Edinburgh] A dot map was a novel methodology for analysing epidemics in the 1830s.

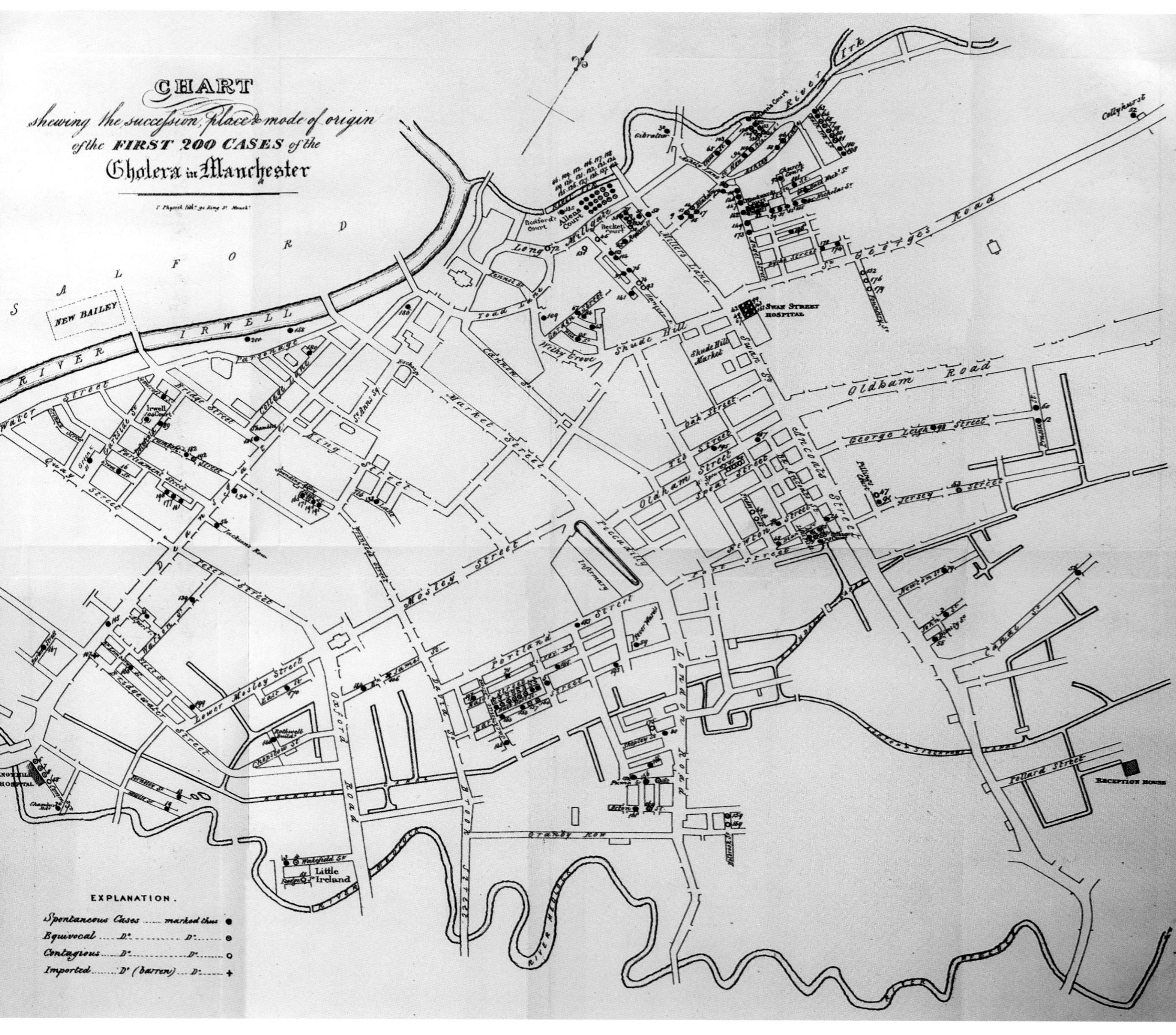
CHART
shewing the succession, place & mode of origin
of the FIRST 200 CASES of the
Cholera in Manchester
SALFORD
NEW BAILEY
RIVER IRWELL
River Irk
Collyhurst
Gibraltar
Allens Court
Becket Court
Long Millgate
Todd Lane
Withy Grove
Cannon St
Millers Lane
St Georges Road
Shude Hill
Shude Hill Market
Swan St
SWAN STREET HOSPITAL
Oldham Road
George Leigh Street
Oak Street
Tib Street
Spear Street
Oldham Street
Ancoats Street
Jersey Street
Newton Street
Port Street
Newton St
Canal St
Piccadilly
Infirmary
Market Street
Exchange
St Ann's Sq
King Street
Parsonage
Bridge Street
Water Street
Quay Street
Irwell Court
Jacksons Row
Peter Street
Mosley Street
Princess Street
Portland Street
Lower Mosley Street
Bridgewater Street
Watson St
Oxford Road
London Road
Granby Row
Pollard Street
RECEPTION HOUSE
KNOTT MILL HOSPITAL
Wakefield St
Little Ireland
RIVER MEDLOCK
EXPLANATION.
Spontaneous Cases ...... marked thus
Equivocal ...... D° ...... D°
Contagious ...... D° ...... D°
Imported ...... D° (barren) ...... D°

S^T. ANNS WARD

S^T. JAMES PART 1

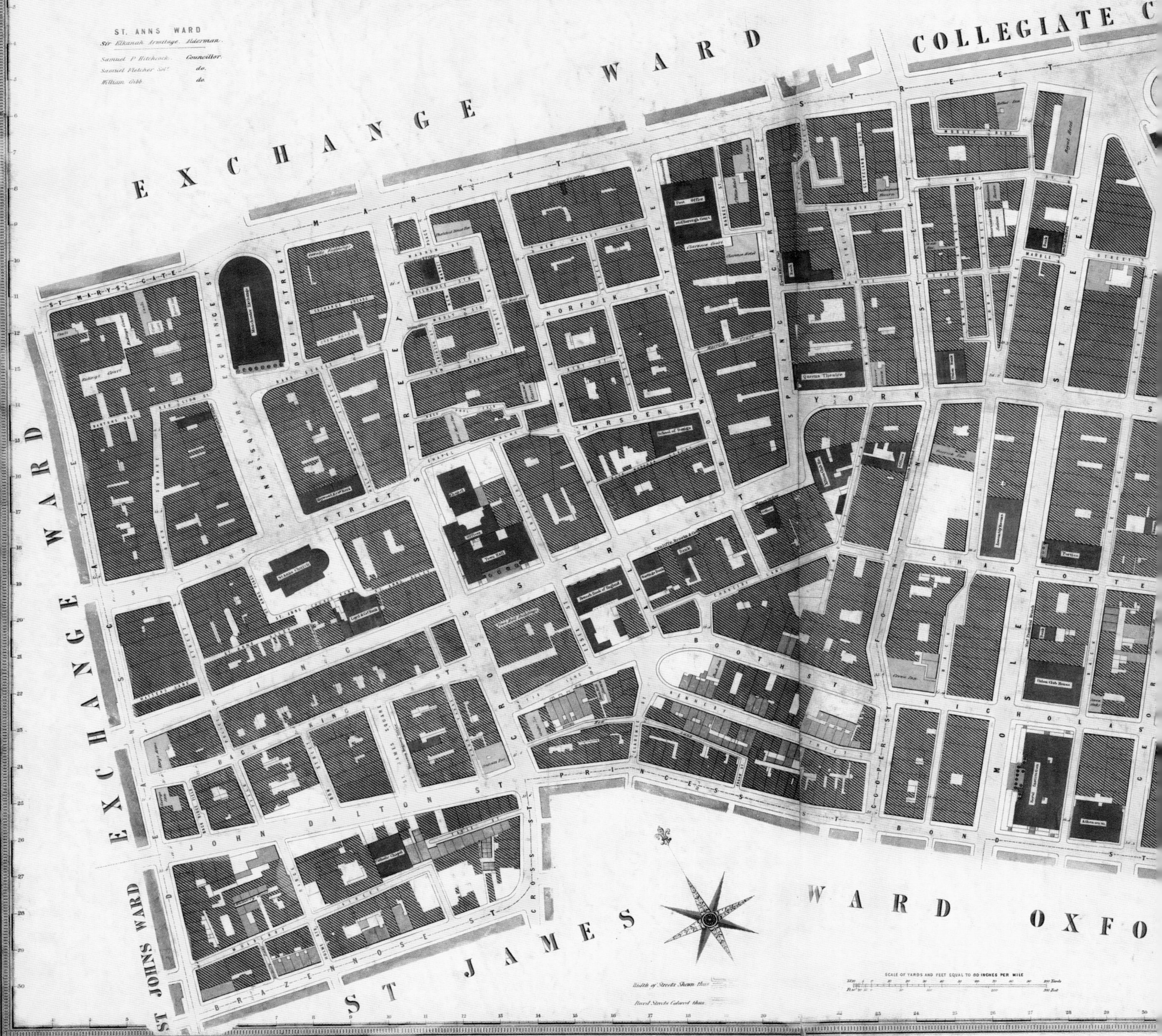

Published by Joseph Adshead, N.o 45, George Street, Manchester 1850 Entered at Stationers Hall

1851a

Adshead's maps: a flawed masterpiece

Adshead's plan of the township of Manchester is a quite extraordinary piece of work: on one hand. it provides superbly detailed and handsome cartography of the town; on the other, a curious mish-mash of sheets that are far from easy to piece together. The map was published at the huge scale of 80 inches to the mile (1:792) and was produced on 23 sheets together with a reference map of the whole township. Completed in 1850 and published in 1851, it was the work of two local men. Richard Thornton completed a survey of the township in 1849, and this was revised and published by Joseph Adshead in 1851. Adshead was an unlikely man to have got involved. Born in 1800, he settled in Manchester around 1820 and played a significant role in local affairs as a hosiery merchant, reformer and pamphleteer. He was part of the consortium that developed Victoria Park, joined the Council of the Anti-Corn Law League, wrote and campaigned on a wide variety of social issues, most notably lobbying for prison reform and for a convalescent hospital in Manchester. He served as alderman for St Ann's Ward. This seems a curious background to lead him into map publication, but he deserves immense gratitude for having done so since the Thornton/Adshead plan is a singular contribution to the cartography of the town. The printing of the map was by two further local men, George Bradshaw (of railway timetable fame) and his printing partner William Blacklock.

The map uses the nine municipal wards as its organising focus so that each sheet stops at the boundaries of a ward. The detail is impressive. Five types of building are distinguished, each demarcated either with hachuring or colour: public buildings, typically churches, schools and hospitals; warehouses and businesses; mills and works, virtually all being specified as cotton-mills, or dye-works, or manufactories, or ironworks, and many shown with the owner's name; hotels, inns and

OPPOSITE. J. Adshead, *Adshead's twenty four illustrated maps of the Township of Manchester* . . . (1851) Sheet 1 [Digital Archives] Mosley Street is on the right.

public houses; and private houses. The sheets name virtually every street (and often the names of terraces of houses), record the width of some streets, demarcate pavements, and list the names of the aldermen and councillors in each ward. They provide great detail of the railway stations, for example showing the departure and arrival platforms at London Road Station. They depict details of some of the larger parks and even outline the layout of some private gardens and ponds, and include examples of street lamps such as at the junction of Market Street and Victoria Street. The map also notes the location of the homeopathic hospital in Bloom Street, which will have appealed to Adshead since homeopathic medicine was one of his causes. Above all, with their colour and confident engraving and calligraphy, the 23 sheets are handsome and seductive works of art.

They reveal much of what the township had become by mid-century. For example, Sheet 1, shown above, covers much of the town centre including the Exchange, St Ann's and the Infirmary. By mid-century, it was virtually wholly given over to commerce and business. Mosley Street, which had earlier been a desirable residential location, was now overwhelmed by public and commercial buildings, including the Portico Library, the Royal Institution, two banks, the Royal Hotel, the Union Club House and the Assembly Rooms. A lone private house sat on the corner of Marble Street. It also shows some of the most significant public buildings that were later to be replaced – the old Town Hall on King Street and the Manchester Exchange. It includes two notable banks: in St Ann's Street, Heywood & Co.'s Bank (established by Benjamin Heywood in 1788 and long run by members of the Heywood family until it was acquired by the Manchester & Salford Bank, which in turn became Williams Deacon's Bank and was acquired by the Royal Bank of Scotland in 1930); and in a Georgian house in King Street, Lloyd & Co., which eventually became a constituent part of the National Westminster Bank.

Sheet 11 covers the medieval core of the town and includes the Cathedral, the Corn Exchange, Chetham's College, the Free Grammar School and Victoria Station and the earliest Jewish Synagogue in the town. Sheet 10 covers the array of specialist markets in the old shambles – including fruit, fish, poultry and meat, flour meal and cheese. Both sheets show the profusion of inns in the old town centre, including the 'Rowland Hill Tavern', no doubt reflecting the popularity of Hill's 1840s' reform of the postal system, which introduced prepayment using postage stamps and the universal penny post.

Sheets 14 and 15 cover much of Ancoats and show that the town's manufacturing economy was based on much more than textiles alone. As well as the huge textile mills fronting the Rochdale Canal (notably McConnel & Co. and Murray's Mill), there were flint glass and bottle glass works, ironworks and chemical works. Another sheet – number 18, which covers the districts of Newton, Bradford and Beswick – shows numerous small coal pits, some of which have old tramway lines linking the pits to the Ashton Canal.

The combination of a wide range of manufacturing, the often-enormous mills and the coal through which machines were generally powered provided the mixture that generated the smoke, noise and general squalor that so astounded visitors to the town. What is apparent from Adshead's plan is that there was really no part of the then township of Manchester that did not benefit – or suffer – from the impact of manufacturing. Even in its outskirts there were factories, quarries and mines. For example, close to the township border Queen's Park (where Sheet 23 shows details of the layout of gardens and lakes, and also the cheek-by-jowl provision of a playground for females and a separate playground for males), there were brick-works, a brewery, cotton-mill and dye-works. And in nearby Collyhurst, Sheets 20 and 21 show the huge Collyhurst Sandstone Quarry and Delph Pits, which provided the reddish stone with which many of the town's important buildings were built (including some of its historic edifices such as the Cathedral and St Ann's in Manchester, and Holy Trinity in Salford).

So not only is the plan a handsome piece of art and cartography; it also has much from which we can learn about the mid-century town. However, it also has infuriating limitations. Since it covers only the township of Manchester, it omits Salford, and to the south it stops at the meandering line of the

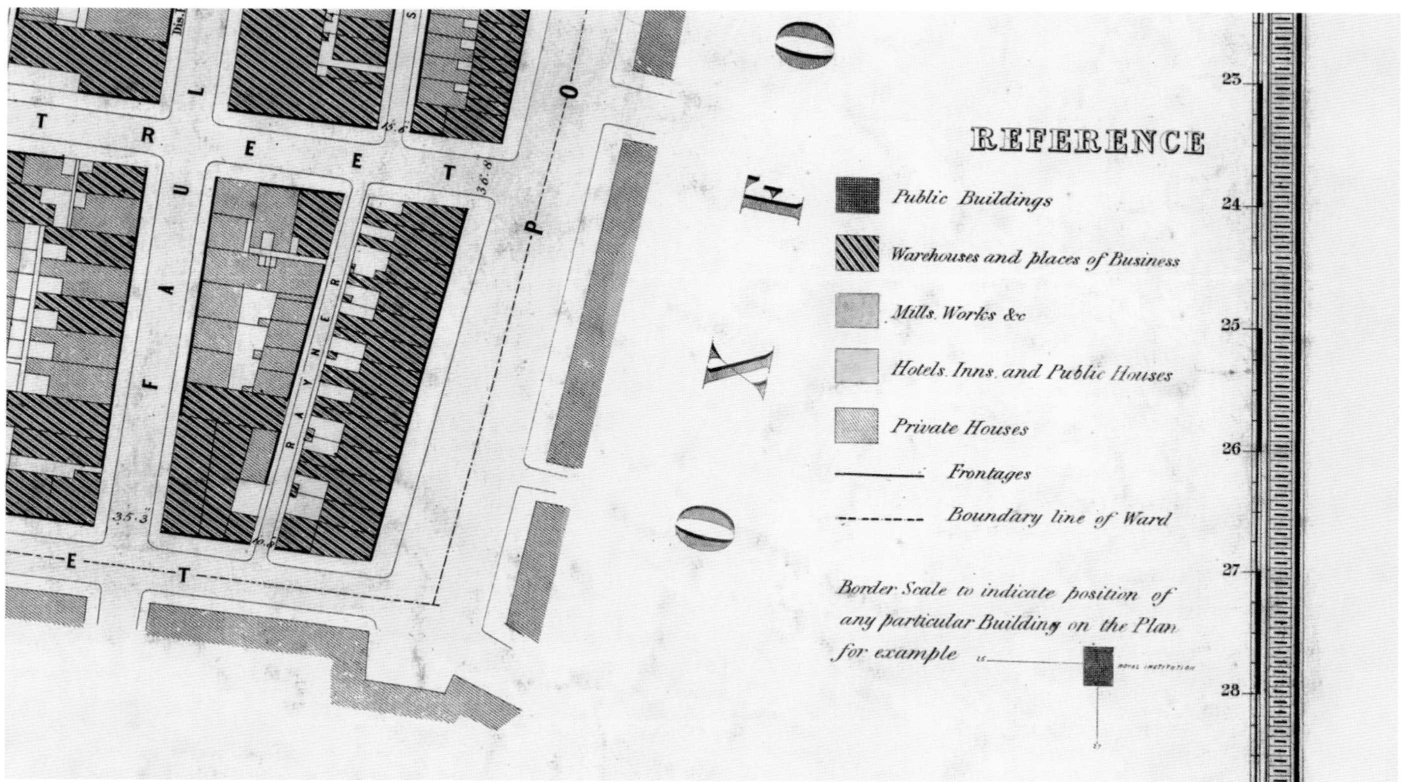

Adshead's key to categories of buildings

Medlock, thereby excluding the fast-expanding townships of Chorlton-on-Medlock and Hulme. The original idea had been to expand the coverage beyond the Manchester township, but that aspiration was doubtless dashed by the fact that Thornton died in 1851 and the Ordnance Survey's 60-inch plan was published in that same year. However, the comparison between Adshead and the OS is rather telling. The OS plots the town in a series of 49 sheets, each of which abuts unambiguously onto its neighbours to form a continuous plan with each sheet oriented identically to the north. Adshead, by contrast, plots each ward separately and, with the single exception of St Ann's, on more than one sheet. St Michael's Ward needs five sheets and New Cross needs no fewer than six. St James's Ward is not only shown on two sheets, but one of these is forced to split off the most southerly part of the Ward and show it on the right-hand side of the sheet. Furthermore, there is no consistency in the orientation of the 23 sheets: north-north-east is at the top of eight sheets; north-west is at the top of six; north-east at the top of three; north-north-west at the top of two; west-north-west at the top of a further two; and west and east-north-east at the top of the final pair. This is the inevitable consequence of having decided to show wards rather than simply divide the whole plan into a grid of contiguous rectangles. It means that, in spite of the key map, trying to put together adjacent wards is likely to bamboozle most users of the map.

OVERLEAF.
Sheet 11 [Digital Archives] The highly mixed land uses of the old core of the town.

TRINITY
RIVER
HUNTS BANK
WALKERS
Palatine Hotel
Chetham College & Library
College Grounds
Palatine Livery Stables
Free Grammar School
College Gates
Old Swan
Sun Tav.
Bucks Co.
Prices Co.
Spread Eagle
Black Moors Head
Old Pack Horse
Head Masters Ho.
Free Grammar Sch.
Apple Market
LONG MILL
Pheasant Tav.
Bakehouse Co.
White Lion
Heltons Co.
Farriers Court
Marsden Court
Cathedral
FENNEL STREET
Lamp
Victoria Bridge
Victoria Terrace
CATHEDRAL CHURCH YARD
Hanging Bridge
Commercial Hall
Crown & Anchor
Church Gates
CATEATON ST
HANGING DITCH
HALF STREET
Crown & Thistle
Hope & Anchor
Hope & Anchor Co.
No. 2 Court
No. 1 Court
Wheat Sheaf
Corn Exchange
White Horse
Dog & Patridge
TODD
Old Boars Head
Stables
Hydes Cross
WITHY
Huntsmans Co.
British Queen
Fleece Tav.
EXCHANGE

ST MICHAELS
HILL
MILLER STREET
HANOVER STREET
STREET
CHARTER STREET
MAYES STREET
REDFERN STREET
EDWARD STREET
NORTH STREET
CROSS ST
NEW STREET
BRADSHAW ST
GATE
IRK
CROFT
Irk Cotton Mill
Collier & Tiddens Smallwares Mill
Union Tav.
Crown and Shuttle Vault
Bedfords Co.
Grammar School Corn Mill
Cathedral Schools
Jews Synagogue
Manchester Arms
Chapel
Royal Oak
Old Balloon
Burton & Fullerton's Timber Yard
C. Steadmans Saw Mill
Timber Yd.
Wilton Arms
Cock Inn
Smallwares Mill
Old George & Dragon
Castle & Falcon Tav.
Houghtons Ct.
Back Cross Street
Chapel
Nags Head Tav.
Duke of York
J. Williamson's Saw Mill
Corn Mill
Higher Turks Head
Back Redfern Street
Back Balloon Street
Back Hanover Street
Balloon Street
Timber Street
Garden Street
Back Garden Street
Chapel Place
Snow Hill
Riga Street
Mayes Street
Clock Alley
Back Clock Alley
Howarths Gates
Bodens Buildings
Pilgrim Street
Carter Street
Parsons Court
New Gates
School Co.
Hill & Co's Dye Works
Bay Horse Tav.
Wool Pack
Dew
Back Mayes St.

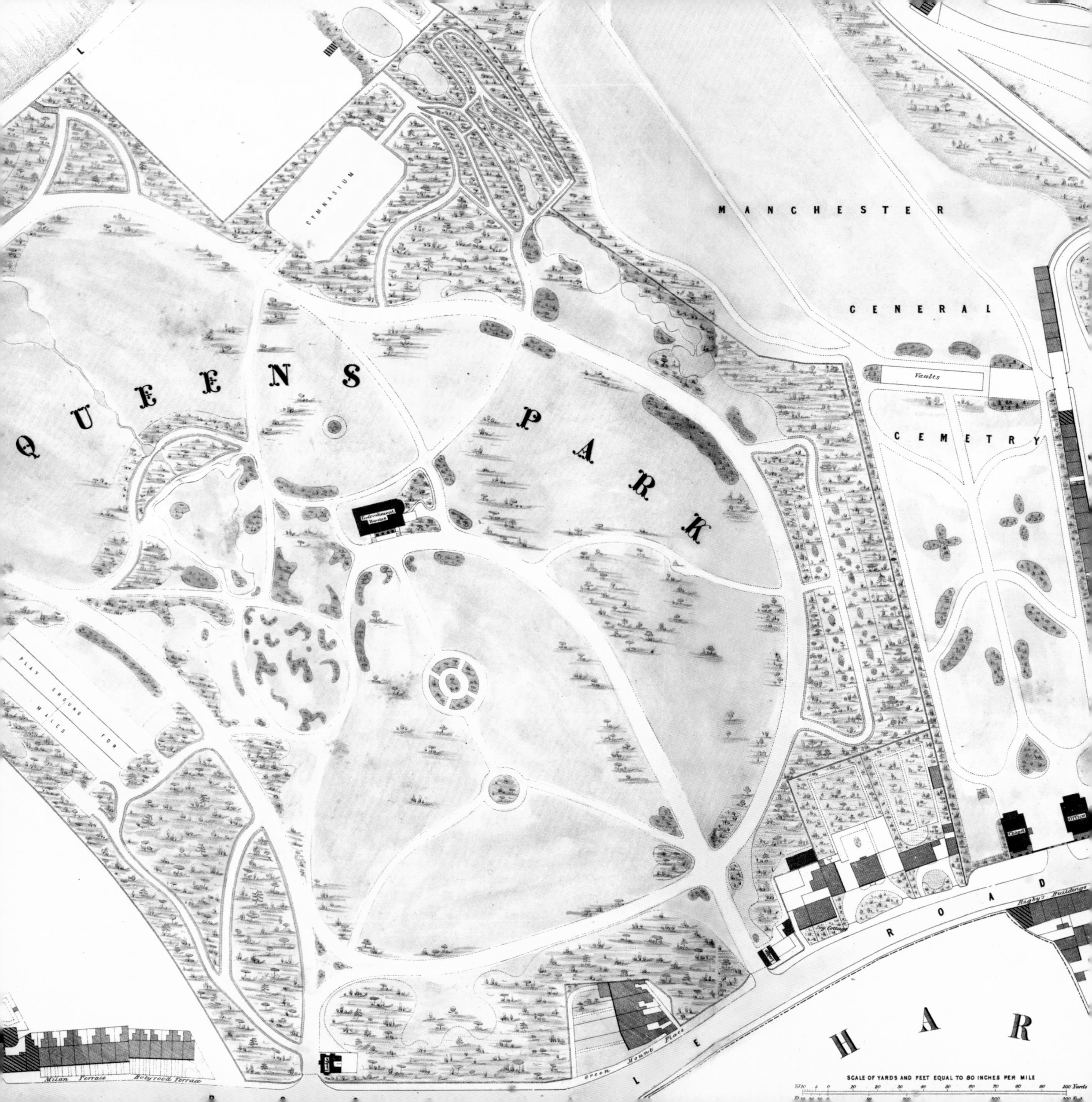

GYMNASIUM
MANCHESTER GENERAL CEMETERY
Vaults
QUEENS PARK
Refreshment Rooms
PLAY GROUND FOR MALES
Chapel
Office
Rigby's Buildings
Ivy Cottage
ROAD
Green Mount Place
Milan Terrace
Holyrood Terrace
HAR
SCALE OF YARDS AND FEET EQUAL TO 80 INCHES PER MILE

1851b

Municipal parks: health and civic pride

By 1914, the presence of the local state was impossible to ignore. To walk around any of the major towns was to pass through a municipal landscape that included market halls, fire stations, gasworks, schools, libraries, art galleries, tram sheds, abattoirs and, of course, town halls. Many of these buildings announced their ownership by displaying the municipal coat of arms, carved in stone or terracotta, and foundation stones that disclosed details of forgotten public ceremonies. Public parks were to be counted as one of the earliest features of this landscape of civic progress.

Edwardian Manchester had some 70 parks and recreation grounds covering an area of 1,480 acres, almost 7 per cent of the land in the borough. These were cherished and celebrated amenities in a city which, moreover, could claim a special place in establishing the idea that parks were part of a civilised urban world. Witnesses from Manchester had been prominent among those giving evidence to the *Select Committee on Public Walks* (1833), the first official inquiry into the shortage of publicly accessible open spaces in towns that had 'grown in a hurry' during the industrial revolution. James Phillips Kay and Joseph Brotherton were among those who argued that the provision of open spaces would do much to improve urban life, especially by providing alternative recreation spaces for the working classes.

The 1840s saw Manchester and Salford emerge as leaders in the movement to establish public parks. In 1844 a campaign, supported by charitable subscriptions, was launched to provide public parks. Funds were raised to purchase three blocks of land, which became Peel Park (32 acres) located on land close to the centre of Salford, Philips Park (30 acres) located in Bradford, on the eastern boundary of the working-class district of Ancoats, one of the most densely populated parts of the town, and Queens Park (31 acres) located in Harpurhey, which at that date was not part of the municipal borough. Their opening on 22 August 1846 was a red-letter day, regarded as evidence that Manchester was taking action against the problems that some observers felt

---

OPPOSITE. J. Adshead, *Adshead's twenty four illustrated maps* . . . (1851) Sheet 23 [Digital Archives]

were likely to overwhelm the community. Joshua Major, a Leeds-based landscape gardener, had won the competition to design all three parks. His plans provided formal areas of flower beds and serpentine walks, as well as spaces for children's games.

Public parks were new and intricate physical and social spaces. The rules introduced to govern the behaviour of users resulted in them being seen as part of a wider strategy of rational recreation promoted by an urban middle class alarmed at the behaviour of a large and potentially dangerous working class. In the more fanciful rhetoric of the public platform, parks were agents of civilisation encouraging public responsibility and social cohesion, and offering amenities that provided the working classes in particular with an alternative to pub-based leisure. Such ideals seemed to be borne out when in 1857 Queen Victoria visited Peel Park to be greeted by 50,000 well-behaved Sunday school children who sang the national anthem.

Queen's Park in Harpurhey was the most remote of the original parks, 2 miles from the city centre. It was situated between the Rochdale Road and the River Irk, adjoining the Manchester General Cemetery, one of the commercial cemeteries opened in response to the burial crisis in the city. The park's 30 acres of undulating land included a lake, and Adshead's map shows the serpentine walks that were a feature of Major's parks. New landscaping was to be introduced over the years, but the most obvious alteration to the park came in 1884 with the opening of what was the city's first purpose-built art museum in the centre of the park, on the site of the original Hendham Hall.

Manchester stayed in the vanguard of the parks movement. Municipal parks proved to be dynamic institutions, and over time the initial assumptions about their social purpose were to be modified. Evidence of this change was seen in the larger parks providing amenities such as bowling greens, swimming pools and football pitches. Given the city's notorious pollution, keeping parks flourishing and attractive was a considerable challenge, not least in Philips Park, which by the early twentieth century could best be located by looking for the chimneys of the huge adjacent Bradford Gas Works. Land continued to be acquired for new parks, but the cost of land in or close to the city centre meant that the larger parks were located in the suburbs. In some cases, such as Alexandra Park (opened 1874), this meant establishing parks beyond the municipal boundary. The incorporation of Moston and Blackley in 1890 was to result in the wild landscape of Boggart Hole Clough becoming a municipal park. By far the largest of these acquisitions was in the very north of the city where Heaton Park was purchased from the aristocratic Wilton family for £230,000 in 1902, the Council having failed a few years previously to purchase Trafford Park from Sir Humphrey de Trafford. However, this growth in parks did not mean that all districts had easy access to a major park; for example, Hulme, whose population exceeded 70,000 by the 1880s, did not have its own public park.

By the beginning of the twentieth century, progress was also made in providing open spaces in the city centre. The closing of churchyards led to the creation of what a later generation of landscape designers would refer to as pocket parks. The demolition of buildings also offered the opportunity for open spaces; for example, in the case of Piccadilly the removal of the Infirmary created a temporary space that was eventually to become permanent. One of the most expensive parks to be opened in Manchester in the early twentieth century was at the junction of Sackville Street and Whitworth Street, a 1-acre area that cost £26,000.

During the Victorian period, the public park became one of the defining institutions in a rapidly changing urban landscape. Before the Victorians, there were no municipal parks in any provincial town; by the time of the Edwardians, most towns boasted at least one major landscaped park with a familiar array of features including conservatories, bandstands, fountains and statues. They represented a remarkable change in land use. In Manchester and elsewhere, they had become regarded as essential amenities that it was the duty of the Corporation to provide and maintain. For many, they were much more than this; they were outstanding symbols of civic pride.

OPPOSITE. Manchester Corporation (1904) [MLA] Philips Park. The Medlock flows though the centre of the park, and the Rochdale and Ashton canals are respectively above and below it.

MANCHESTER CORPORATION.

Plan to accompany Parks Committee's Report of 22nd July, 190

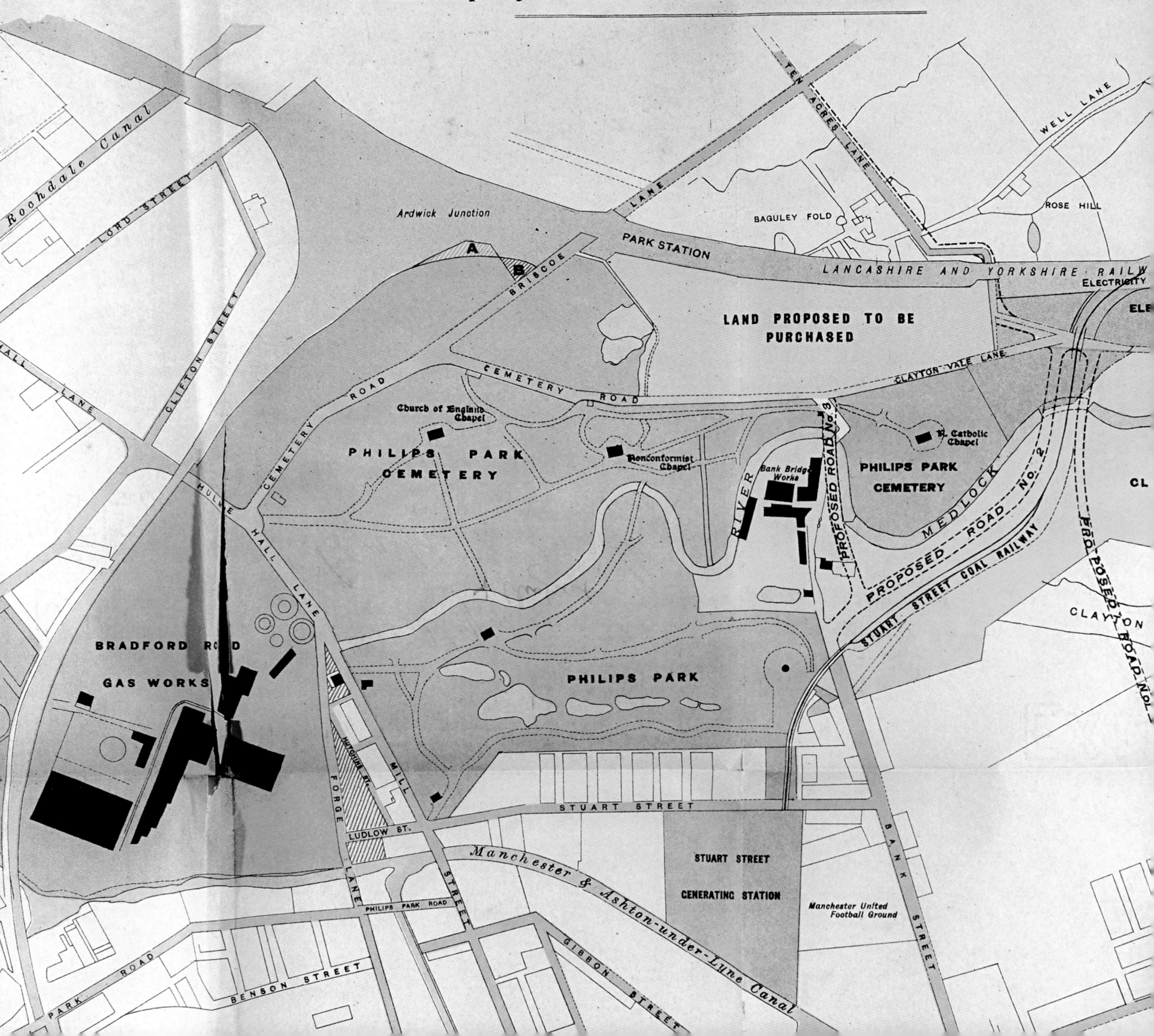

SALFORD
HULME
CHORLTON
UPON
MEDLOCK
BROUGHTON ROAD
GREEN GATE
CANNON STREET
ADELPHI STREET
KING STREET
BURY STREET
ST STEPHENS STREET
CHAPEL STREET
BATHS
ST PHILIPS
ST JOHNS
TOWN HALL
PRESBYTERIAN
POST OFFICE
EAST LANCASHIRE
NEW BAILEY
NEW BAILEY ST
STANLEY STREET
LOWER KING STREET
BRIDGE STREET
PARSONAGE
ST MARYS
BLACKFRIARS STR
VICTORIA BR STR
VICTORIA STATION
WORKHOUSE
UNITARIAN
BRIDGE STREET
CHETHAM COLLEGE
FENNEL ST
CATHEDRAL
CORN EXCHANGE
HANGING DITCH
RIVER
MILLER STREET
BATHS
SWAN ST
SHUDE HILL
SMITHFIELD MARKET
WITHY GROVE
ST GEORGE'S ROAD
TO HARPURHEY
ST MICHAELS
ST GEORGES CHURCH
OLDHAM ROAD
OLDHAM STREET
GREAT ANCOATS
POST OFFICE
CANNON STREET
CHURCH ST
HIGH STREET
ST PAULS CH
MARKET STREET
STEVENSON SQUARE
ST CLEMENTS
DALE STREET
PORT STREET
CHAPEL
DEANSGATE
KING STREET
TOWN HALL
POST OFFICE
SPRING GARDENS
CROSS ST
JOHN DALTON ST
BANK OF ENGLAND
PRINCESS STREET
FOUNTAIN ST
MOSLEY STREET
ROYAL INFIRMARY
PICCADILLY
YORK STR
PORTLAND STREET
DALE ST
POST OFFICE
STORE STREET
LONDON ROAD
CANAL
BOND STREET
RIVER
WATER STREET
QUAY STREET
GARTSIDE STREET
ST MARY'S HOSPITAL
LOWER BYRON STR
ST JOHNS
PETER STREET
FREE TRADE HALL
THEATRE ROYAL
MOUNT ST
COOPER STREET
LOWER MOSLEY ST
NEW MECHANICS INSTITUTION
DAVID STREET
SACKVILLE STREET
ORDSALL LANE STATION
LIVERPOOL ROAD
FREE LIBRARY
ST MATTHEWS
ROAD
BRIDGEWATER STREET
OXFORD STREET
GRANBY ROW
KNOTT MILL STATION
BOWDON & ALTRINCHAM STATION
OXFORD ROAD STATION
POST OFFICE
KNOTT MILL
BRIDGEWATER VIADUCT
DAWSON STR
EGERTON STREET
CHARLES STR
TEMPLE STREET
DOWNING ST
ST GEORGES
CHESTER ROAD
GREAT JACKSON
GEORGE STREET
MEDLOCK STREET
GROSVENOR STREET
GROSVENOR ROAD
RUSHOLME
CEMETERY
CAVENDISH STR
POST OFFICE
POST MONEY ORDER OFFICE
UPPER BROOK
CLIFFORD STR
POST MONEY ORDER OFFICE
CAVALRY BARRACKS
BEDFORD STREET
ST WILFRIDS C. CH
CITY ROAD
STRETFORD NEW ROAD
ROSAMOND STREET WEST
BRUNSWICK ST
OXFORD

1857

The Art Treasures Exhibition and illustrated town guides

The second half of the nineteenth century prompted a demand for easily read street maps as improved transport encouraged greater travel, both for business and a growing 'tourist' market. Map-makers were quick to spot the potential and produced a flood of simplified portable maps to cater for the demand. Typically, their maps were included as illustration for guidebooks, with titles such as *The Strangers' Guide to . . .*

Hale and Roworth's map was published both as a pocket map folded into card covers and as an accompaniment to Joseph Perrin's 1857 *Manchester Handbook*. Its timing is significant since that was the year of the Art Treasures Exhibition, which attracted huge numbers of visitors to Manchester. Sales of the official catalogue exceeded 167,000 copies, and an unknown number of visitors purchased such publications as Abel Heywood's penny guide, Edward Bellhouse's *Operative's Guide to the Art Treasures Exhibition* and dialect skits such as Tom Treddlehoyle's *Peep at t' Manchstir Art Treasures Exhebeshan.*

The 1851 Great Exhibition in London had shown the commercial scope for spin-offs for travel agents, makers of gimcrack souvenirs and publishers of guidebooks and town plans. Manchester followed the London model. Like Crystal Palace, the Exhibition was mounted in a building of cast iron and glass, erected in Stretford on the south-west flank of the city and with a specially built station at Old Trafford. As in London, cheap excursion trains helped to boost attendances. Thomas Cook, for example, organised a number of such trips, including a 'moonlight excursion' leaving Newcastle-on-Tyne at midnight, arriving in Manchester at 7 a.m. and returning at 6 p.m. The Exhibition showed that Manchester had the confidence and expertise to plan and execute a huge and highly public project. In doing so, it challenged common perceptions of Manchester as 'Coketown', a cultural backwater that knew much about making money but little about the arts.

Perrin's guidebook set out to correct the perception that

OPPOSITE. Hale & Roworth, *The picture map of Manchester* (1857) [UML]

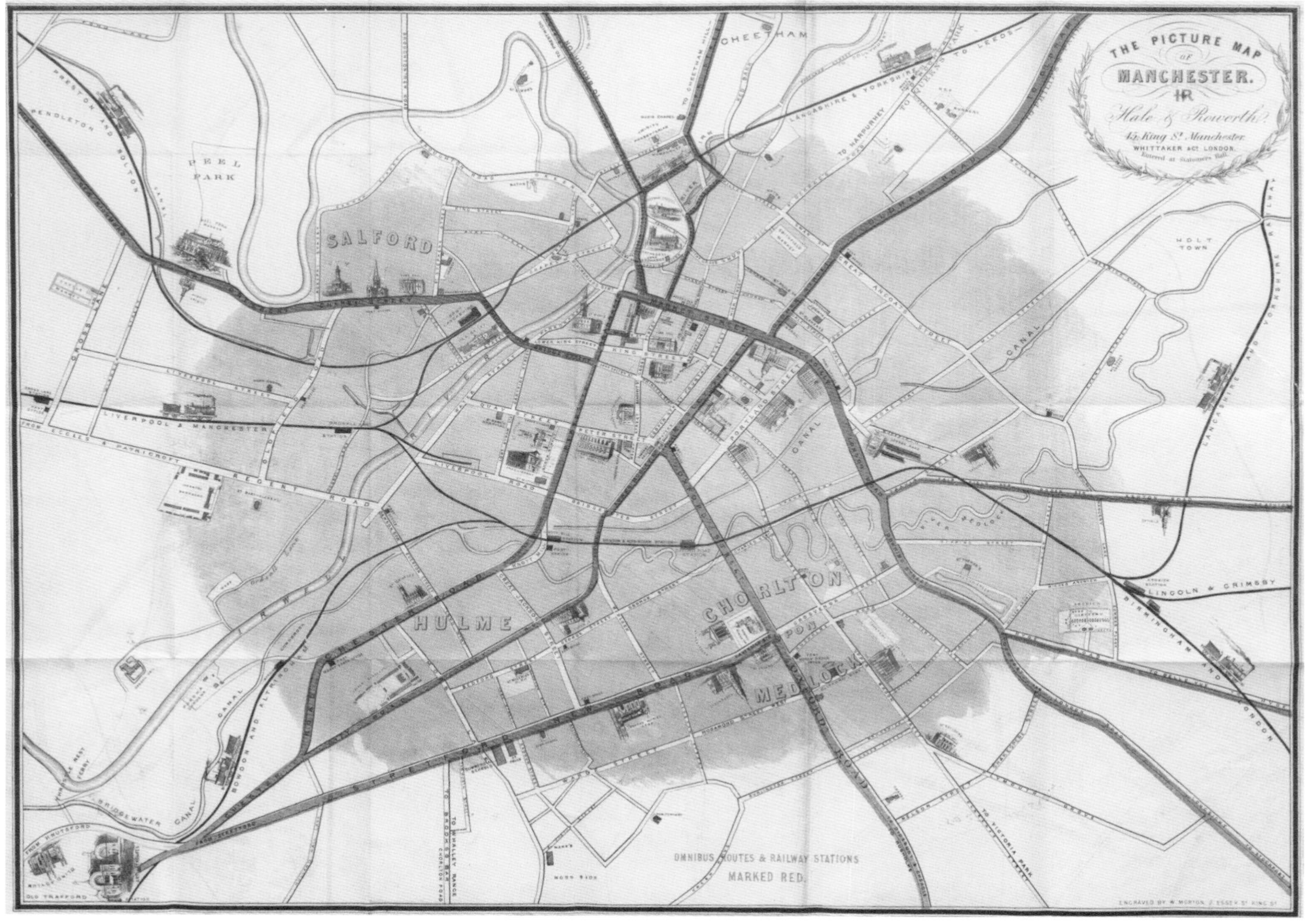

Manchester lacked the attributes of a great city. It included Hale and Roworth's map, which has a strong emphasis on transport routes to the Exhibition at Old Trafford. Red 'omnibus' routes outline the network of roads, and heavy black lines demarcate railway lines to stress the overall spider's web of communications. Railways, while hardly a novelty in 1857, still warranted the rather charming views of engines drawn on the principal lines and the striking view of the long frontage of Victoria Station. A novel but rather less successful element is the way in which views of buildings are displayed, showing them as they would be seen from the street. In theory, this makes sense, showing on which side of a street a building lay, but it has the not inconsiderable disadvantage that to see both the views and the text that names them the map has to be turned around 90 or 180 degrees.

The map's main interest is which buildings are illustrated

OPPOSITE. J. Mennie, *Ernst & Co.'s illustrated plan of Manchester and Salford* (1857) [MLA] Architecturally distinctive warehouses became an integral part of the city's identity.

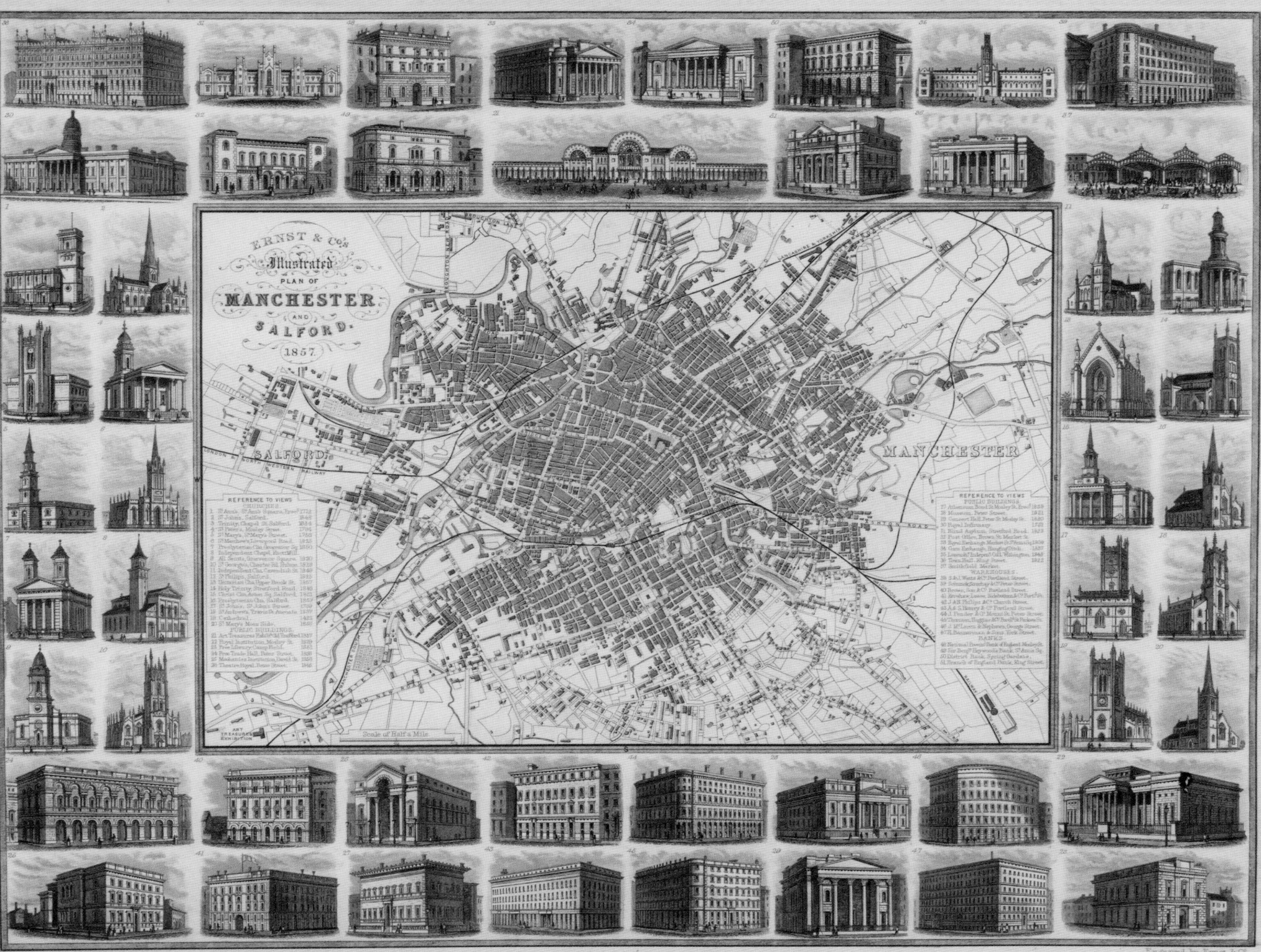

ERNST & Co's
Illustrated
PLAN OF
MANCHESTER
AND
SALFORD.
1857.
SALFORD
MANCHESTER
LONDON & NORTH WESTERN RAILWAY
REFERENCE TO VIEWS
REFERENCE TO VIEWS
Scale of Half a Mile
ART TREASURES EXHIBITION
Designed & Drawn by J. Mennie.
Engraved by Ernst & Co.

ERNST & Co's
Illustrated
PLAN OF
MANCHESTER
AND
SALFORD.
1857.
REFERENCE TO VIEWS
CHURCHES.
1 St. Ann's, St. Ann's Square. Erecd. 1712
2 St. John's, Salford. 1848
3 Trinity, Chapel St. Salford. 1834
4 St. Peter's, Mosley Street. 1794
5 St. Mary's, St. Mary's Street. 1756
6 St. Matthew's, Liverpool Road. 1825
7 Presbyterian Chn. Grosvenor Sq. 1850
8 Independent Chapel, Knott Mill.
9 All Saints, Grosvenor Square. 1820
10 St. George's, Chester Rd. Hulme. 1828
11 Independent Cha. Cavendish St. 1849
12 St. Philip's, Salford. 1825
13 Unitarian Cha. Upper Brook St. 1857
14 Holy Trinity, Stretford Road. 1846
15 Christ Chn. Acton Sq. Salford. 1831
16 Presbyterian Chn. Salford. 1847
SALFORD
THE CRESCENT
LIVERPOOL STREET
LONDON & NORTH WESTERN RAILWAY
BROUGHTON NEW ROAD
BROUGHTON LANE
REGENT ROAD
ALTRINCHAM

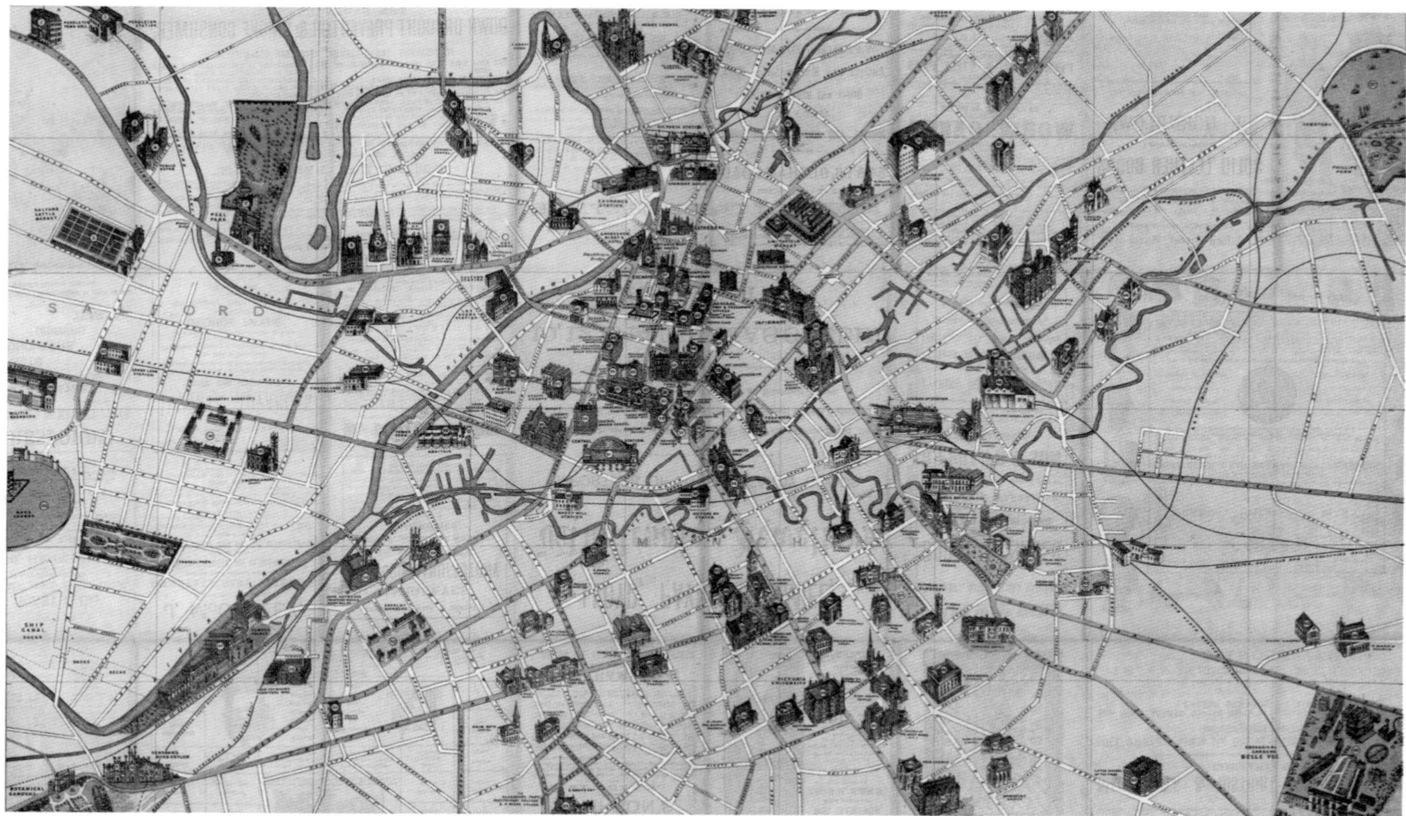

J. Heywood, *John Heywood's pictorial map of Manchester and Salford . . .* (1886) [NLS]

or simply left named but without a profile. 'Tourists' (if that is not too premature a term) were clearly well catered for, with the depiction of many of the town's more handsome buildings and with the view of the Art Treasures Exhibition (flanked by a picture of the Blind Asylum housed in a handsome building that, like the recently rebuilt Royal Infirmary, represented the public face of charitable Manchester). Tourist and business visitors would also have welcomed the depiction of the Queen's Hotel in Piccadilly – the only hotel to be shown. A variety of entertainments are also displayed, such as the Free Trade Hall, Theatre Royal and Peel Park Museum in Salford.

For visiting business people, the location of some of the trading centres would have been important, hence the use of views such as those of the magnificent Watts Warehouse, the Royal Exchange and Corn Exchange. That said, the map's most striking feature is that it excludes any indication of the mills and industrial premises on which the town's fortunes depended. Ancoats for example – centre of much of the textile industry – has not a single building depicted, and the same is true of Castlefield. If the map aimed to attract sales from the business world, the absence of any manufacturing sites must cause some surprise. On the other hand, showing the location of the Bank of England, money-order offices and the numerous post offices would have been valuable commercial aspects of the map.

The decisions about which buildings to illustrate in part reflect the space available to incorporate views. There is only one instance where a view cuts across the street pattern – the handsome tall tower of St Mary's Church, which cuts rather crudely across both The Parsonage and the Irwell. Elsewhere, views are fitted into the available space. This explains the cursory treatment of St Ann's, which is not explicitly named and whose outline is purely nominal. The same lack of space explains the absence of a profile of Holy Trinity, despite it being by far Salford's most important church. Equally, a number of the major churches in the centre of the town are not profiled, for example St Paul's and St Clement's. There is also a tendency to focus on Anglican churches. Catholic churches are identified, and Salford's Catholic Cathedral is illustrated, but the great majority of chapels tend to be named but not drawn. The Friends' Meeting House and the Scottish Presbyterian church in Chorlton-on-Medlock are given portraits, and there is a large and striking view of the Unitarian Chapel on Upper Brook Street. These, however, are the exception.

What is clear is that the map was aimed not just as a wayfinder but as a puff for the town as a place of civilised and elegant modernity. Here was a town that could offer a plethora of handsome buildings, a profusion of churches and chapels, a wide variety of entertainments, modern transport as well as the public services on which its inhabitants relied.

A more conventional map, clearly intended as a souvenir of the Exhibition, was published by Ernst & Co. in 1857. It comprised a plan of the city framed by 51 drawings of its principal buildings – the work of James Mennie. Among the featured buildings were churches (including St Peter's, St James's and All Saints), warehouses (including the recently completed Watts Warehouse on Portland Street), banks (including the Bank of England in King Street) and public buildings (including the Royal Infirmary and Royal Manchester Institution). These were identified and numbered in the key, but, oddly, their location was not indicated on the map itself. Advetised as 'Views in Manchester', it was in the tradition of pictorial urban maps, a form that had been seen most recently in John Tallis's 'Manchester and its environs', which included coloured vignettes of the Royal Infirmary, Royal Manchester Institution and the Town Hall.

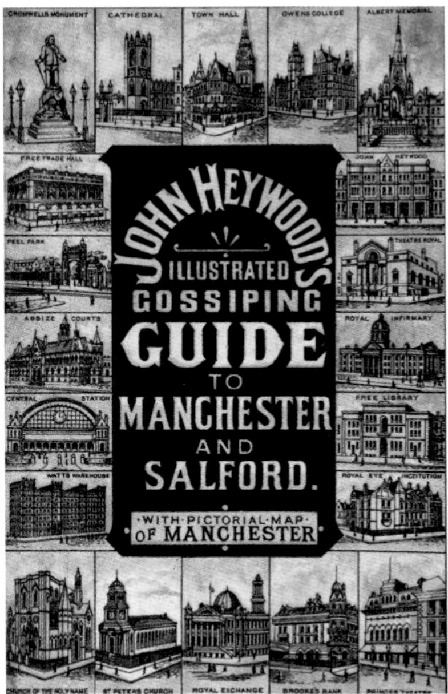

Heywood's *Gossiping Guide* (1887) [CHETS] cover.

Thirty years later, the design and thrust of John Heywood's 1886 map bears a striking resemblance to the Hale and Roworth map. Like Hale and Roworth, Heywood highlights streets served by tramways, covers virtually the same area of the town, incorporates an array of views of 'public' buildings, identifies the stations serving the now more numerous railway lines and also notes the post offices. Neither map has a scale bar, which allows both to show the town centre at a somewhat larger scale than the peripheral areas. However, inevitably there are differences between the two, reflecting Manchester's continuing growth, the most striking of which are the appearance of Waterhouse's new Town Hall and of Central Station. Interestingly, Heywood adopts a somewhat different approach to depicting buildings by orienting them to show their façades even though many in fact faced a different direction. So for example Central Station is shown with its façade facing south not north and the Town Hall as facing south rather than west. Cheekily, if understandably, the only commercial buildings that Heywood shows are the Heywood stationer's shop and furniture showroom on Deansgate. The only suggestions of industry are the two views of the Heywood furniture factory and printing works, both shown with smoking chimneys. The only other smoking chimneys are from three public baths. So, like Hale and Roworth, his view of Manchester is bereft of industry and the map basically extols the town as a place of culture and refinement.

Blackfriars Bridge
EXCHANGE
S. ANN'S SQUARE
ST MARY'S CHURCH
GUARDIAN
EXAMINER
LLOYD, PAYNE & AMIEL
HYAM
CANNON STR.
CHURCH ST
MARKET STR
HIGH STREET
ALBERT
T. ARMSTRONG & BROTHER
KING STREET
QUEEN'S THEATRE
DEANS GATE
S. ANN'S STREET
S. ANN'S CHURCH
CARLTON CLUB
PALL MALL
BROWN ST
SPRING GARDENS
GENERAL POST & TELEGRAPH OFFICES
MANC. & LIV. DISTRICT BANK
REFORM CLUB
INFIRMARY
REFERENCE LIBRARY
BANK OF ENGLAND
YORK ST
QUEENS HOTEL
JNO DALTON ST.
JOHN HEYWOOD STATIONER WAREHOUSE
JOHN HEYWOOD CH & SCHOOL FURNITURE SHOW ROOMS
MASONIC HALL
MAN. & SAL. BANK
CROSS ST
REVENUE OFFICES
COOPER ST
ART GALLERY
ATHENÆUM
MOSLEY STREET
JACKSONS ROW
CHIEF FIRE STATN
GRAND CIRCUS
TOWN HALL ALBERT SQ.
COMEDY THEATRE
CITY POLICE COURTS.
PETER
THE FOLLY
FREE TRADE HALL
THEATRE ROYAL
YOUNG MENS CHRIS. ASSN
S. PETER'S CHURCH
S. PETER'S SQ.
SACKVILLE
TECHNICAL SCHOOL
CENTRAL BOARD SCHOOL.
CONCERT HALL (Private)
TURKISH BATHS
PORTLAND STREET
PRINCES THEATRE
CENTRAL STATION
LOWR MOSLEY STREET
OXFORD
DAVID ST
S. JAMES'S HALL
THEATRE
GREAT BRIDGEWATER ST
ROCHDALE CANAL

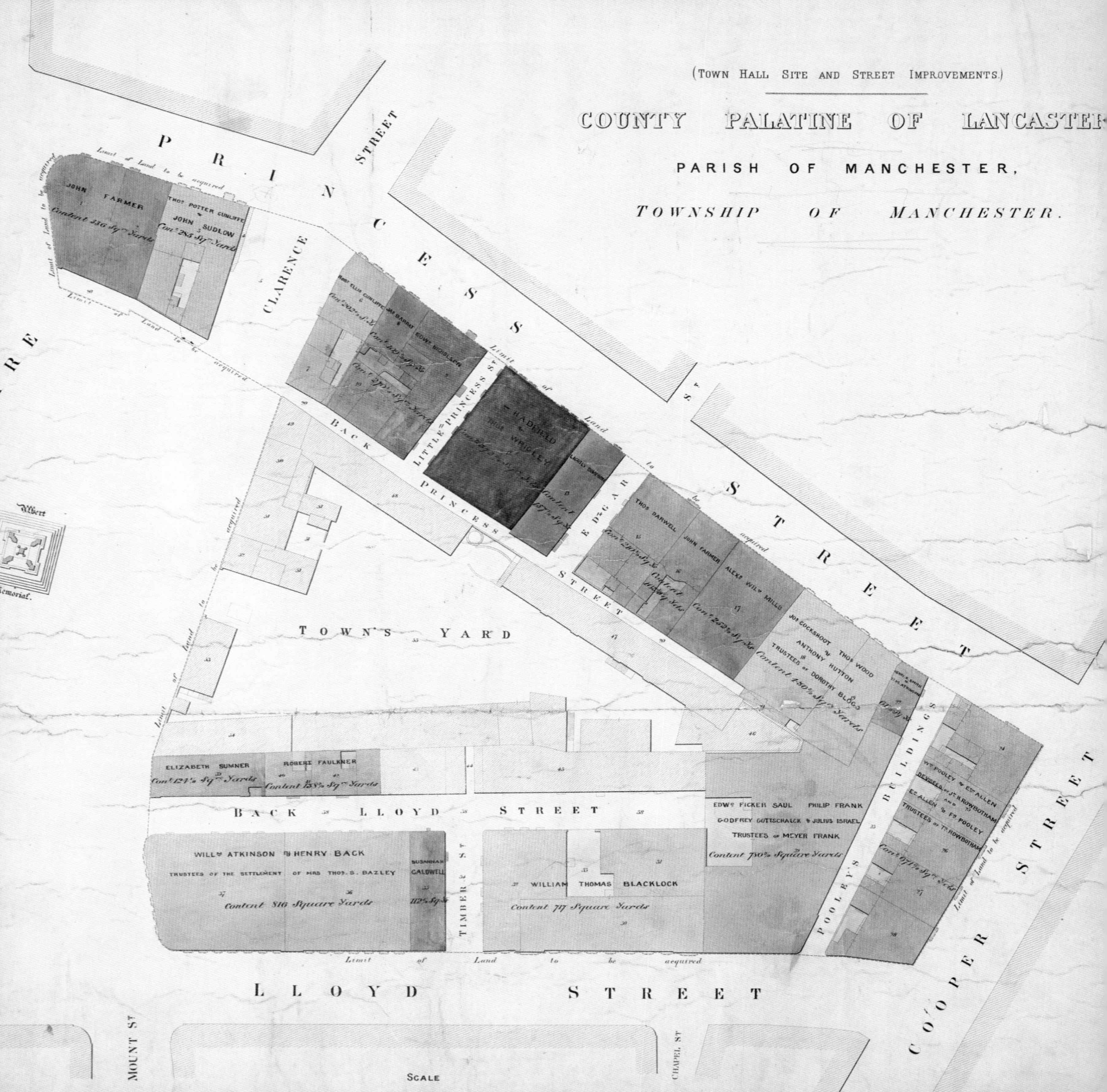
(TOWN HALL SITE AND STREET IMPROVEMENTS.)
COUNTY PALATINE OF LANCASTER
PARISH OF MANCHESTER,
TOWNSHIP OF MANCHESTER.
PRINCESS STREET
CLARENCE STREET
Limit of Land to be acquired
JOHN FARMER
Content 236 Sq. Yards
THOS POTTER GUNLIFFE & JOHN SUDLOW
Cont 285 Sq. Yards
LITTLE PRINCESS ST
BACK PRINCESS STREET
EDGAR ST
THOS DARWELL
JOHN FARMER
ALEXR WILM MILLS
Con 252 Sq. Yds
JOS COCKSHOOT & THOS WOOD
ANTHONY HUTTON
TRUSTEES OF DOROTHY BLOGG
Content 450 Sq. Yards
Albert Memorial
TOWN'S YARD
ELIZABETH SUMNER
Con 124 Sq. Yards
ROBERT FAULKNER
Content 158 Sq. Yards
BACK LLOYD STREET
WILLM ATKINSON & HENRY BACK
TRUSTEES OF THE SETTLEMENT OF MRS THOS. S. BAZLEY
Content 816 Square Yards
SUSANNAH CALDWELL
112 Sq. Yds
TIMBER ST
WILLIAM THOMAS BLACKLOCK
Content 717 Square Yards
EDWD FICKER SAUL PHILIP FRANK
GODFREY GOTTSCHALCK & JULIUS ISRAEL
TRUSTEES OF MEYER FRANK
Content 730 Square Yards
POOLEY'S BUILDINGS
WM POOLEY & ED ALLEN
DEVISEES OF JN ROWBOTHAM
ED ALLEN AND FD POOLEY
TRUSTEES OF TH ROWBOTHAM
Con 474 Sq. Yds
COOPER STREET
LLOYD STREET
MOUNT ST
CHAPEL ST
SCALE

1866

A municipal palace: Alfred Waterhouse's Town Hall

Manchester's first Town Hall was constructed in the mid 1820s on King Street. For its time it was a significant building, styled elegantly along Greek-revival lines by the architect Francis Goodwin. However, as the city prospered through the Victorian age, it seemed too modest in scale. Partly in a response to more monumental town halls being built in competitor cities such as Birmingham, Leeds and Bradford, there were moves to construct a more impressive seat of civic government and symbol of the power of Manchester.

A public competition was launched for the design of a new town hall. It needed to be suitably imposing but also functional and had to fit within a constrained site on an irregular triangle of land used by the Council for street cleansing equipment and fire engines. The surrounding narrow streets and courts contained small businesses and miscellaneous workshops that were seen as having little value. The conditions of the competition required a five-storey building with a 'grand front and entrance' on Albert Square, which had been created in the early 1860s to mark the death of Prince Albert.

Alfred Waterhouse was one of 123 architects in the competition and he won, with judges noting that his plan for the building had 'such great merits, is so admirably and simply disposed, and so well lighted that we cannot but feel it is thoroughly entitled to the first place'. While its overall conception and interior layout was superior, Waterhouse faced significant criticism for its aesthetics, with many people pointing to what they saw as flaws in its visual appearance. The *Manchester Guardian* savagely remarked that 'the main façade is wretchedly composed; the principal entrance is like a hole into a beehive'. The squat shape of the clock tower and spire was particularly derided, with *Building News* calling it horribly ugly and having 'no connection with the building from which it springs'.

OPPOSITE. G. Falkner, *Town Hall site and street improvements* (1866) [MLA]
The land and buildings to be cleared for the triangular site are shown in colour.

During the time between the architectural competition and the start of construction, Waterhouse was able to revise and improve the design, particularly the clock tower and main entrance onto Albert Square. The design would continue to be refined even as the lengthy construction progressed.

Waterhouse was born in 1830 into a wealthy Quaker family in Aigburth, near Liverpool. He eventually became one of the leading British architects in the second half of the nineteenth century. His best work is associated with the Gothic revival style, but many of his designs were more nuanced and diverse in their influences. He had close connections with Manchester as he received his early architectural training in the city in the 1850s. He subsequently set himself up in a practice in the Cross Street Chambers. His first major commission was Manchester's Assize Courts in Strangeways, which won him national recognition as an architect able to tackle large-scale projects in an effective manner. Later he was responsible for the design of the original Owens College buildings on Oxford Road.

Construction of the new Town Hall took around five years, with a workforce at its peak of about 1,000 men, most of whom were stonemasons. Much time and money was spent on the interior decor and fittings, over which Waterhouse had a continuing influence. There was copious decorative carved stone, mosaic flooring, grand staircases, polychrome tiling and vaulted ceilings along the major public corridors. The roof of the Great Hall was lavishly decorated with gold and adorned with the coats of arms of the nations that traded with Manchester. Around its walls were newly commissioned murals by the local artist, Ford Madox Brown, recounting episodes in the history of the city. Also installed was a huge hydraulic-powered organ in the public hall and a great peal of bells in the tower. Eventually, the Council got cold feet over the cost of fittings and not all the murals that Waterhouse had wanted were installed.

The building was formally opened in 1877 with one of the leaders of the Free Trade movement, John Bright, asserting emotively that 'it is a municipal palace . . . Whether you look at its great proportions outside or its internal decoration . . . there is nothing like it . . . in any part of the United Kingdom and I doubt whether there is in any of the great, famous old cities of the Continent of Europe'. The effect of the neo-Gothic design made the building appear older than it was – granting some gravitas to a modern industrial metropolis. However, it is interesting that the Council continued to call it 'Town Hall' rather than 'City Hall'.

The completed building had hundreds of working offices, an imposing council chamber with seating for 16 aldermen and 48 councillors, and a 100-foot-long Great Hall. However, within a couple of decades, partly as a result of the growing responsibilities of the Corporation – including electricity generation, education and the provision of council housing – there was a serious lack of spaces for officials and committees. By the 1930s, another major building was mooted to provide additional office space. The Central Library and Town Hall Extension departed markedly from the neo-Gothic of the main Town Hall, but their design was widely regarded as a success when they opened in 1934 and 1938 respectively.

Yet the Town Hall was not universally appreciated in the mid twentieth century. The soot-stained Gothic detailing was seen by some as antiquated and even ugly. The building lacked the clean lines favoured by interwar modernist architecture, and the 1945 *City of Manchester Plan* proposed that it should be replaced with a more modest and contemporary building.

As a building, cleaned of soot in the late 1960s and Grade I Listed, the Town Hall sits proudly at the heart of a civic complex encompassing the circular Central Library, public squares and various public monuments including Lutyens's Cenotaph. It has been central to the political and wider cultural life of the city for well over a century. Cognitively, it is a focal point – the *centre* of the city centre. It is the point to which people gravitate to celebrate, commemorate and protest. It remains the most iconic and recognisable building in Manchester, better known than older compatriots such as the Cathedral and Chethams. It is one of a handful of sites that have come to symbolise the spirit of Manchester in the public consciousness.

OPPOSITE. A. Waterhouse, *Manchester Town Hall plan of first floor* (*c.*1875) [MLA] The small attached plan at lower left shows the outline of the Lord Mayor's apartment on a mezzanine floor.

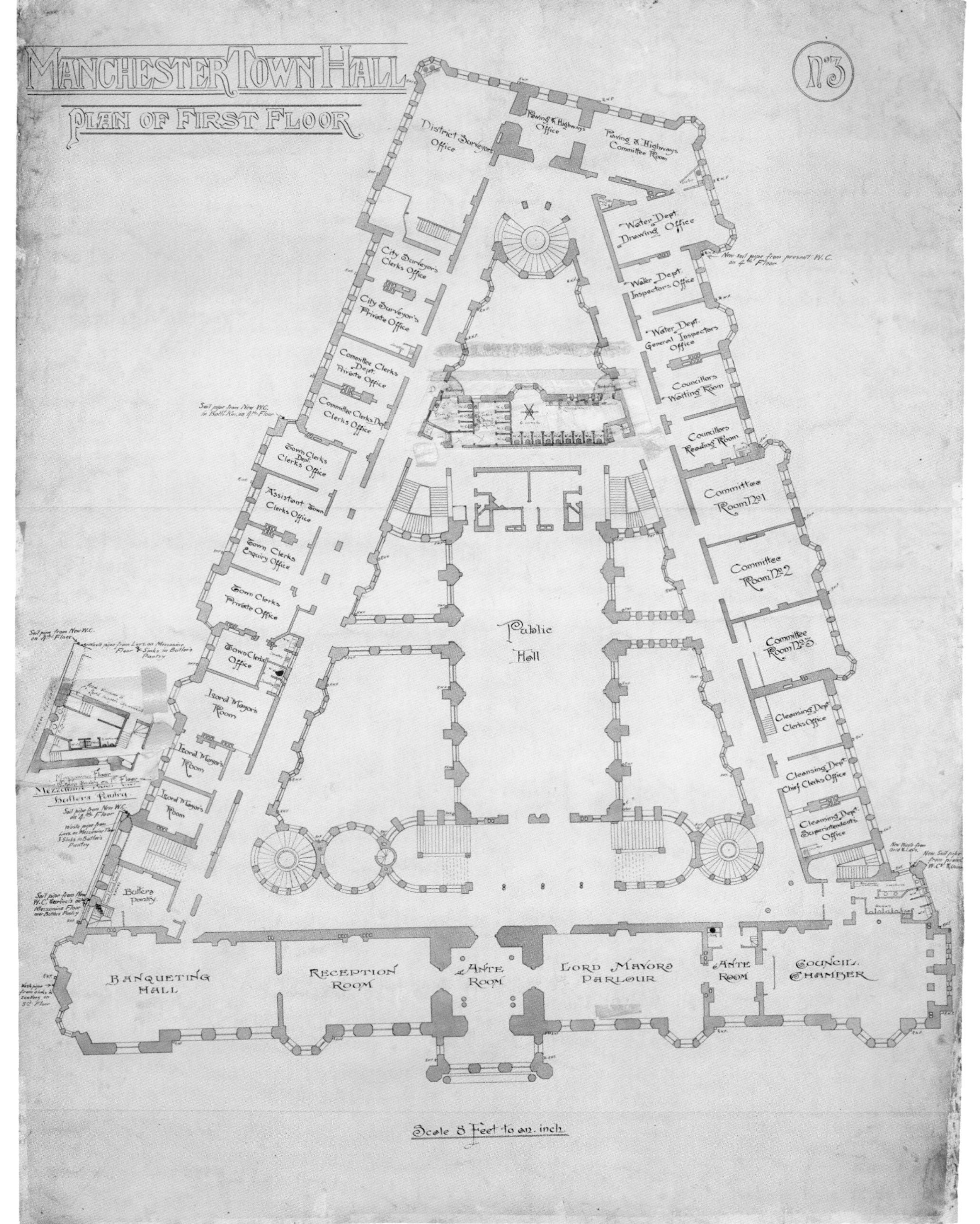

MANCHESTER TOWN HALL
PLAN OF FIRST FLOOR
Nº3
District Surveyors Office
Paving & Highways Office
Paving & Highways Committee Room
Water Dept. Drawing Office
Water Dept. Inspectors Office
Water Dept. General Inspectors Office
Councillors Waiting Room
Councillors Reading Room
Committee Room Nº1
Committee Room Nº2
Committee Room Nº3
Cleansing Dept. Clerks Office
Cleansing Dept. Chief Clerks Office
Cleansing Dept. Superintendents Office
City Surveyor's Clerks Office
City Surveyor's Private Office
Committee Clerks Dept. Private Office
Committee Clerks Dept. Clerks Office
Town Clerks Dept. Clerks Office
Assistant Town Clerks Office
Town Clerks Enquiry Office
Town Clerks Private Office
Town Clerks Office
Lord Mayor's Room
Lord Mayor's Room
Lord Mayor's Room
Butlers Pantry
Public Hall
BANQUETING HALL
RECEPTION ROOM
ANTE ROOM
LORD MAYORS PARLOUR
ANTE ROOM
COUNCIL CHAMBER
Scale 8 Feet to an inch

LOWER BROUGHTON
CLARENCE ST.
CUMBERLAND ST.
GROSVENOR SQUARE
BROUGHTON
BROUGHTON ROAD
NEW ROAD
NDLETON
PEEL PARK
ST. STEPHENS
WARD
TRINITY
WARD
BLACKFRIARS
WARD
CRESCENT
SALFORD
WARD
ST. JOHNS
BANK PARADE
CHAPEL ST.
BURY STREET
GREEN GATE
QUEEN ST.
KING ST.
BLACKFRIARS ST.
NEW BAILEY ST.
BRIDGE ST.
QUAY ST.
STANLEY ST.
HAMPSON ST.
CROSS LANE
BOLTON ROAD
BURY & YORKSHIRE CANAL
L & N.W. Ry (LIVERPOOL & MANCHESTER)
DUNCAN ST.
UNION ST.
CHARLES ST.
ST. JOHN ST.
CUMBERLAND ST.
HARDMANS ST.
IRWELL ST.
ISLINGTON
STIVENSON ST.
CLEMINSON ST.
BANK ST.
MARKET ST.
CHURCH ST.
FLAX ST.
HOPE ST.
CANNON ST.
BLACKBURN ST.
ADELPHI ST.
SANDON ST.
ST. SIMON ST.
RICHMOND ST.
RAVALD ST.
ALBERT ST.
PARSONAGE
LR KING ST.
JOHN DALTON
QUEEN ST.
BK. QUEEN ST.
JACKSONS ROW
BOOTLE ST.
CHE
1
2
3
4
5
11
12
13
14
15
16
22
23
24
25
26
32
33
34
35
36
37
41
44
45
46
53
55
56
57
58

1868

Paying the cabbie

The 1868 cab-fare map of Manchester and Salford was the work of the Liverpool-based publisher Llewellyn Syers who produced two similar maps, one for Liverpool (1868), and another covering Birkenhead and much of the Wirral Peninsula (1869). They were brave – although ultimately rather unsuccessful – attempts to provide a handy *vade mecum* to check the mileage charged by horse-drawn cabs.

Two-wheel hansom cabs, which began to appear in Britain from the 1830s, offered a smaller and cheaper alternative to the four-wheel Hackney carriages. They were named after Joseph Hansom who not only developed them but was also an architect whose designs included Manchester's Church of the Holy Name on Oxford Road. Carriages in Manchester were overseen by the Hackney Coach Committee. In 1854, it licensed 52 owners who operated some 220 vehicles, including 116 double-seater coaches, 60 broughams and 25 hansom cabs. These licensed carriages had to display information on fares on the outside. In the absence of taximeters, which calculated the distance travelled but were only introduced from the 1890s, there was inevitably a potential process of bargaining over fares between cabbie and customer, and indeed in London there are examples of formal court cases to resolve disputes. Unsurprisingly, London had many more carriages than did Manchester. Numerous London cab-fare maps were produced by publishers such as William Orr, George Cruchley and William Mogg, consisting either of plans on which straight-line distances could be calculated or with various circles superimposed to aid the approximate calculation of distances from frequently used origins such as major railway stations. Printed lists of fares between various points in the capital were also compiled, an example being Mogg's *Ten Thousand Cab Fares*. Such lists of specific fares do not seem to have been produced for provincial cities, hence the potential value of maps that could help in calculating distances. The appearance of cab-fare

OPPOSITE. L. Syers, *Syers's ready reference & cab fare plan of Manchester & Salford* (1868) [BL]

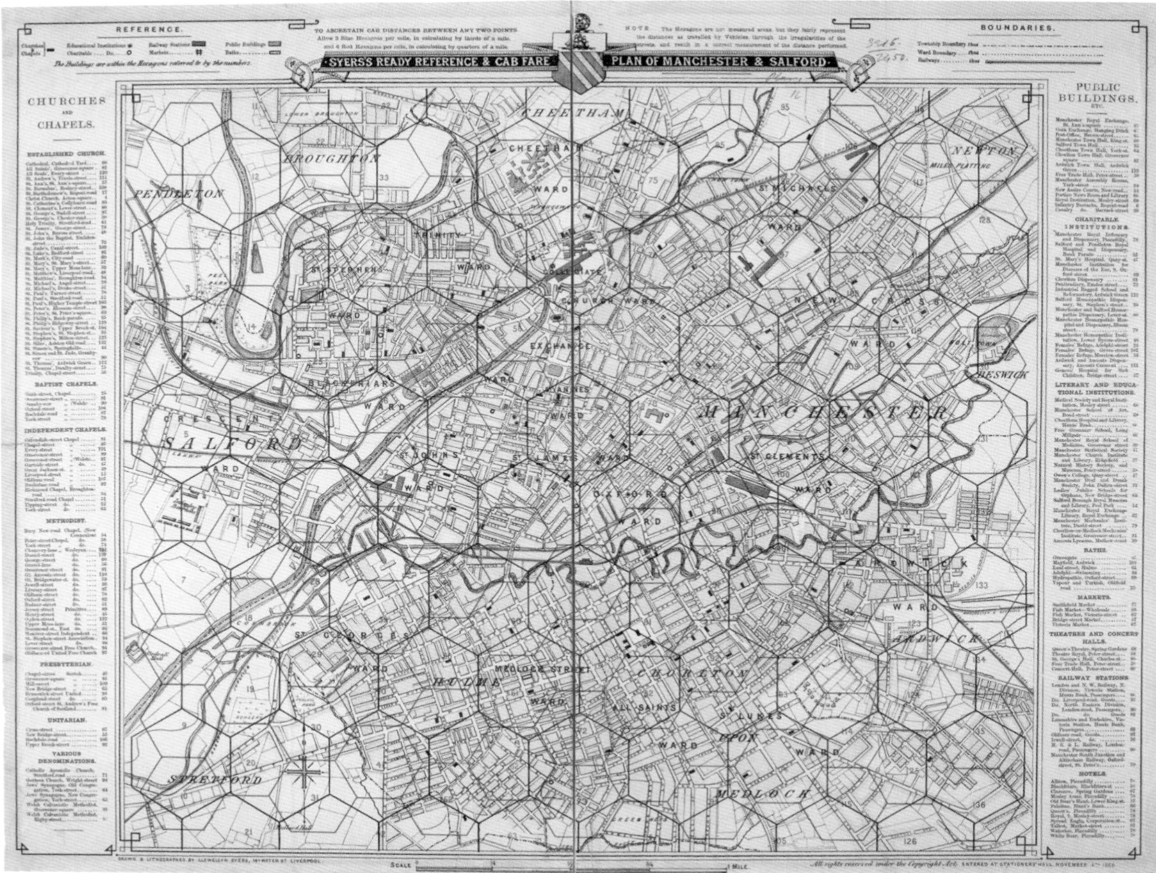

maps for Manchester and Liverpool suggests that by the 1860s the physical growth of some major provincial cities had begun to warrant a wider use of cabs in cities outside the metropolis.

Syers's map is a highly detailed plan of the town with most roads named, some buildings depicted in outline (as with the Strangeways and New Bailey prisons and the principal railway stations), numerous other buildings highlighted with small symbols distinguishing churches and chapels, educational and charitable institutions, markets, and public baths. Township and ward boundaries are shown with their respective names. The map also depicts the lines of railways, canals and rivers. This is a great deal of detail, probably too much given the relatively small size of the map. It certainly makes for eye strain. The map is accompanied by a highly detailed listing of buildings in the side margins. The left margin lists 41 Established churches, 6 Baptist chapels, 14 Independent chapels, 25 Methodist chapels, 7 Presbyterian chapels, 4 Unitarian and 6 other denominations. The right margin lists 15 public buildings such as town halls, 16 charitable institutions, 16 literary and educational institutions, 6 baths, 5 theatres and concert halls, 9 railway stations and 12 hotels.

Given such detail, this is in itself a potentially valuable map. However, as its title suggests, it ostensibly aims to provide a means of calculating distances between any two points, presumably as a means of checking the fares charged by cabbies. To achieve this, Syers superimposed two sets of hexagons across the face of the map, one in blue with a diagonal of one-third of a mile and a second in red with a diagonal of one-quarter of a

Two scales of hexagon, with diameters of one-third and one-quarter of a mile.

An earlier version with only a single set of hexagons with a diameter of one-quarter of a mile. [NLS]

mile (an earlier version used only the smaller red hexagons). Syers claims that by taking the number of hexagons (using three blue hexagons per mile or four red hexagons) an accurate calculation of the distance between two points could then be made. This, of course, could only be true for journeys passing between the mid-points of the hexagons – and even then the configuration of the street network would be likely to increase the distance in comparison to the crow's-fly measure. He has a rather disarming disclaimer at the head of the map saying that the hexagons 'fairly represent the distances as travelled by Vehicles through the irregularities of the streets, and result in a correct measurement of the distance performed'. Ironically, his map includes a scale bar, which would make a more accurate straight-line measure of distance between any two points, although if a journey followed a non-linear route, the hexagons could offer a more accurate calculation.

However, were someone to use the map as a means of calculating distances, they would find it a cumbersome object to consult while sitting in a cab. The British Library example shown here has only a single central fold, rather than being dissected or folded into a more convenient size. So the map would have proved both awkward to hold and, given a jolting cab, not easy to read given its excessive detail. Moreover, were it to come to a disagreement over what was charged as a fare, one can imagine the bemusement of a cabbie when shown the hexagonal logic claimed by the map. It may therefore seem that Syers's marketing strategy to encourage buyers was as much to rely on its novelty element as its role as distance-calculator.

Nevertheless, the map has a quite different potential merit in its use of the red hexagons as a reference system for buildings. Syers uses the 136 red hexagons as the network through which to reference the 182 buildings listed in the map margins. He does not name the buildings on the map, showing them only as one of seven different symbols. This proves to be a reasonably effective method of identification for most of the sites and buildings since the hexagons are sufficiently small that few include more than a single example of a particular category of building, other than churches and chapels. However, the latter are so numerous that inevitably there are many hexagons that include more than a single instance. An example is the area covered by the wards of St Stephens and Blackfriars in Salford where hexagons 25, 34 and 35 each contain three churches/chapels, and hexagon 36 contains two. Moreover, while it would be relatively easy for a map user to locate buildings whose names they knew, it would be relatively difficult to identify in the marginal lists buildings shown by a symbol in any particular hexagon.

Syers's map is an intriguing object – in part novelty, in part an interesting and creative attempt to provide a method for calculating distances without resort to tape measure and scale bar, and in part a different approach to providing a reference system to locate specific buildings. Ultimately, it has to be seen as mere novelty. It is certainly difficult to imagine it helping to resolve disputes between a cabbie and a passenger.

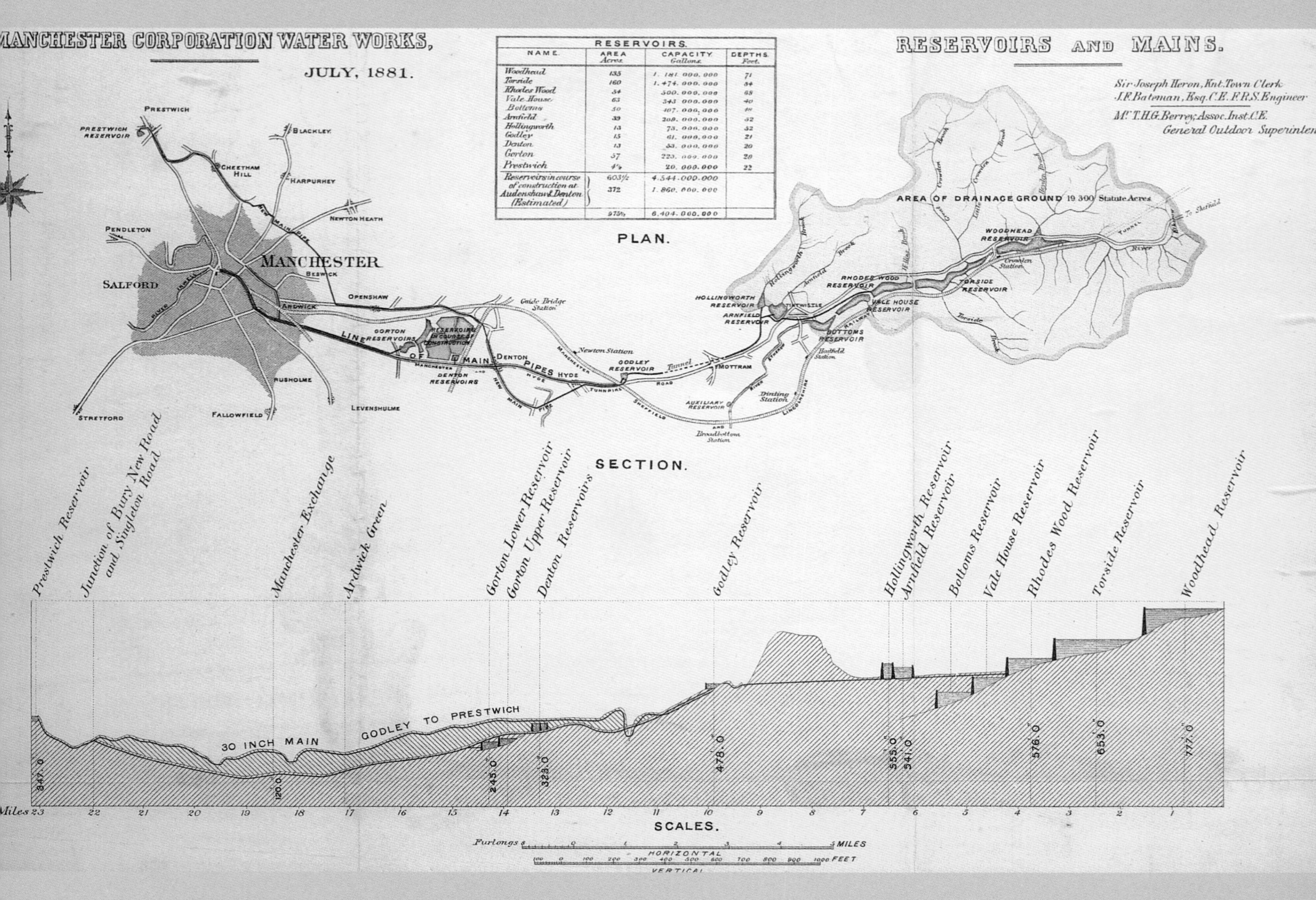

MANCHESTER CORPORATION WATER WORKS,
JULY, 1881.
RESERVOIRS AND MAINS.
Sir Joseph Heron, Knt. Town Clerk
J.F. Bateman, Esq. C.E. F.R.S. Engineer
Mr. T.H.G. Berrey, Assoc. Inst. C.E.
General Outdoor Superinten
RESERVOIRS.
NAME.	AREA Acres.	CAPACITY. Gallons.	DEPTHS. Feet.
Woodhead	135	1.181.000.000	71
Torside	160	1.474.000.000	84
Rhodes Wood	54	500.000.000	68
Vale House	63	343.000.000	40
Bottoms	50	407.000.000	48
Arnfield	39	208.000.000	52
Hollingworth	13	73.000.000	52
Godley	15	61.000.000	21
Denton	13	53.000.000	20
Gorton	57	223.000.000	28
Prestwich	4½	20.000.000	22
	603½	4.544.000.000	
Reservoirs in course of construction at Audenshaw & Denton (Estimated)	372	1.860.000.000	
	975½	6.404.000.000	
PLAN.
AREA OF DRAINAGE GROUND 19 300 Statute Acres.
PRESTWICH
PRESTWICH RESERVOIR
BLACKLEY
CHEETHAM HILL
HARPURHEY
NEWTON HEATH
PENDLETON
MANCHESTER
BESWICK
SALFORD
OPENSHAW
ARDWICK
GORTON RESERVOIRS
RESERVOIRS IN COURSE OF CONSTRUCTION
LINE OF MAIN PIPES
DENTON
HYDE
DENTON RESERVOIRS
RUSHOLME
FALLOWFIELD
LEVENSHULME
STRETFORD
Guide Bridge Station
Newton Station
GODLEY RESERVOIR
Tunnel
MOTTRAM
AUXILIARY RESERVOIR
Dinting Station
Broadbottom Station
HOLLINGWORTH RESERVOIR
ARNFIELD RESERVOIR
TINTWISTLE
RHODES WOOD RESERVOIR
VALE HOUSE RESERVOIR
BOTTOMS RESERVOIR
TORSIDE RESERVOIR
WOODHEAD RESERVOIR
Crowden Station
Hadfield Station
TUNNEL
To Sheffield
SECTION.
Prestwich Reservoir
Junction of Bury New Road and Singleton Road
Manchester Exchange
Ardwick Green
Gorton Lower Reservoir
Gorton Upper Reservoir
Denton Reservoirs
Godley Reservoir
Hollingworth Reservoir
Arnfield Reservoir
Bottoms Reservoir
Vale House Reservoir
Rhodes Wood Reservoir
Torside Reservoir
Woodhead Reservoir
30 INCH MAIN GODLEY TO PRESTWICH
347.0
120.0
245.0
323.0
478.0
555.0
541.0
576.0
653.0
777.0
Miles 23 22 21 20 19 18 17 16 15 14 13 12 11 10 9 8 7 6 5 4 3 2 1
SCALES.
Furlongs MILES
HORIZONTAL
FEET
VERTICAL

1881

Delivering plentiful clean water to the city

The second half of the nineteenth century saw a sustained burst of large-scale civil engineering by Manchester Corporation to supply vastly increased amounts of clean drinking water, as well as culverting unruly rivers to reduce flooding and building a large sewer network to remove effluent. These municipal interventions were a vital contribution to the making of a sanitary and prosperous city.

In the early years of industrial expansion, there was only limited engineering intervention to try to pipe clean water to the city centre, and until the middle of the nineteenth century most residents relied upon local sources, from wells, street pumps, rainwater collection systems, or nearby streams and rivers. However, dramatic environmental degeneration in the nineteenth century, with rapidly increasing sewage and pollution from factories, meant that the quality of river water declined to such an extent that by the 1830s de Tocqueville wrote of 'Manchester's fetid muddy waters, stained with a thousand colours' and likened the whole city to a cesspool. The outbreaks of cholera in Manchester in the 1830s and 1840s were also powerful indications of poor sanitation, although at the time doctors and officials were unaware that the disease was spread via a waterborne bacterium.

Yet the privately run Manchester and Salford Waterworks Company conspicuously failed to improve this situation throughout its 35-year existence, with demand for water frequently outstripping the unreliable supply. By the 1840s, its piping only provided sporadic clean water to around a quarter of households in the city. In response, an 1847 Act of Parliament authorised control of the company by Manchester Corporation and plans for a much larger-scale drinking water supply system were drawn up.

The plan was an example of bold Victorian engineering

OPPOSITE. Manchester Corporation, *Manchester Corporation water works, reservoirs and mains* (1881) [MLA]

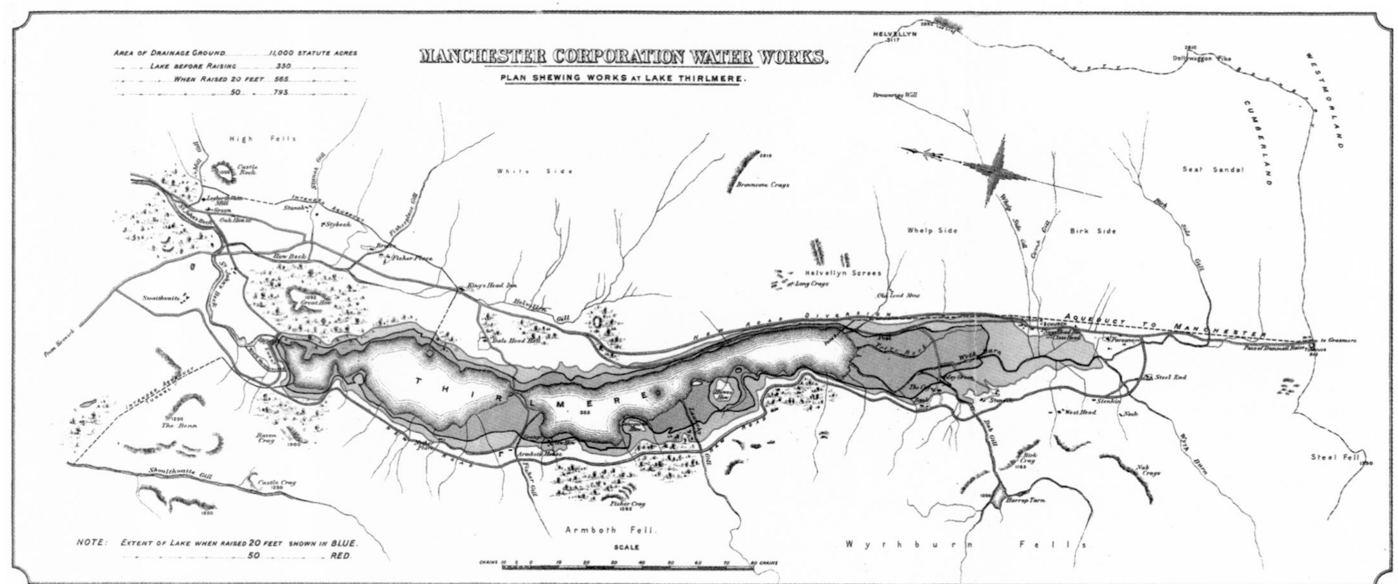

Manchester Corporation, *Plan shewing works at Lake Thirlmere* (*c.*1884) [AUTH]

intended for civic benefit. It involved the city reaching out well beyond its formal administrative boundaries to gather water from the uplands in the Peak District. Under the leadership of the civil engineer John Frederic La Trobe Bateman, a chain of seven reservoirs was constructed up the Longdendale valley, with a connection to the city through an 18-mile aqueduct. The first of these reservoirs, at the Woodhead, proved particularly tricky to construct as its dam wall was initially positioned on geologically faulty ground, but persistence paid off and the first Pennine water was flowing into Manchester by 1851. As the system grew and more reservoirs were completed, it was widely celebrated as an engineering triumph and a source of pride for the Corporation. An important feature of the system, as displayed on the cross-sectional view in the map, is that water could simply flow downhill under the power of gravity and feed into the homes and factories in the city. By the 1860s it was claimed that the amount of artificial storage operated by Manchester Water Works was then the largest in the world.

Manchester's reservoir system in Longdendale covered over 500 acres of surface area and was fed by a drainage ground of more than 19,000 acres. The Corporation had become a major force in the Peak District through its land ownership. There was a series of smaller storage reservoirs closer to the city in which water could settle and be delivered quickly to meet fluctuations in demand. Additional storage reservoirs were also under construction in Audenshaw at the time the map was produced, demonstrating how demand for water kept expanding, usually faster than the rate of population growth. Industrial processes also needed prodigious amounts of clean water.

While the last reservoirs in the Longdendale chain were being built, it became evident that demand for fresh water could only be satisfied for a few years from this scheme. More homes were being connected to a piped supply, and industrial use of clean water had increased. Sources available to Manchester from the Peak District would soon be inadequate, and the Corporation looked even further afield for drinking water supplies.

Manchester's sights alighted on Thirlmere, a natural lake in the Lake District, as a second major source of fresh water; it was potentially much larger than the Longdendale system but also much further from the city. The level of annual rainfall in the area was also much higher than in the Peak District, thereby providing a considerably larger volume of water from a smaller catchment area. In the face of considerable opposition from a nascent conservation movement, the Parliamentary Act authorising the scheme passed in 1879. Construction on the project started five years later, with a fairly small dam on the northern end of the lake, which significantly raised the level and storage capacity of Thirlmere. Several properties were lost to the rising water and roads had to be altered.

In terms of engineering construction, the 96-mile aqueduct to the city was much more challenging and costly than the work at Thirlmere itself. Much of the aqueduct was in buried pipes, but 13 miles of tunnelling was also required. It was completed by 1894 with major civic celebrations to mark the flow of fresh drinking water from the Lake District to consumers across Manchester. Subsequent schemes in the Lake District saw Haweswater dammed, raising its water levels significantly, and the construction of another very long aqueduct south through Cumbria and Lancashire to the city. It was planned early in the twentieth century but would not be completed until the 1950s.

These were bold, civic-minded, expensive, ambitious and, at the time, controversial schemes, but the building of extensive infrastructure to collect and distribute millions of gallons of clean water by Manchester during the Victorian and Edwardian periods was essential for sustaining the city. Having a reliable and safe water supply is easy to take for granted, but it was a remarkable technical, financial and political achievement, and one that remains essential for urban life today.

Equally remarkable is that these elegantly engineered gravity-feed systems still work pretty much as they were initially designed. Although they are now owned and operated by a private utility company rather than Manchester, the chain of large reservoirs built by the Corporation in Longdendale still collects rainwater for the city from the moors. The reservoirs were a dramatic man-made addition to the valley but today look largely natural and have become popular sites for recreation as well as being important wildlife habitats.

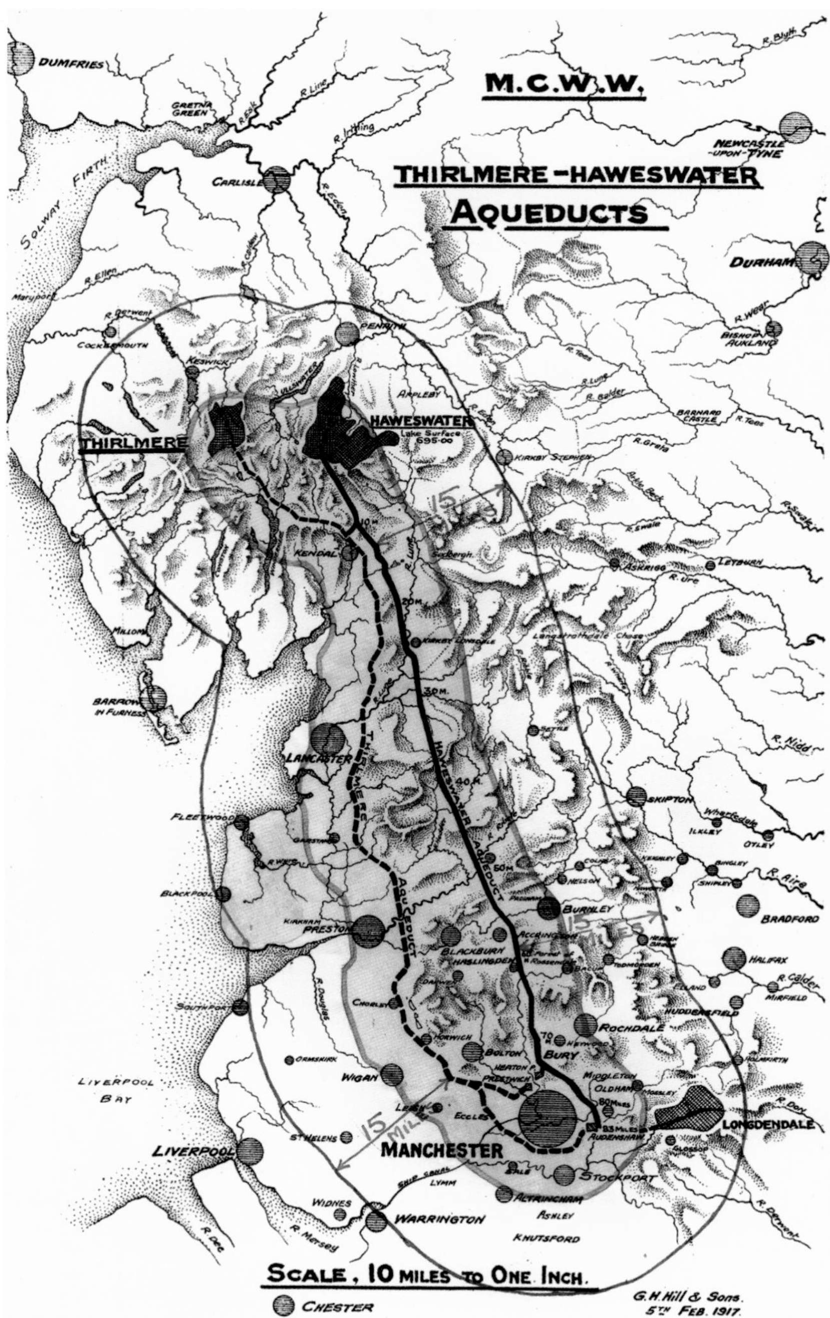

Manchester Corporation, *Thirlmere-Haweswater aqueducts* (1917) [MLA] The two long aqueducts that carried water to Manchester (and to some Lancashire towns along the routes).

Owens College
All Saints Church
Catholic Ch.
Union Chapel
General Burial Ground
POLYGON
PRINCESS

1889a

'The demon drink'

One of the most conspicuous fault-lines in Victorian society was that which separated those who consumed alcohol from those who had renounced it. Temperance supporters viewed alcohol as the root cause of poverty, crime and many other social problems. Initially, temperance societies had focused on persuading individuals to change their attitudes and behaviour, pointing to the personal misery and wider problems caused by the 'demon drink'. This was no easy task given the centrality of alcohol in everyday life. Pubs were the foci around which much community life revolved: places in which to meet, talk and escape from harsh working and living conditions; meeting places for political societies, trade unions, friendly societies, freemasons and sporting clubs; as well as information centres for those seeking work. To swim against these bright currents was incredibly difficult. Most people found it hard to resist the warmth and conviviality of the pub, not least given the cold and cramped housing in which many lived. Temperance societies worked hard to provide an alternative culture, and their supporters were prominent in the campaigns to establish public parks, museums and libraries as counter-attractions to the pub, while over time temperance buildings – hotels, hospitals, coffee shops and billiard halls – and facilities such as drinking fountains were to become part of the streetscape of the Victorian city. Temperance ideals also helped to shape other institutions: Salford was the birthplace of what became the world's largest temperance friendly society – the Independent Order of Rechabites.

From the early 1830s when the temperance movement first put down its roots in northern England, Manchester was one of the centres of the movement. Its importance was further strengthened in 1853 when the United Kingdom Alliance for the Suppression of the Traffic in all Intoxicating Liquors was

---

OPPOSITE. H. Blacklock & Co., *Map of Manchester indicating the licensed victuallers, beer, wine and other licensed houses* (1889) [CHETS]

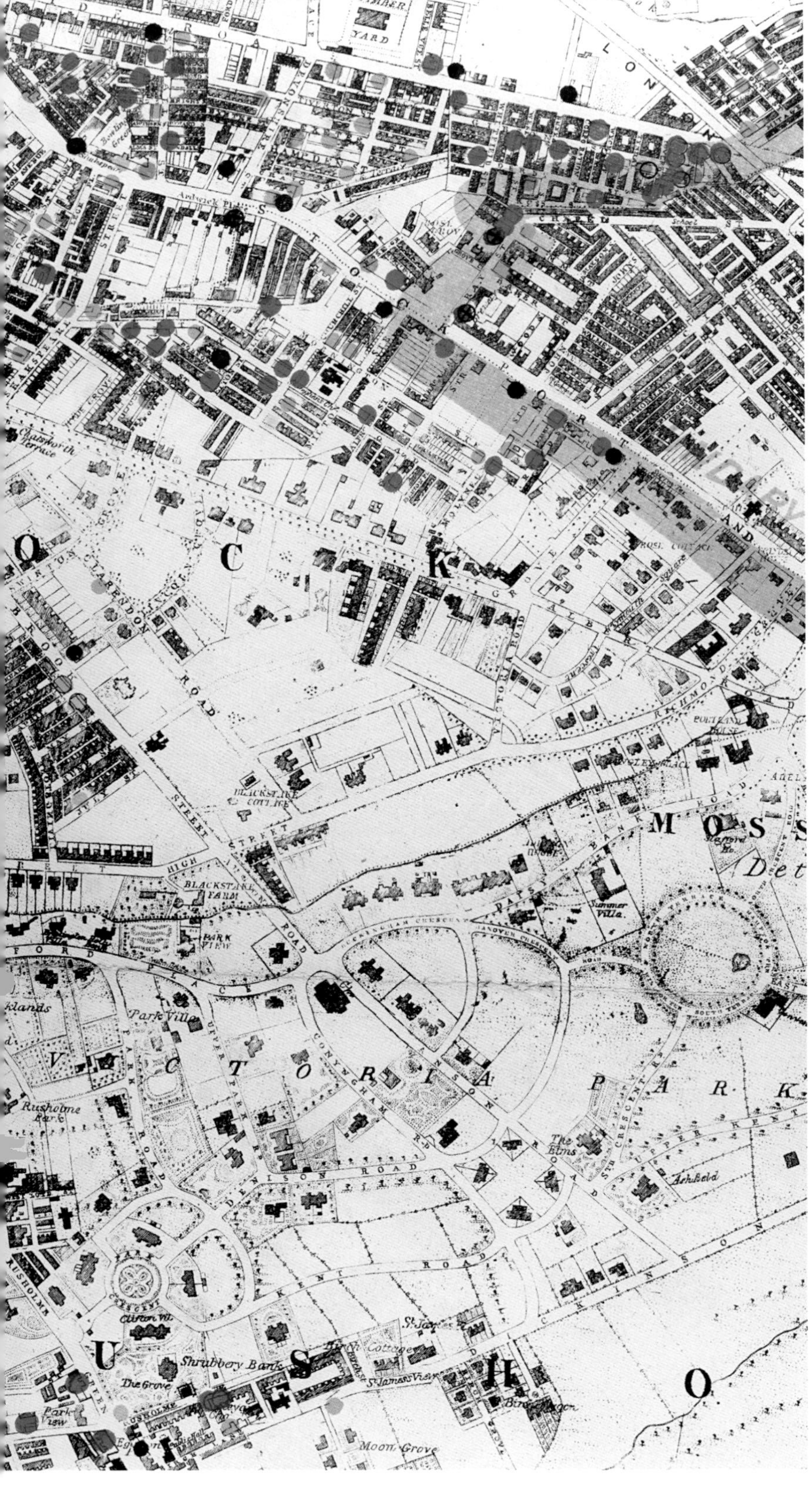

founded in the town. It argued that the policy of moral suasion pursued by most temperance societies had achieved only limited success and that the most effective way of tackling the drink trade was through legislation. It campaigned for the introduction of the local option whereby local ratepayers would vote to decide whether they wanted to prohibit the sale of alcohol in their communities.

Temperance societies relied heavily on the printing press to spread their message. They published books, magazines, tracts, sermons, reciters, novels and biographies of reformed drunkards, all of which aimed to disseminate their arguments, converting individuals to the cause and, of course, strengthening the resolve of those who had taken the pledge. Visual images were also used to spread the gospel of temperance, and, as with George Cruikshank's 'The Bottle' (1847) and 'The Drunkard's Children' (1848), they could have a powerful impact. Maps were to be added to this armoury of temperance propaganda. They were a singular way of visualising the presence of the drink trade in urban communities. Drink maps, which identified the location of licensed premises, were published for a number of towns, including Birmingham (1876), Southampton (1878), Oxford (1883) and Glasgow (1884). The National Temperance Publication Depot's 'The Modern Plague of London' (1886) overprinted licensed premises on Bacon's map of London and suburbs.

Manchester's earliest drink map was commissioned by the Mayor in 1888 for the use of the local magistrates responsible for issuing drink licences. It was a hand-drawn map showing the main roads and streets in the borough on which the location of 'pubs' was marked. Six categories of licensed premises were identified: licensed victuallers; beer to be consumed on the premises; beer to be consumed off the premises; beer and wine to be consumed on the premises; 'sweets' licences, covering the sale of British wines; and other licences. The map was brought to the attention of the wider public when it was published in

There is a striking contrast between Victoria Park and adjacent poorer areas in north Rusholme.

Map key.

the *Manchester Guardian* in January 1889. It occupied a full page of the newspaper and was accompanied by an editorial and an article about the drink trade in the city. It generated discussion. The *Catholic Times* considered the map to be 'more instructive than a volume of sermons' as it demonstrated the extent of the drink trade more clearly than the statistics repeatedly used by temperance groups. More pessimistically, it drew attention to the Herculean task facing the temperance movement after over 50 years of campaigning. It appears to have made an impact as it was subsequently printed as a single sheet by the United Kingdom Alliance, with an explanatory commentary which ended, predictably, with the call that ratepayers be given the local option.

Another version of this map entitled *Map of Manchester indicating the Licensed Victuallers, Beer, Wine and other Licensed Houses* was published in 1889. It met some of the difficulties of using the Alliance's map by marking the licensed premises on a larger-scale map with the different types of licences distinguished by colour symbols. Breweries were also identified. The map was printed by Henry Blacklock, the long-established Manchester printing company more widely known as the publishers of Bradshaw's railway guides. The larger scale and use of colour symbols on the map made it easier to identify the location of the licensed premises. It showed the spatial distribution of pubs in the city with the heaviest concentration of beer houses and pubs in working-class districts such as Ancoats and Hulme. Pubs also lined the main transport routes – Stretford Road, Manchester Road, Great Ancoats Road – and clustered around the city's markets. In contrast, licensed premises of all types were less numerous in the suburbs populated by the middle classes.

In 1888 the total number of licensed premises in Manchester was 2,585, a small increase since 1878, although the borough had increased in size with the incorporation of Rusholme, Bradford and Harpurhey. Beer houses, brought into existence by the easing of the licensing system in 1830 in an attempt to reduce spirit consumption, were the dominant type, especially in working-class districts. Arrests and prosecutions for drink-related offences were in decline but still remained the largest category recorded in the annual police reports. Alcohol still remained a major problem in the city, and for some of its inhabitants drink remained 'the shortest way out of Manchester'.

1889b

A bird's-eye view

Manchester was firmly fixed in the mental map of late Victorian Britain as one of its great provincial cities. Established writers as well as penny-a-line journalists visited the city to describe its businesses, institutions and buildings for the readers of magazines and newspapers. Artists followed, creating their own visual record of the city. One of the most compelling and original images was the aerial view published in the weekly illustrated magazine, *The Graphic*, in November 1889. *The Graphic's* popularity was largely due to its images, which had been made possible by advances in engraving and lithography. Manchester featured frequently in its pages – for example, during the holding of its Royal Jubilee Exhibition in 1887.

The Graphic employed Henry William Brewer (1836?–1903) to provide a large bird's-eye image of the city centre. Brewer was an experienced artist with a talent for architectural perspectives. His interest and appreciation of architecture – he was a member of the Royal Institute of British Architects – was evident in many of his drawings. His aerial view of Manchester followed similar perspectives of Liverpool and Birmingham that he had drawn for the magazine. Brewer would have been aware that he was working in a long tradition that could be traced back to the panoramas and aerial views of European towns from the sixteenth century and, more immediately, to the work of artists such as Thaddeus Fowler and Oakley Bailey who provided numerous views of towns in the United States, commissioned as exercises in city boosterism.

Brewer chose to view Manchester from the Salford side of the Irwell, more precisely from above the recently opened Exchange railway station. As earlier topographical artists of Manchester had discovered, there was no single scenic viewing

---

OPPOSITE. H.W. Brewer, *A bird's-eye view of Manchester in 1889* (1889) [AUTH]

OVERLEAF. The view from Salford with the Irwell and Manchester Cathedral in the foreground.

RAWN BY H. W. BREWER

England

A BIRD'S-EYE VIEW OF

1. Cathedral
2. Exchange Station Approach
3. Victoria Bridge
4. Victoria Terrace
5. Victoria Street
6. Salford
7. Blackfriars Bridge
8. St. Mary's Church
9. Grosvenor Buildings and Hotel
10. Deansgate
11. St. Mary's Gate
12. Victoria Buildings and Hotel
13. Exchange
14. Market Place
15. *Courier* Office
16. Corporation Street
17. Civil Service Stores—Wholesale Department
18. Fennel Street
19. Grammar School
20. Old College
21. Smithfield
22. Great Ancoats Street
23. Ancoats Hall
24. St. Andrew's Church
25. Nicholl's Hospital
26. Ardwick Green
27. London Road Station
28. Portland Place
29. City Police Courts
30. Rochdale Canal
31. Infirmary
32. Piccadilly
33. Lewis's
34. Watts' Warehouse
35. Victoria Park
36. Church of the Holy Name

MANCHESTER IN 1889

37. Eye Hospital
38. {Victoria University / Owens College}
39. Chorlton Town Hall
40. School of Art
41. Independent Church
42. Grosvenor Square
43. All Saints' Church
44. Presbyterian Church
45. St. James's Church
46. Post Office
47. Reform Club
48. Royal Institution
49. Athenæum
50. Town Hall
51. St. James's Hall
52. St. Peter's Church
53. Reference Library
54. Albert Memorial
55. St. Ann's Church
56. St. Mary's Roman Catholic Church
57. Free Trade Hall
58. Theatre Royal
59. Young Men's Christian Association
60. Central Station
61. St. Philip's Church
62. Holy Trinity Church
63. St. Mary's, Moss Lane
64. Alexandra Park
65. Bridgwater Viaduct
66. Bridgwater Canal and Knott Mill Station
67. St. Matthew's Church
68. Tonman Market
69. Dean's Gate Market
70. St. John's Church

point, but the view across the Irwell did provide a reasonably wide sweep of the city and its principal buildings. No account of Brewer's working methods survives, but it can be surmised that he would have started by consulting published maps, establishing a grid on which would be marked the major streets. A reconnaissance on foot would have followed, identifying individual buildings, taking particular note of taller structures visible on the skyline. There is no record of him sketching the city from a balloon as he did when making an aerial view of London, but it can be assumed that he made use of church towers and other tall buildings in observing and making preliminary sketches.

His bird's-eye view of Manchester showed an intensively developed man-made world with few open spaces. Except for the Irwell, which separated Manchester from Salford, the natural features of the landscape had been almost entirely obliterated by the galloping urbanisation of the industrial revolution. The view offered a positive image of a bustling commercial city, not a city of slums. Also largely absent was the city's much-discussed smoke, responsible for that patina of soot, which led some visitors to think that Manchester had been built from blocks of coal. Producing an aerial view demanded artistic skills of a high order, but unlike photographic panoramas – such as Squire Knott's 1876 view of the cotton-spinning town of Oldham – no artist could capture the complexity and detail of the townscape. Manchester was, of course, far too large for the entire city to be represented in detail, a problem that necessitated the generalised treatment of particular areas, the inner suburbs eventually becoming a mass of hazy lines.

In focusing on the centre, it was the city's warehouses rather than its factories and industrial works that were dominant. A key feature of the bird's-eye view was the emphasis given to specific buildings, which were drawn in greater detail and on a larger scale. Brewer drew some 20 buildings, all numbered and identified in a key. The important public buildings illustrated included Waterhouse's Town Hall, in front of which stood Manchester's own Albert Memorial, and major commercial buildings such as the Royal Exchange – the parliament of the cotton lords. Philanthropic Manchester was represented by buildings such as the Royal Infirmary in Piccadilly. The city's importance as a retail centre was indicated by the inclusion of buildings such as the store recently opened by David Lewis in Market Street. Of the city's railways stations, only Central Station, opened in 1880, is easy to identify. The perspective adopted also meant that some buildings, notably the Cathedral with its rebuilt tower and the recently extended buildings housing Manchester Grammar School, were given a greater prominence, indeed one that drew attention away from the modern to the older medieval centre of the town. The Irwell itself in the foreground of the drawing also appears more prominent and certainly cleaner than most Mancunians would have recognised.

As in all such constructed townscapes, Brewer's aerial perspective cannot be viewed as a literal transcription of reality, although when the street layout and his representations of the major buildings are checked against other evidence there is a high degree of accuracy. Nevertheless, the perspective looking across the Irwell from Salford and the prominence given to buildings close to the river is a reminder that other faces of this vibrant and bustling city would have been revealed had Brewer chosen to draw it from other points of the compass.

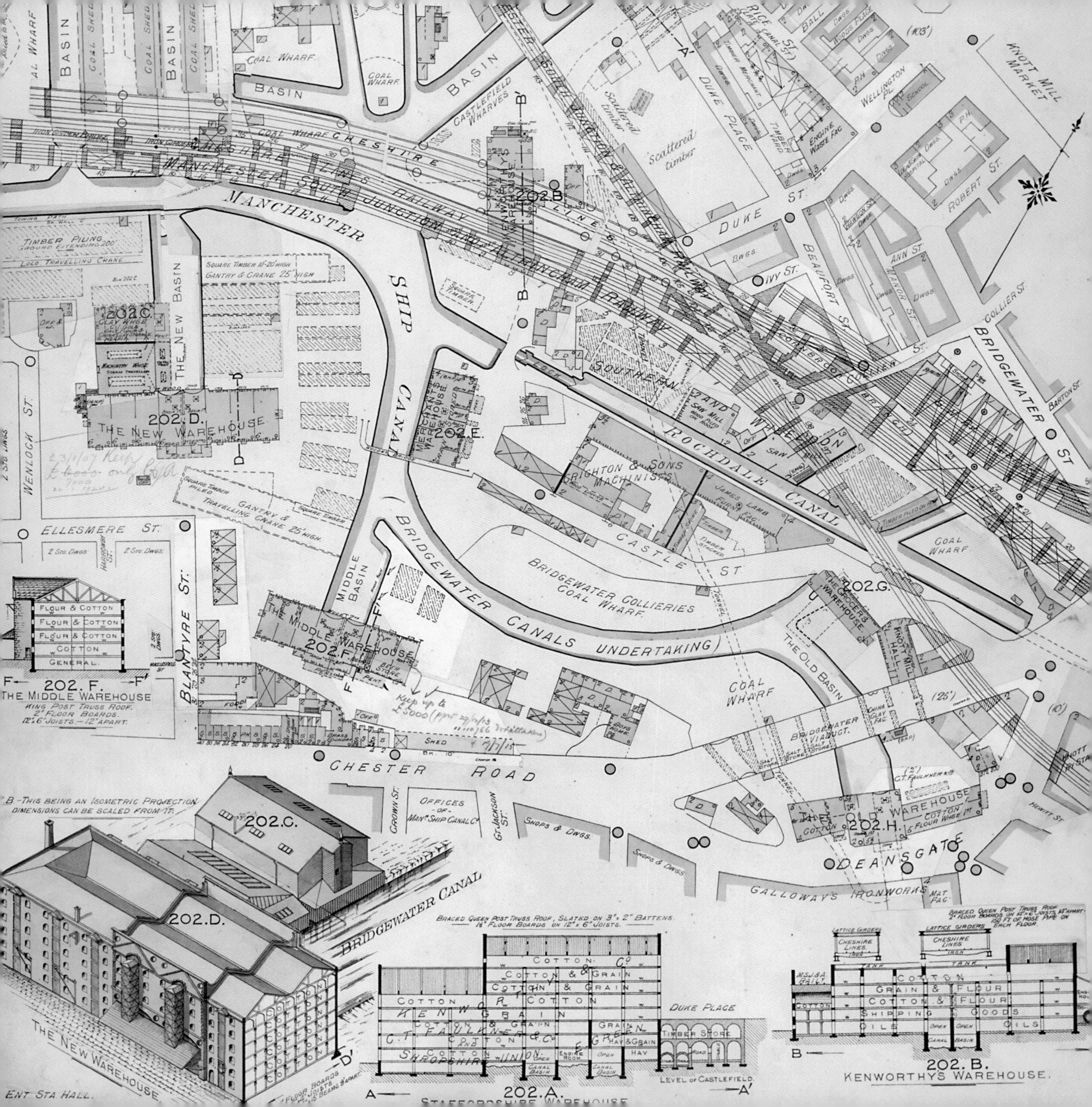

BASIN
COAL SHED
COAL WHARF
CASTLEFIELD WHARVES
CHESHIRE LINES
MANCHESTER SOUTH JUNCTION AND ALTRINCHAM RAILWAY
MANCHESTER SHIP CANAL
TIMBER PILING
LOCO TRAVELLING CRANE
THE NEW BASIN
SQUARE TIMBER 10'-20' HIGH
GANTRY & CRANE 25' HIGH
202.B.
202.C.
202.D.
THE NEW WAREHOUSE
202.E.
MERCHANTS WAREHOUSE
202.F.
THE MIDDLE WAREHOUSE
MIDDLE BASIN
202.G.
THE GROCERS WAREHOUSE
202.H.
THE OLD WAREHOUSE
THE OLD BASIN
KNOTT MILL MARKET
KNOTT MILL HALL
DUKE PLACE
DUKE ST.
IVY ST.
BEAUFORT ST.
ANN ST.
ROBERT ST.
WELLINGTON PL.
COLLIER ST.
BRIDGEWATER ST.
BARTON ST.
SOUTHERN
ROCHDALE CANAL
CRIGHTON & SONS MACHINISTS
JAMES LAMB FURN. FAC.
CASTLE ST
BRIDGEWATER COLLIERIES COAL WHARF
BRIDGEWATER CANALS UNDERTAKING
COAL WHARF
BRIDGEWATER VIADUCT
WENLOCK ST.
ELLESMERE ST.
BLANTYRE ST.
CHESTER ROAD
GROWN ST.
OFFICES OF MAN. SHIP CANAL Co
GT. JACKSON ST.
SHOPS & DWGS.
DEANSGATE
GALLOWAY'S IRONWORKS
HEWITT ST.
FLOUR & COTTON
COTTON
GENERAL
202. F.
THE MIDDLE WAREHOUSE
KING POST TRUSS ROOF.
2" FLOOR BOARDS.
12 x 6 JOISTS - 12" APART.
B - THIS BEING AN ISOMETRIC PROJECTION DIMENSIONS CAN BE SCALED FROM IT.
BRIDGEWATER CANAL
THE NEW WAREHOUSE
BRACED QUEEN POST TRUSS ROOF, SLATED ON 3" x 2" BATTENS.
1½" FLOOR BOARDS ON 12" x 6" JOISTS.
COTTON & GRAIN
KENWORTHY & Co
C. T. FAULKNER & Co
SHROPSHIRE UNION
E. GREEN
DUKE PLACE
TIMBER STORE
LEVEL OF CASTLEFIELD
202. A.
STAFFORDSHIRE WAREHOUSE
LATTICE GIRDERS
CHESHIRE LINES
GRAIN & FLOUR
COTTON & FLOUR
SHIPPING GOODS
OILS
CANAL BASIN
202. B.
KENWORTHYS WAREHOUSE.

1889c

The fear of fire: Goad's insurance plans

Before the industrial revolution, fire ranked alongside epidemics as one of the major hazards of the urban world, a force that was capable of stalling the growth of towns, large or small. Urban development increased the risks associated with fire. By the early nineteenth century, the destruction of factories and warehouses by fire had become a common occurrence in Manchester and its satellite textile towns. The fear of fire encouraged communities to provide themselves with more effective firefighting services, a response that also highlighted the necessity of a reliable water supply. At the same time, concerns over fire encouraged the development of fireproof buildings, using materials such as cast iron and fire-resistant brick.

Fire insurance companies came of age in the industrial revolution as businesses and householders recognised the advantages of insuring their buildings, goods and possessions against loss from fire. London insurance companies led the way, opening agencies in the main provincial towns in the eighteenth century. The fear of a small domestic fire becoming a conflagration engulfing neighbouring houses, and indeed whole streets, was a spectre haunting fire insurance companies. The companies became increasingly sophisticated in estimating risk, but they did not generally invest in producing their own maps, preferring to rely on the available commercial town maps and, from the 1840s, on the large-scale town plans (1:1056) of the Ordnance Survey. Lancashire towns were among the earliest to be surveyed at this increased scale.

This reliance on published maps remained largely unchanged until the late 1880s when Charles Goad established a company to produce plans of British towns, exclusively for the use of fire insurance companies. Goad had gained experience in surveying and publishing fire insurance plans in Canada, a business that was closely modelled on that founded by Daniel Sanborn in the United States.

OPPOSITE. C. Goad, *Castlefield* (1889) [Digital Archives]

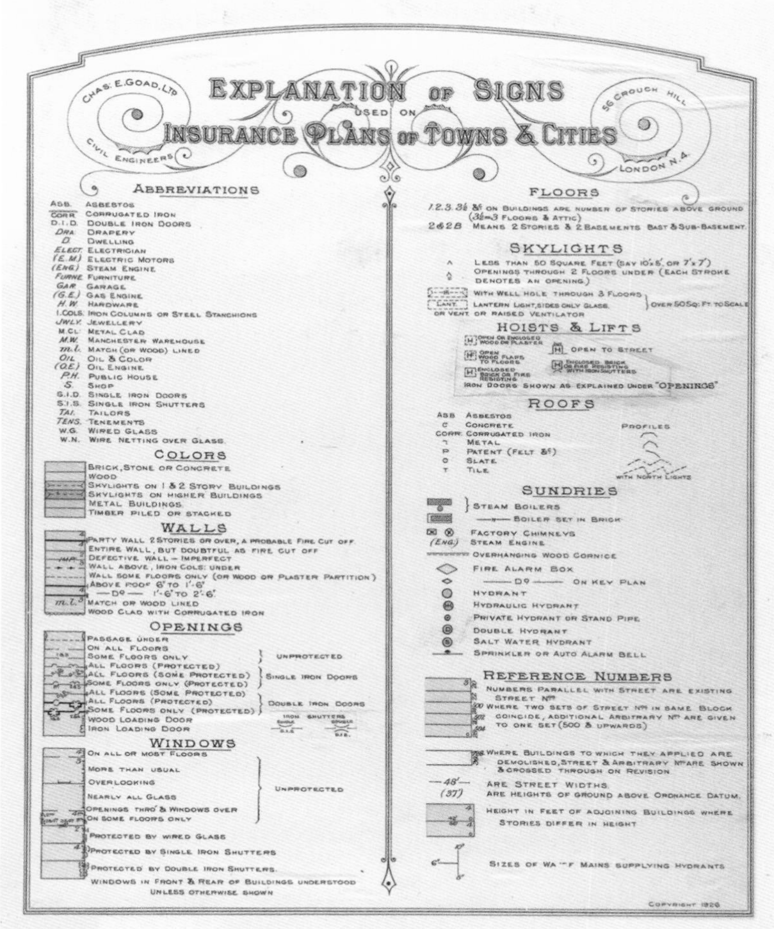

Goad (1926) [Digital Archives] Map key.

Goad's fire insurance plans differed from the Ordnance Survey town plans. They were produced to meet the specific needs of the insurance companies, providing a range of information that would enable them to assess more accurately the risks associated with insuring properties. To that end the plans were on a scale of 40 feet to the inch and tried to cover all properties in a town or city centre. The materials used in the construction of buildings were recorded, each being given a distinctive colour. Details were provided of the internal layout of buildings, particular attention being paid to the construction of roofs, party walls and doorways.

The accuracy of the maps depended on the painstaking work undertaken by the company's surveyors who were required to liaise with local fire departments and, wherever possible, to obtain access to the interiors of building. Once the survey of a town had been completed, the data were returned to Goad's offices in London where the maps were produced. The finished maps were bound in atlas volumes, the number of volumes for any one town being dependent on its size.

Goad did not sell but leased each volume of plans to the insurance companies, reminders being pasted into each volume that they were to be kept safe and under no circumstances copied: 'Every company should bear its share in the expense of the common benefit'.

Goad's plans became known for their scale, detail, legibility and accuracy. The plans were revised regularly. Importantly, a new survey did not usually require the printing of a new map, revisions being added to the original sheets by simply pasting in correction slips, each volume thereby becoming a palimpsest. Thus, in consulting a sheet that is dated, for instance, 1901 but first surveyed and printed ten years earlier, exactly how much of that original sheet remains visible will depend on the number of correction slips added between 1891 and 1901. Identifying both the extent and date of the corrections is difficult even when examining the original volumes.

The sheet covering the Castlefield canal basin provides a sense of the scale and detail of the Goad's plan. It shows the basin at the beginning of the twentieth century. It was a landscape dominated by warehouses, among the earliest of which was the Grocers' and the Staffordshire. Given their commercial importance and their vulnerability to fire, additional information was provided on the plans in the form of isometric drawings allowing further detail to be included, including the types of goods stored. Among the warehouses is the Duke of Bridgewater's, which was located at the Deansgate end of the basin. It was the first warehouse to be built at Castlefield, constructed of stone with shipping holes that allowed barges to be loaded and unloaded inside the building. Unfortunately, it has not survived, being almost completely destroyed by fire in 1919. The sheet covering the medieval part of the town is similarly detailed, showing the lines of original streets and the Market Place as well as more recent developments such as the Victoria Hotel and the new Corn Exchange, both of which occupied distinctive triangular plots.

Goad's business depended on it remaining sensitive to the changing needs of the fire insurance companies and on ensuring that his plans provided more relevant information than could be obtained from the cheaper Ordnance Survey town plans. Goad focused on the business, commercial, retail and industrial districts of towns, since it made little financial sense to extend the coverage to residential districts. However, in addition to the maps covering Manchester city centre, a number of volumes detailing transport warehouses in the wider region were produced. These were known as the Carriers' Warehouses volumes, and they eventually covered the cotton district as far north as Preston and as far south as Stockport, and in the west and east the plans extended to Wigan and Sowerby Bridge respectively.

What were once the working tools of the fire insurance companies have become important historical documents, providing often unique evidence of urban land use and allowing one to trace piecemeal changes in the use of buildings in different parts of the city centre during the late nineteenth and twentieth centuries.

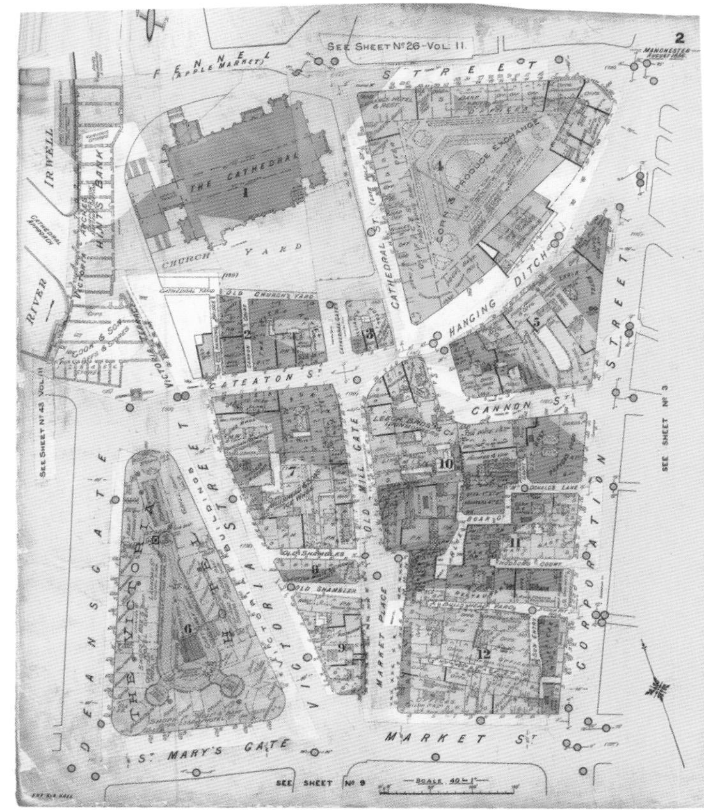

ABOVE AND OVERLEAF.
Showing the wide-ranging commerce of the old core.

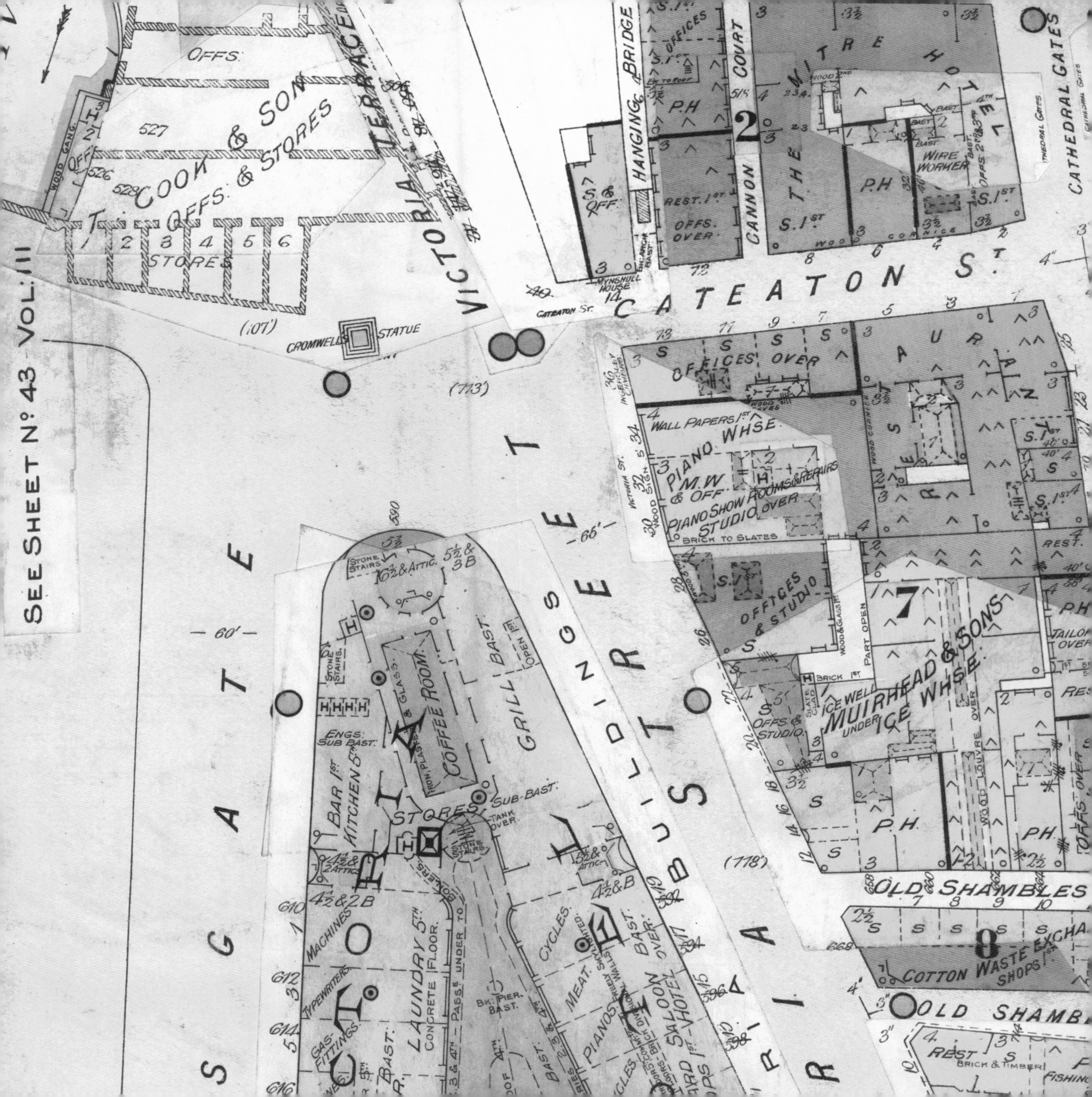

SEE SHEET Nº 43 VOL: III

T. COOK & SON
OFFS. & STORES
OFFS.
STORES
527
526
528
1 2 3 4 5 6
WOOD GANG.

VICTORIA TERRACE

CROMWELLS STATUE
(107')
(113')

HANGING BRIDGE
CANNON COURT
THE MITRE HOTEL
WIRE WORKER
OFFS. OVER
REST. 1ST
P.H
S. 1ST
MYNSHULL HOUSE
CATHEDRAL GATES

CATEATON ST.

RESTAURANT
OFFICES OVER
WALL PAPERS 1ST
PIANO WHSE.
PIANO M.W. & OFF.
PIANO SHOW ROOMS & REPAIRS
STUDIO OVER
BRICK TO SLATES
OFFICES & STUDIO
OFFS. & STUDIO
MUIRHEAD & SONS
ICE WHSE.
ICE WELL UNDER
WOOD LOUVRE OVER
TAILORS OVER
OFFS. OVER

VICTORIA STREET
- 60' -
- 65' -

GRILL
COFFEE ROOM
STORES
SUB-BAST.
KITCHEN 5TH
BAR 1ST
ENGS. SUB BAST.
STONE STAIRS
MACHINES
TYPEWRITERS
GAS FITTINGS
LAUNDRY 5TH
CONCRETE FLOOR
CYCLES
MEAT
PIANOS
SALOON
HOTEL

BUILDINGS

OLD SHAMBLES
COTTON WASTE EXCHA.
SHOPS 1ST
REST.
BRICK & TIMBER
FISHING
(118')

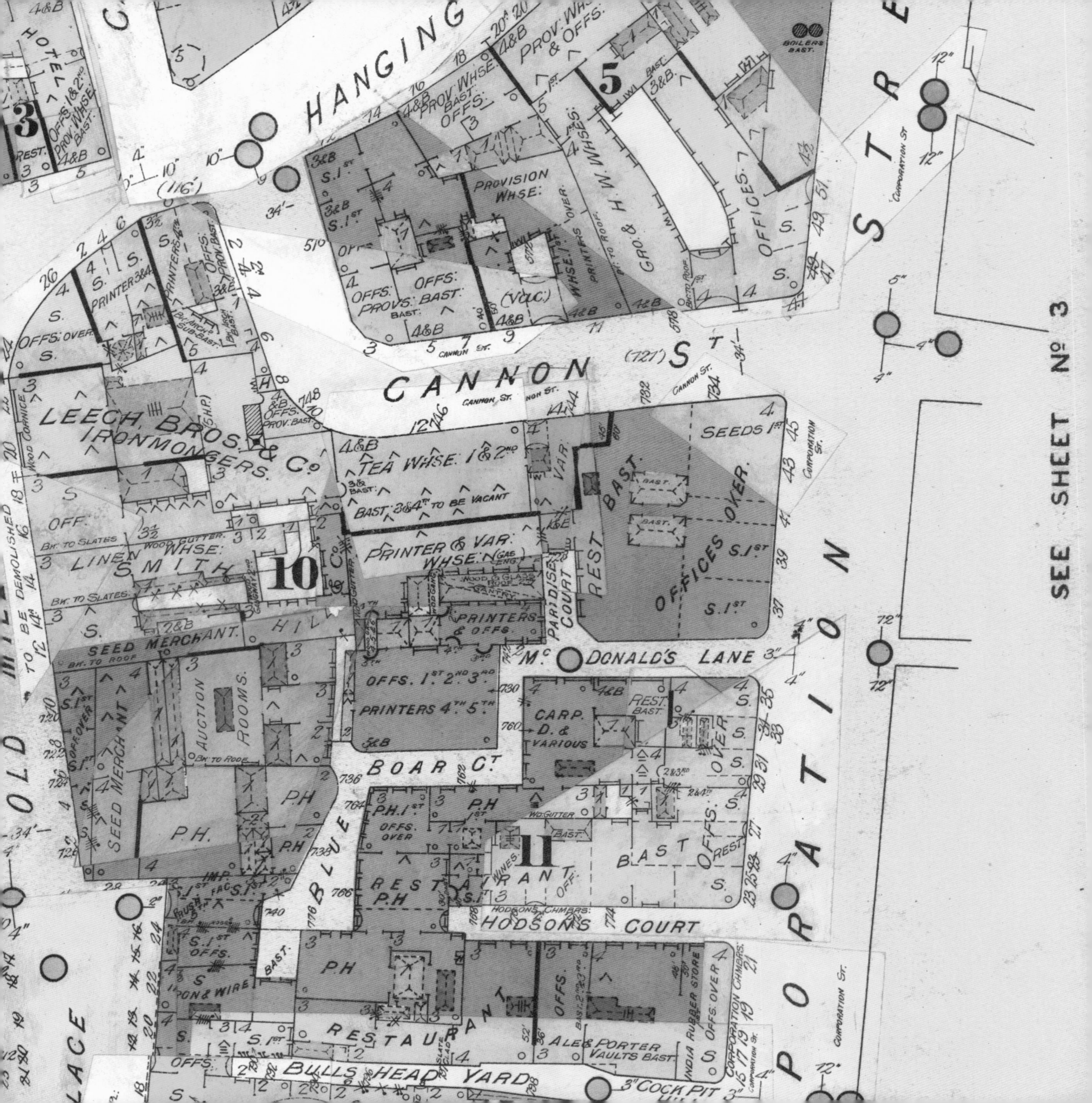

HANGING
CANNON ST
(721')
CANNON ST.
3
5
10
11
HOTEL
PROV. WHSE. & OFFS.
PROVISION WHSE:
GRO. & H.W. WHSE:
OFFICES
BOILERS BAST.
LEECH, BROS & Co
IRONMONGERS
TEA WHSE. 1&2ND
BAST: 3&4TH TO BE VACANT
PRINTER & VAR. WHSE.
LINEN WHSE:
SMITH
SEED MERCHANT
TO BE DEMOLISHED
PARIDISE COURT
REST BAST.
OFFICES OVER
SEEDS 1ST
PRINTERS & OFFS.
McDONALD'S LANE
OFFS. 1ST 2ND 3RD
PRINTERS 4TH 5TH
AUCTION ROOMS
CARP. D. & VARIOUS
BOAR CT.
BLUE
P.H.
REST P.H
HODSONS CHMBRS.
HODSONS COURT
RESTAURANT
OFFS.
ALE & PORTER VAULTS BAST.
INDIA RUBBER STORE
CORPORATION CHMBRS.
BULLS HEAD YARD
COCK PIT
OLD
LACE
CORPORATION ST.
STREET
CORPORATION
SEE SHEET No 3

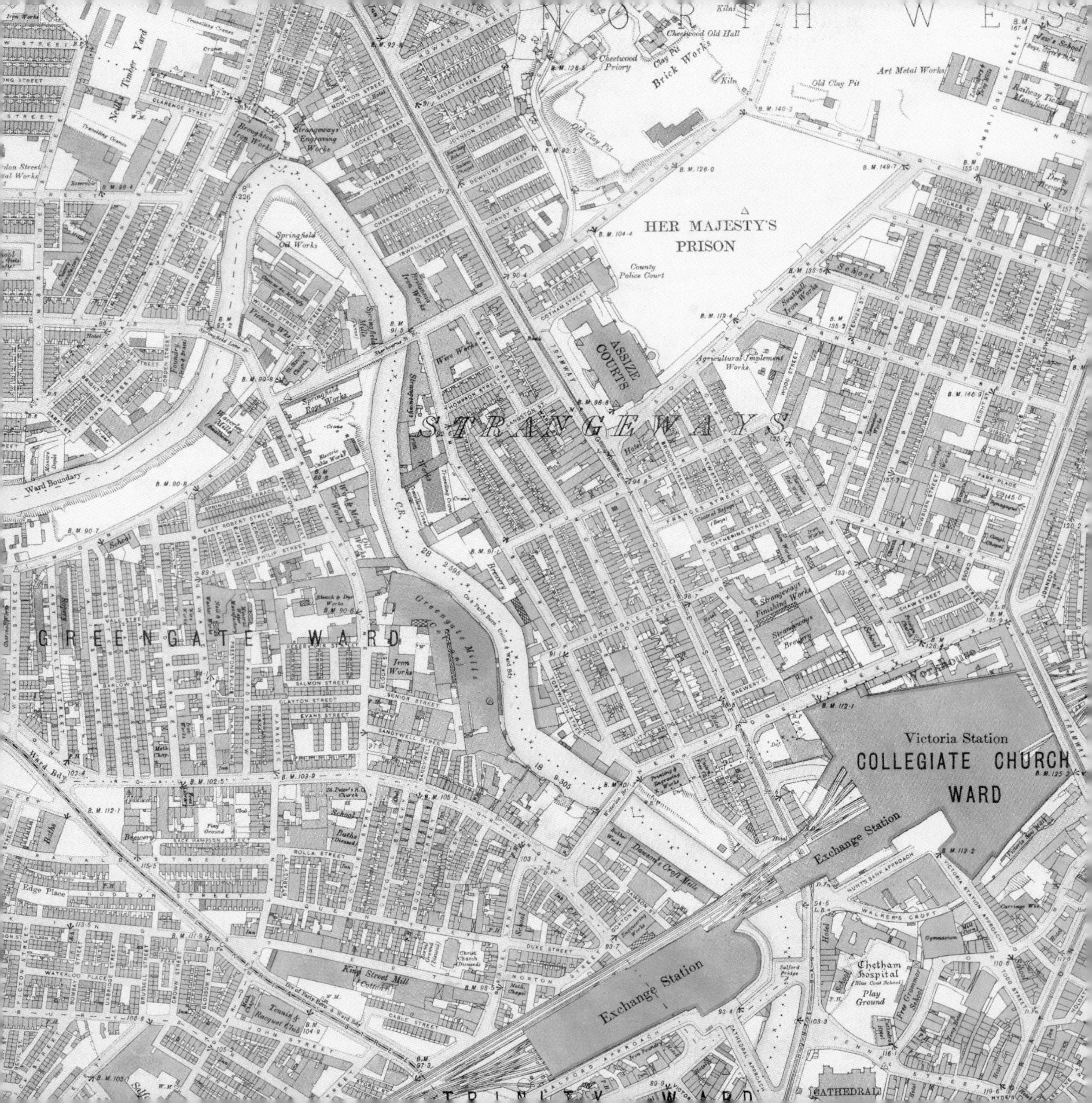

NORTH WEST
HER MAJESTY'S PRISON
County Police Court
ASSIZE COURTS
STRANGEWAYS
GREENGATE WARD
COLLEGIATE CHURCH WARD
Victoria Station
Exchange Station
Cheetwood Old Hall
Cheetwood Priory
Clay Pit
Brick Works
Old Clay Pit
Art Metal Works
Railway Ticket Manufactory
Jews' School
Derby Brewery
Timber Yard
Travelling Cranes
Broughton Iron Works
Strangeways Engraving Works
Springfield Oil Works
Britannia Iron Works
Victoria Wks.
Springfield Mills
Springfield Rope Works
Wire Works
Strangeways Iron Works
Electric Cable Works
Waverley Mills
Ward Boundary
Southall Iron Works
Agricultural Implement Works
School
Central Refuge (Boys)
Strangeways Finishing Works
Strangeways Brewery
Workhouse
Greengate Mills
Iron Works
Bleach & Dye Works
Brewery
Play Ground
Baths
Dawson's Croft Mills
King Street Mill
Tennis & Racquet Club
Christ Church
Chetham Hospital (Blue Coat School)
Free Grammar School
Gymnasium
CATHEDRAL
Salford Bridge
Salford Approach
Cathedral Approach
Victoria Station Approach
Hunts Bank Approach
Walker's Croft
Printing & Engraving Works
Edge Place
Waterloo Place
Salmon Street
Clayton Street
Evans Street
Sandywell Street
Senior Street
Barlow St.
Duke Street
Norton
Cable Street
John Street
Rolla Street
Harris Street
Lockett Street
Moulton Street
Sagar Street
Joynson Street
Dewhurst Street
Cheetwood Street
Irwell Street
Cotham Street
Barker Street
Thompson Street
Langston St.
Frances Street
Catherine Street
Nightingale Street
Brewery St.
Shaw Street
Foulkes St.
Park Place
Wood Street
Springfield Terrace
East Robert Street
East Philip Street
Clarence Street
Catlow St.
Ellison Street
Wilfred Street
Bright St.
Long St.
Carey St.
Springfield Lane
Sherbourne St.
Tramway
C.R.
TRINITY WARD

1892

Salford at the end of Victorian industrialisation

Salford is the sister city to Manchester but has always played a secondary role to its neighbour across the Irwell. It was a separate township, and was granted borough status in 1844 and city status in 1926. Understandably, given the prominence of Manchester, many visitors are simply unaware of the distinct identity of Salford; indeed, even many local people would struggle to clearly define its geographical boundary.

The Ordnance Survey County Series plan from 1892 captures in great detail the development of central Salford at the end of the Victorian era. It was printed at 25.344 inches to the mile which, as Brian Harley and Chris Phillips note, means it:

> . . . delineates the landscape with great detail and accuracy. In fact practically all the significant man made features to be found on the ground are depicted. The series is thus a standard topographical authority.

For most of the country, this was the most detailed cartography available in the nineteenth century. Some sheets could be purchased with hand-colouring, as with the Salford sheet shown here. Most buildings were shaded in carmine, indicating brick or masonry construction; the odd grey structures were made of timber; the presence of glass roofs is indicated by hatched blue shading.

The sheer density of building is striking as is the defining meander of the Irwell that divides the two cities. The original settlement of Salford centred around Holy Trinity Church. Most of the medieval remnants were obliterated by new building in the nineteenth century, not least the opening of Exchange Station in 1884. Its footprint on the OS map shows that it was a sizeable and busy station, famous for being linked to Victoria Station via a shared platform nearly 700 metres long that spanned the river.

By the time of the OS survey, virtually all land within the

OPPOSITE. Ordnance Survey 25-inch [Salford] (1892) [UML]

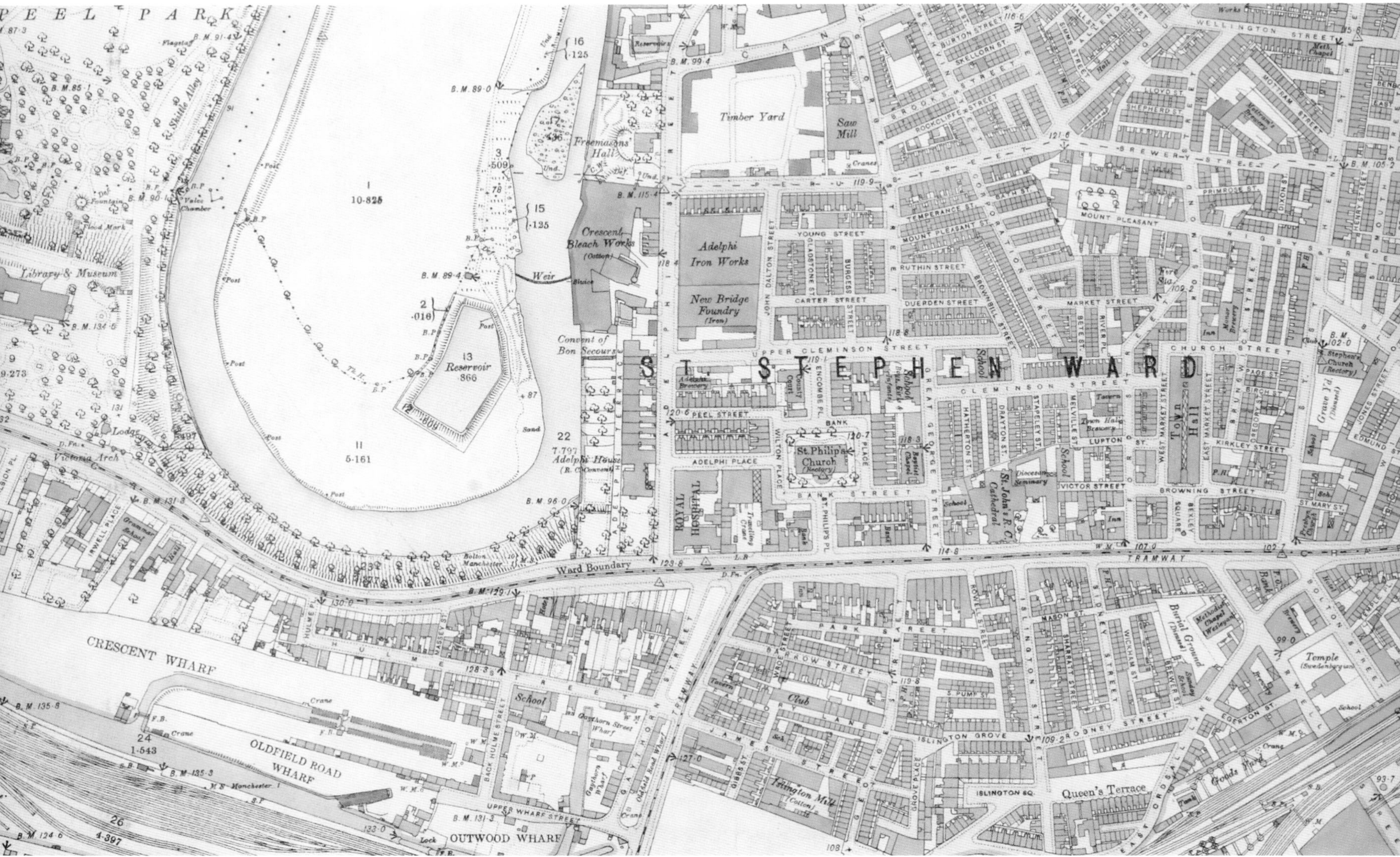

Dense development in the core of Salford contrasts with the Crescent and the Library and Museum to the left.

defining loop of the Irwell was densely built-up with acres of terraced housing and extensive industrial activity, including large ironworks, sprawling cotton-mills and textile manufacturing such as bleaching and dyeing. In the middle of the town were the Salford Gas Works with three distinctive circular gasometers on Bloom Street. The map shows how the banks of the Irwell were favoured for industrial activity, exploiting the river as a source of water and a means of disposing noxious wastes. It was a forbiddingly urban setting, with no recreation grounds or public parks in the central area (although there was a large Racquet Club on Blackfriars Street, which included an indoor court for real tennis). Salford had become the quintessential 'dirty old town' of Ewan MacColl's famous folk song.

Chapel Street was then – and remains today – the main thoroughfare through Salford. Fronting onto Chapel Street near The Crescent were the major civic institutions: the Town Hall, built with private money, and opened in 1827 on the elegantly laid-out Bexley Square; St John's, the Catholic Cathedral consecrated in 1848; and the Royal Hospital, which had originally opened in the 1820s and was significantly expanded by the 1880s. Chapel Street leads to Peel Park, and on to Pendleton and more spacious suburban housing beyond. Salford's Library and Museum on Lark Hill was one of the first municipally-run free libraries in Britain.

Given the prominence of the Irwell, with buildings crowded along both banks, it was not surprising that there were many serious floods, including particularly damaging incidents in 1866, 1881 and 1886. Their impact was amplified as the river channel had been narrowed by development and the dumping of rubbish, and the natural floodplains had been built on. In response, there were proposals to better control the river, including radical re-engineering such as digging large tunnels to bypass the meander through the city centre. None were realised. However, one advantage of the relative narrowness of the Irwell channel was that it could easily be spanned and, as the map shows, by the 1890s there were nine road bridges in the space of a couple of miles.

One of the largest industrial concerns in Salford by the 1890s, and enjoying an extensive river-fronted site, was Greengate Mills, a substantial cotton manufacturing complex operated by the brothers Edward and George Langworthy. They had opened a much smaller mill on the site in the 1840s and it expanded over subsequent decades, spinning yarns and producing khaki drills, flannelettes and varied kinds of printed textiles. As well as enjoying considerable success as a Victorian cotton magnate, Edward Langworthy was also heavily engaged in the civic life of Salford, serving at different times as a councillor, alderman, MP and Mayor. He was also a generous benefactor, giving substantial gifts to cultural institutions including Salford's Library and Museum, funds for school education and an endowment for a professorship in experimental physics at Owens College, now the University of Manchester. Known as the Langworthy Chair, this has enjoyed an impressive lineage being held by a series of Nobel Prize winners (Ernest Rutherford, Lawrence Bragg, Patrick Blackett, Andre Geim and Konstantin Novoselov).

Over the years, there have been proposals to merge the twin cities of Manchester and Salford for administrative efficiencies, given their close proximity and shared economy. Yet many Salfordians passionately defend their separate identity, asserting the cognitive and cultural divide at the Irwell boundary. This tribal allegiance to locality stems from attachment to the place where one is born and grows up. The character of central Salford – captured dramatically on the OS 25-inch map – was one of dense housing, dirty industry and the constraining presence of the River Irwell, but was a place called home by many thousands of people.

The Langworthys' Greengate cotton-mill.

1894

Manchester Ship Canal: 'The Big Ditch'

The 'Big Ditch' stretched for over 36 miles from its terminal docks on the edge of the city centre, shown on this detailed bird's-eye view, to the village of Eastham on the southern bank of the Mersey estuary. It was a hugely ambitious piece of civil engineering aimed to bring ocean-going merchant ships direct to Manchester. It came to typify the audacity of the late Victorian age in transforming the environment. At its peak, some 17,000 workers were involved in construction and it cost a fortune to complete – over £15 million (about £1.65 billion in today's value). It cost the lives of many workers – it is reported that 154 men were killed on site, with a further 186 suffering permanent injuries – yet from its opening in 1894 the Ship Canal was a source of great pride for the city. It was endlessly promoted as a sign of the industriousness and ingenuity of Manchester men.

While it was one of the most audacious projects of the nineteenth century, the aspiration to sail vessels directly to Manchester had started much earlier. Parliamentary powers were granted in 1720 to allow the Mersey and Irwell to be dredged, widened and straightened. However, the volume of traffic along the river in the eighteenth century was limited by the relatively modestly sized flat sailing barges that could move along the channel. Likewise, the barges that could use the Bridgewater Canal were limited in size. The Mersey Irwell Navigation Company considered plans to make it possible for larger vessels to reach Manchester, but no action was taken.

Given the volume of raw imports required to supply Manchester's industries and the range of exports from its warehouses, by the middle of the nineteenth century many industrialists in the city and the surrounding towns felt they were being held captive by Liverpool and its dock dues and the onerous railway charges for moving freight between the two cities. It was widely reported that the high transport costs were a crucial impediment to profits and threatened the very

---

OPPOSITE. H. Blacklock & Co., *Port of Manchester, Manchester Docks* (1894) [Canal Maps Archive]

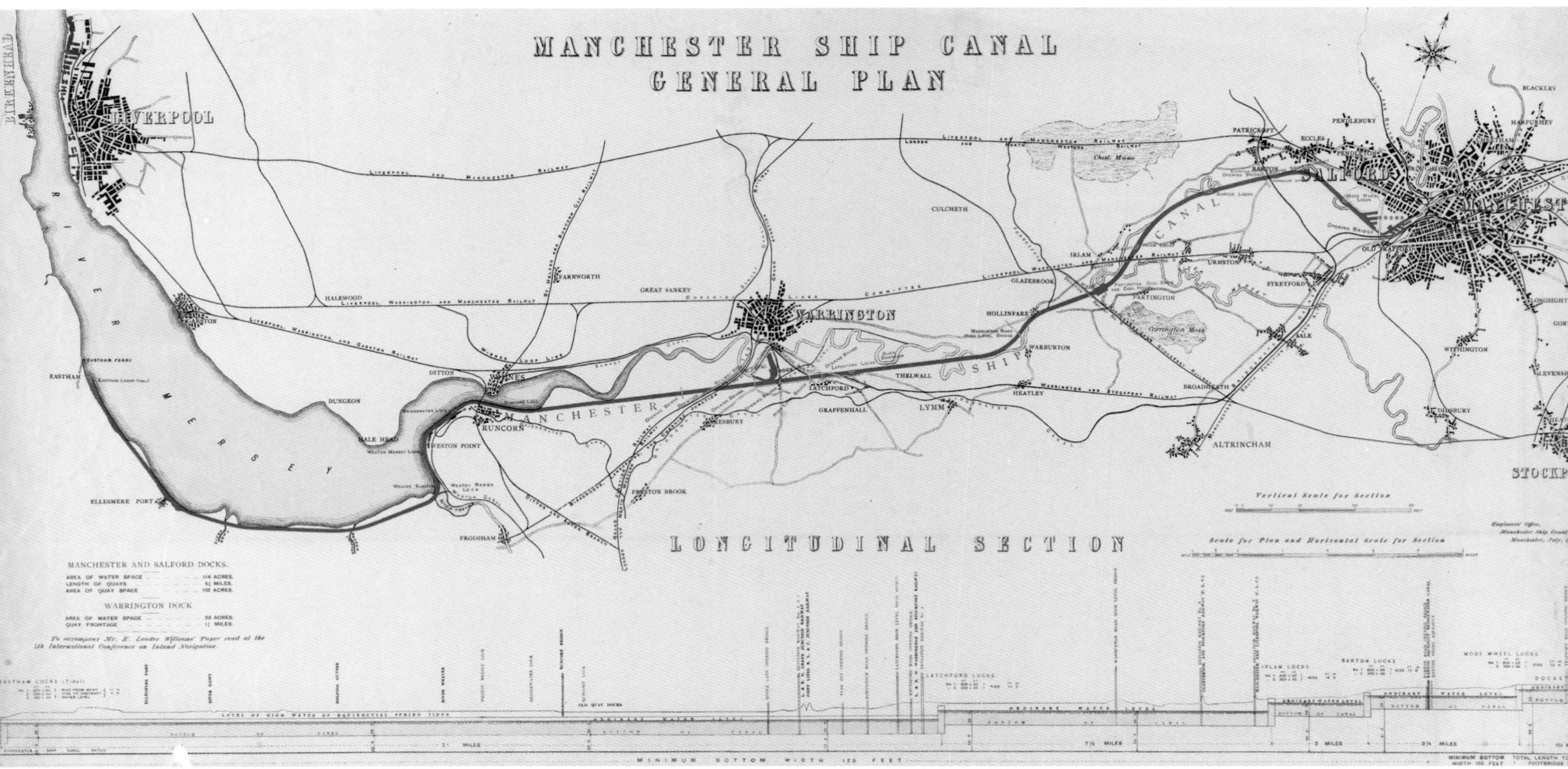

Ship Canal Company, *Manchester Ship Canal, general plan . . .* (1890) [UML] The long extension by the side of the Mersey from Runcorn to Eastham reflects Liverpool's attempt to scupper the canal.

competitiveness of Manchester as a commercial centre.

Various schemes for a ship canal to Manchester were put forward, but the crucial stage was a meeting of wealthy industrialists, local politicians and other influential men organised by the engineer and successful boilermaker Daniel Adamson at his house, The Towers, in Didsbury in 1882. A committee was formed, and detailed engineering proposals were solicited. The plan judged best was by Edward Leader Williams, which included using locks to maintain water levels rather than relying wholly on tidal flow to move the ships to Manchester.

The battle by promoters of the MSC to obtain parliamentary authorisation was protracted and hard fought. There was fierce opposition from factional interests, including the Liverpool harbour authorities, railway operators, the owners of the Bridgewater Canal (who also had the rights of the Mersey and Irwell Navigation) and various large landowners, including the de Trafford family who owned the manorial estate, which would be in the line of the canal as it approached Salford and Manchester. Many people expressed scepticism as to whether the proposal to sail ships along so many miles of narrow channel to a landlocked city was possible in engineering terms and, just as importantly, in terms of economic feasibility. Nevertheless, the third private bill eventually received Royal Assent in 1885, having been scrutinised by six different Select Committees who

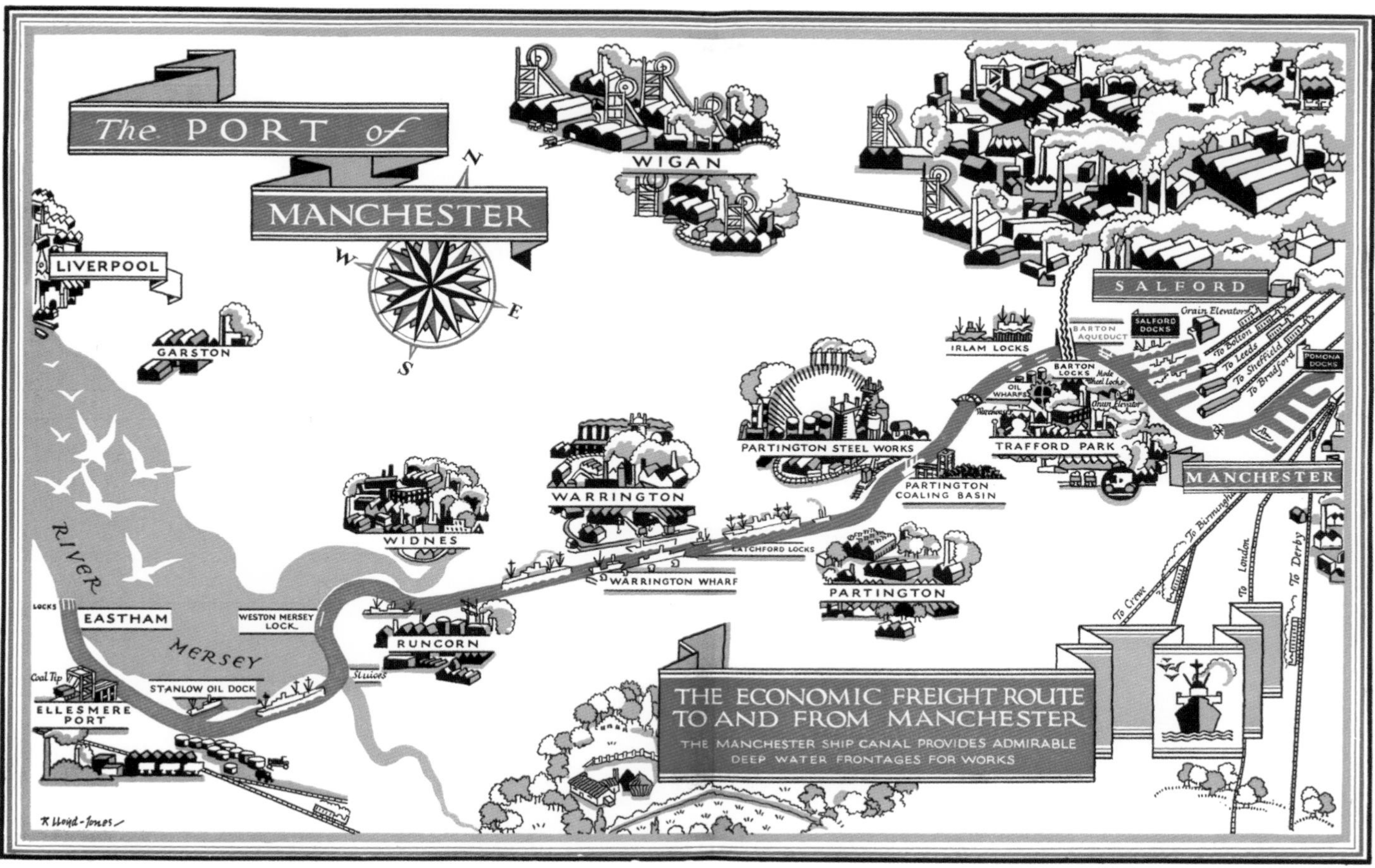

R. Lloyd Jones, *The Port of Manchester* (*c.*1927) [AUTH]

heard evidence from 543 witnesses and asked 87,936 questions.

Importantly, the battles in parliamentary committee rooms had a major consequence for the actual route of the canal. Engineering witnesses called on behalf of the Mersey Docks and Harbour Board opposed the original plan for a dredged channel down the estuary, claiming it would disturb tidal flows and impair the docks at Liverpool. To try to thwart Manchester's scheme, they argued for an expensively engineered extension to the canal along the shore of the Mersey and for entry into the estuary much lower down. Rather than resisting, the MSC engineers acquiesced, and the result is the apparent oddity of an indirect route to the sea whereby the canal hugs the Cheshire shoreline for many miles, extending to Eastham and deep-water access to the estuary.

Construction finally started in 1887. It was a massive piece of geo-engineering, excavating a channel over 35 miles long with a depth of 8.5 metres and bottom width of 36.5 metres, sufficient to allow ships to pass each other. In total, it is estimated that 41 million cubic metres of soil and rock had to be removed, and the vast amounts of spoil raised surrounding farmland by several metres. Some of the spoil was simply piled up to create Mount Manisty near Eastham, an artificial hill some 30 metres high. Many bridges were needed for existing roads, railways and canals, and these had to be at a high level

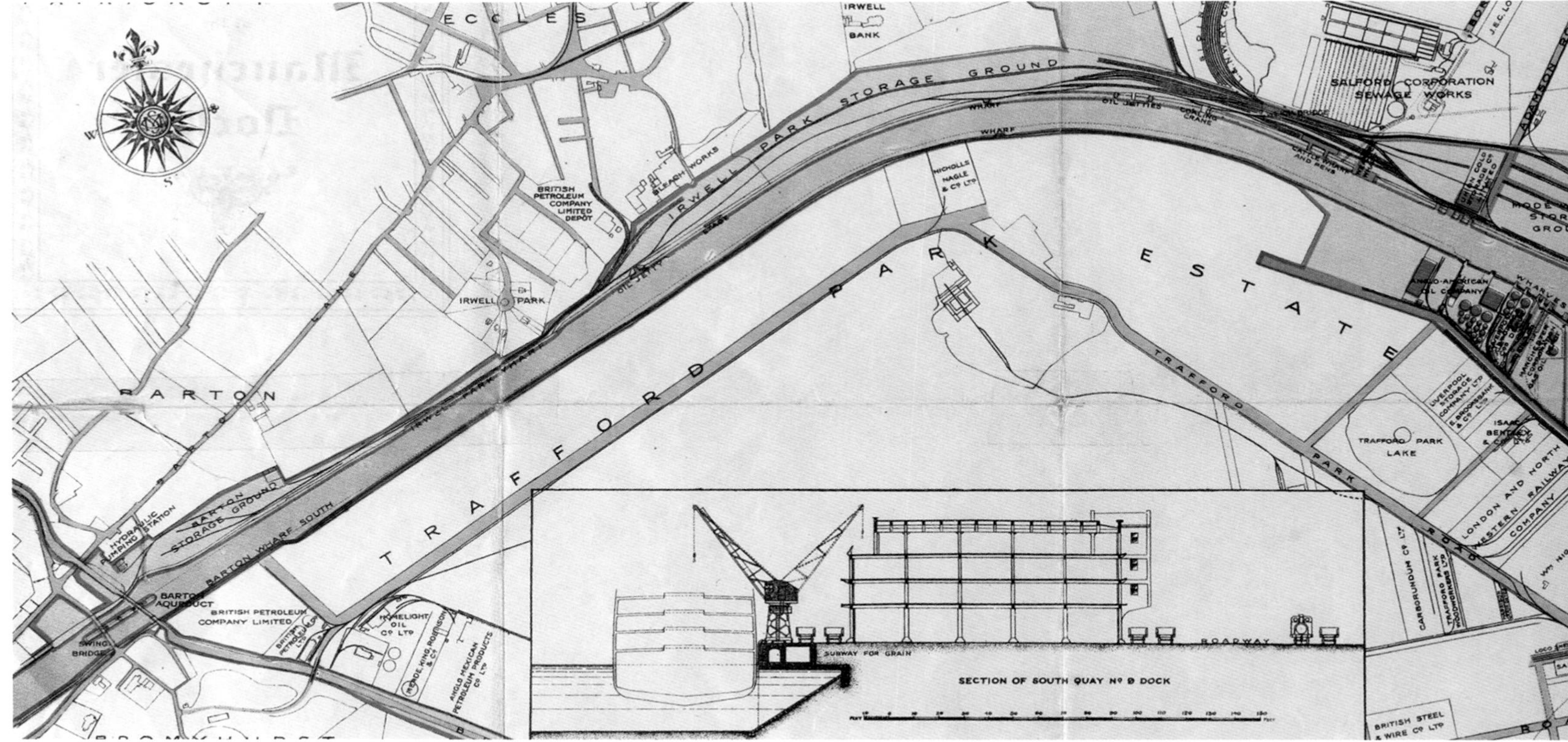

or designed to be swung open to allow large vessels to pass.

The costs also rose markedly as construction was underway as a result of various delays, including winter flooding in 1890 and 1891. Towards the end of the project, it was in danger of failing financially and it was bailed out by Manchester Corporation. So while it started as a private venture, it was only finally achieved with the support of the city. It was in this respect, truly *Manchester's* Ship Canal when it was completed.

The public opening was held in the new year of 1894 when great crowds came to see an official flotilla of some 70 vessels sail up to Salford, led by the steam yacht *Norseman*. The steamship *Pioneer*, owned by the Co-operative Wholesale Society, claimed the honour of being the first vessel registered to make commercial use of the new docks when it sailed up MSC with a cargo of sugar. The city had become a port, a fact marked with official pomp a couple of months later when Queen Victoria ceremonially opened the lock gates at Mode Wheel, to the accompaniment of a 21-gun salute fired from the Manchester racecourse.

The Manchester Docks, the terminus of MSC, were built as close as possible to the city and were designed to accommodate large ocean-going ships. Extensive quayside facilities and warehouses were needed to handle the expected volume of goods arriving and departing annually. The chosen location was an area of low-lying, marshy ground in the Irwell floodplain, land that had remained relatively undeveloped despite a century of industrialisation and urban growth. The port was known as the Manchester Docks even though most of the land was actually in Salford.

The bird's-eye perspective captures in great detail the scale and complexity of the working port. As well as the main docks, there were long warehouses, plus many cranes, steam trains and carts. The prominent structure in the foreground was a large grain store. A strip of land along the Trafford side of

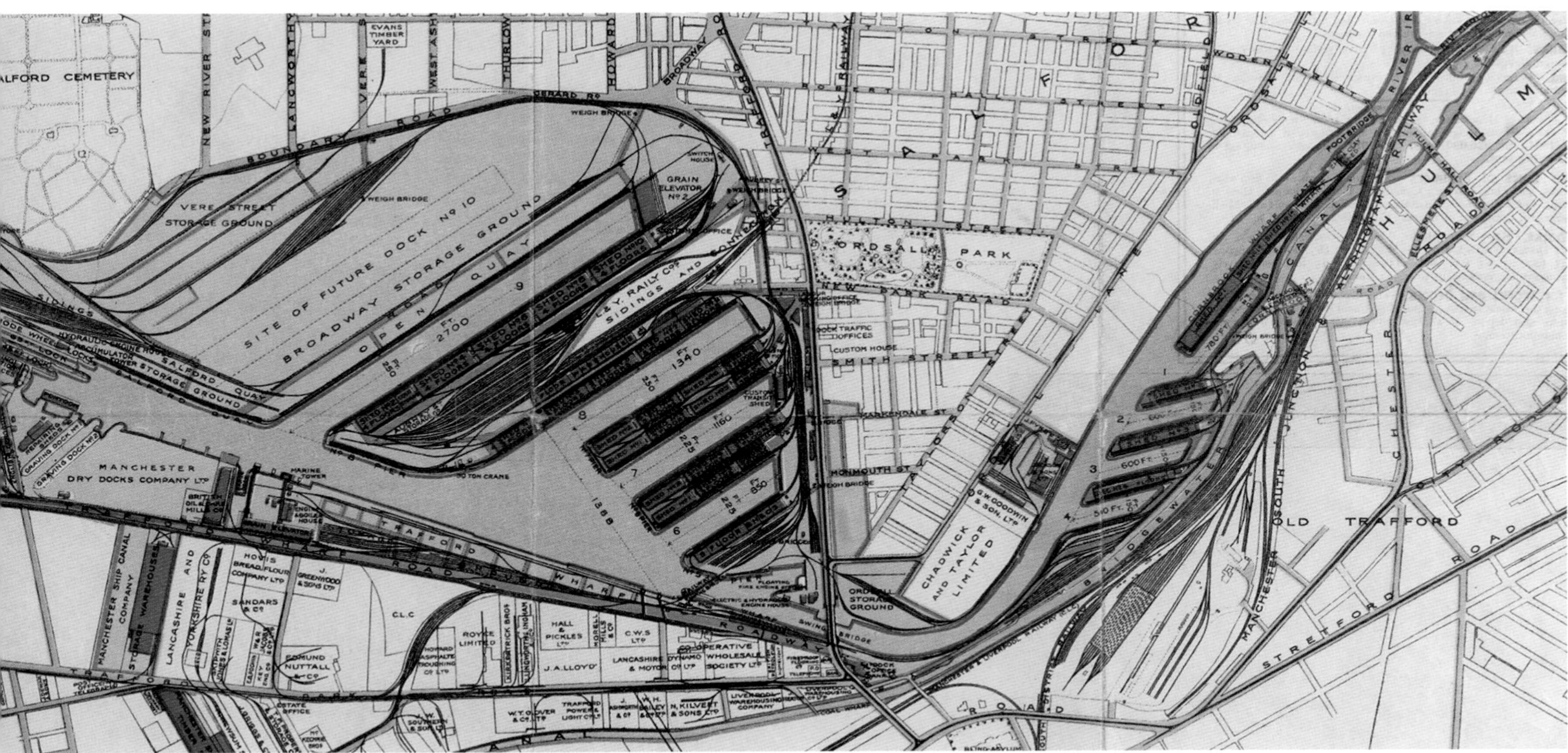

MSC was purchased by the company from the de Trafford's for use as wharfs, prior to the establishment of Trafford Park. The range of port facilities along the half-kilometre-long Trafford Wharf included extensive timber yards, where wood was stacked to season, and stock pens to handle imports of sheep and cattle, along with auction facilities and abattoirs. The Pomona docks catered for canal barges and smaller ships with shallower draft. All the quays and warehouses were interconnected by a network of railway tracks. By the Edwardian era and up to the First World War, traffic grew, with the annual tonnage increasing from 2.9 million in 1901 to over 5.5 million in 1913. The port was also significantly expanded when the Canal Company took over the racecourse site to build Dock No. 9, opened in 1905 by King Edward VII.

The Port of Manchester was particularly busy during the Second World War and enjoyed boom times in the subsequent decade. Peak traffic for the canal was in 1959 when over 18.5 million tonnes were handled. Despite precipitous decline from the 1970s, and the subsequent closure of the docks, the Ship Canal remains open as a working waterway linking Manchester to the sea. It also serves a vital role as a drainage system for the wider region.

Well over 120 years since its construction, most of the system remains in place. It is one of the greatest industrial monuments of the nineteenth century. For Manchester, the 'big ditch' was clearly a bold undertaking, and while one should not overstate its impact on the regional economy, it was clearly a significant infrastructure that shaped the city through much of the twentieth century.

ABOVE. Ship Canal Company, *The Manchester Docks* (1914) [UML] The extensive railway system is evident. 'Future Dock No. 10' was an anticipated extension but was never constructed.

BENTCLIFFE BLEACH WORKS
BARTON LANE FARM
MANCHESTER SHIP CANAL
CRICKET GROUND
STABLES
THE MANSION
DEER PARK
SALFORD
MODE WHEEL LOCKS
SITE OF MANCHESTER CATTLE LAIRS
DRY DOCKS
Boat House
FISH POND
Deer Shed Wood
Old Deer Shed Wood
KEEPERS LODGE
Barton Lodge
BRIDGEWATER
SIR HUMPHREY CANAL
TRAFFORD MOSS HOUSE
Township Boundary
(Barton) (Stretford)
Aviary
Keepers Lodge
Hattons Wood
LAND BELONGING TO SIR HUMPHREY E. DE TRAFFORD, BARONET.
MOSS FARM
MOSS BRIDGE
MANCHESTER MAIN OUTFALL
BRIDGEWATER
CHESHIRE
WATERS MEETING FARM
TAYLORS BRIDGE
WATERS MEETING
Land belonging
To Stretford
Plan of TRAFFORD PARK, MANCHESTER
For Sale by
MESSRS. CHINNOCK, GALSWORTHY, & CHINNOCK.
1896.

1896

Trafford Park: the first industrial estate

The plan from the sale brochure for Trafford Park came at the turning point of its fortunes. In 1896 it was still a handsome rural retreat. Its 1,183 acres were well wooded with a large deer park, a fishpond and three farms. The Hall itself sat in the midst of large gardens with a pond, fountain and aviary. It still had its three entrance lodges, one in the north-west at Barton and two in the east at Throstle Nest and Old Trafford. They gave access to the Hall along tree-lined drives.

The Park was part of the estate of the de Traffords. However, with the opening of the Ship Canal, it sat in what had effectively become an island demarcated by the Bridgewater Canal to the south and the Ship Canal to the north. Sir Humphrey Francis de Trafford had vigorously opposed the construction of the Ship Canal and, while he failed to stop it, he forced the Canal Company to build a 10-foot wall along the 5-mile frontage between the Park and canal and provide a wharf for the exclusive use of the de Traffords. Despite this, the ever-encroaching development of the city prompted him to offer the Park for sale. In 1893, the City Council established a committee to advise on the purchase with a view to creating public gardens. However, faced with the high costs of both the Thirlmere reservoir and the Ship Canal, the city decided against further major expenditure. Sir Humphrey offered the Park for sale in May 1896, but the auction price was never reached. When he readvertised it later in the year, it was bought by Ernest Terah Hooley, a financier who made his money by buying and reselling companies. Fraudulent practices led to his serving two prison terms. He went bankrupt in 1898 and played no further part in the Park's fortunes. However, his purchase of the Park set in train the creation of the world's first industrial estate.

Hooley floated Trafford Park Estates Ltd and his initial concept was to create a high-status residential area, but he was persuaded to use the land for industrial development by Marshall Stevens, who had been General Manager of the Canal

---

OPPOSITE. Chinnock, Galsworthy & Chinnock, *Plan of Trafford Park* . . . (1896) [MLA]

Company but was appointed by Hooley to manage the Park. Stevens led the development of the estate as general manager and chairman for over 30 years and was initially embroiled in tripartite struggles between the Ship Canal Company, railway interests and his beloved estate. The relationship with the Canal Company was complex: the two shared mutual interests in maximising trade, but the company protected the interests of its docks on the north of the canal by opposing the construction of competing docks on the estate on the south side. The railway companies in turn were cautious about supporting what was in effect a rival private company looking to develop rail links to serve the Park.

Stevens, however, proved a doughty fighter. He built a 3-mile tramway and succeeded in getting a railway link with the Canal Company. An initial gas tramway was followed by a light railway in 1901 and an electric tramway in 1903. A vestigial network was beginning to evolve to create increasingly good connectivity. However, in its early years, the estate struggled to attract industrial investment. It operated at a financial loss for its first ten years. The 1902 map shows that industrial development was relatively limited and concentrated on the eastern side, which benefited from its links to Manchester and Salford. The west was given over to sporting and leisure activities. The Hall itself was used as a hotel; the ornamental lake was leased for boating; and in the grounds were two golf courses and a polo field.

However, significant putative developments were already obvious. Not least was the creation of the British Westinghouse Electric Company, a subsidiary of the American Westinghouse firm. It purchased no less than 11 per cent of the Park and started operations in 1902, making turbines and electric generators. By the following year, it had some 6,000 employees, half of the total jobs in the Park. In 1919, it split from its American parent to become Metropolitan Vickers. Its involvement in the

Park had two significant consequences: the attraction of large numbers of American firms and the boost it gave to the creation of a residential 'village' to house employees.

The 'Village' was designed on an American gridiron pattern with 12 numbered streets running east–west crossed by 4 numbered avenues. Its residents were catered for by three churches, schools, a working men's club, a library, wash-house and clinic. It also included an ornate hotel. Much of the estate survived until the final houses were demolished in the 1980s.

The First World War gave a significant boost to the industrial park. The vacancy rate fell from 51 per cent in 1915 to 34 per cent in 1921. The third plan, for 1936, shows this transformation shortly before the start of the Second World War. Not only had most sites been occupied, but the railway network had filled out to create accessibility for almost every part of the estate. Indeed, the railway carried no less than 3 per cent of the total freight on all railways in Great Britain. Second, more than 300 American companies operated in the Park and, in 1938 they were joined by Kelloggs, a highly significant inward investment. The plan shows some of these American companies: the Southern Cotton Oil Company, the British Reinforced Concrete Company, the Rubber Regenerating Company, the English Textilose Manufacturing Company, the Carborundum Company and the Trussed Concrete Steel Company. The major American company not shown on the plan is Ford Motors, which started production in 1911, initially as an assembly plant importing car bodies from America, but soon as a fully operational plant manufacturing its own bodies. It was the first Ford factory outside America, and the success of its right-hand drive Model T Ford made it the largest car manufacturer in Britain. It introduced Ford's revolutionary assembly-line production system at about the same time as it was developed in America. Its Trafford factory had been located in the plot at the junction of Westing-

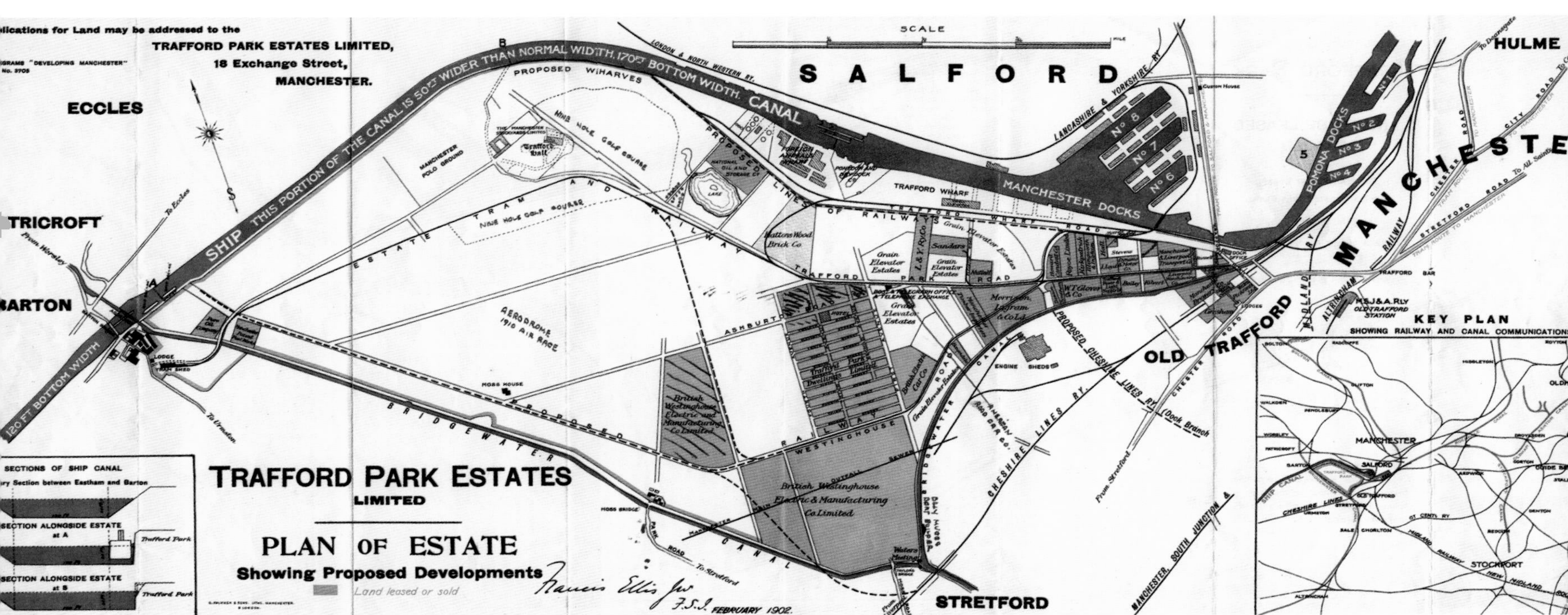

Trafford Park Estates Ltd, *Plan of Estate showing proposed developments* (1902) [MLA]
The contrast between the east and west reflects the initially slow development of the Park.

house Road and First Avenue (square 60/76), but the company moved to Dagenham in 1931 to facilitate access to the European market, hence its absence from the 1936 plan.

Over the years, other significant firms that operated in the Park have included GEC, Ciba-Geigy, CWS, AEI, Procter & Gamble, ICI, Glovers Cables and Kraft Foods. The wide mix of chemicals, light and heavy manufacturing, foodstuffs, fuels and minerals provided a counterpoint to 'Cottonopolis', thereby helping to soften the impact of the Great Depression on the city's economy. A notable name on the map is Arthur Guinness, whose company bought a site in 1913 (square 52). The intention to open his first brewery site outside Ireland was frustrated by the First World War, and in 1932 the company decided that a London base would be preferable and it brewed beer at Park Royal in north London from 1936. Today, the only signifier of the Guinness connection with Trafford Park is the naming of Guinness Road.

The Second World War proved the zenith of the Park's fortunes. War production gave it an enormous boost and employment rose to a peak of 75,000. The most notable feature was the output of Rolls-Royce Merlin engines, manufactured from a redeveloped Ford factory, which opened in the Park in 1941. The Merlin powered the Avro Manchester bomber produced by MetroVicks at Trafford and the much more numerous Avro Lancaster bomber produced at Chadderton. The Park suffered extensive damage in the blitz of December 1940 when the MetroVicks factory was badly damaged. In the later attack of May 1941, the newly reopened Ford factory was bombed, but rapidly recovered and was able to produce over 30,000 Merlin engines by the time it closed at the end of the war. The other symbolically significant casualty was Trafford Hall itself. Having survived the industrialisation of the Park, it was badly bombed in 1940 and was duly demolished, thereby ending the most tangible link between the Park and the de Trafford family.

Trafford Park Estates Ltd, *Trafford Park Manchester* (1936) [MLA] The extensive holdings of Metropolitan Vickers and the fully built 'village' shortly before the Second World War.

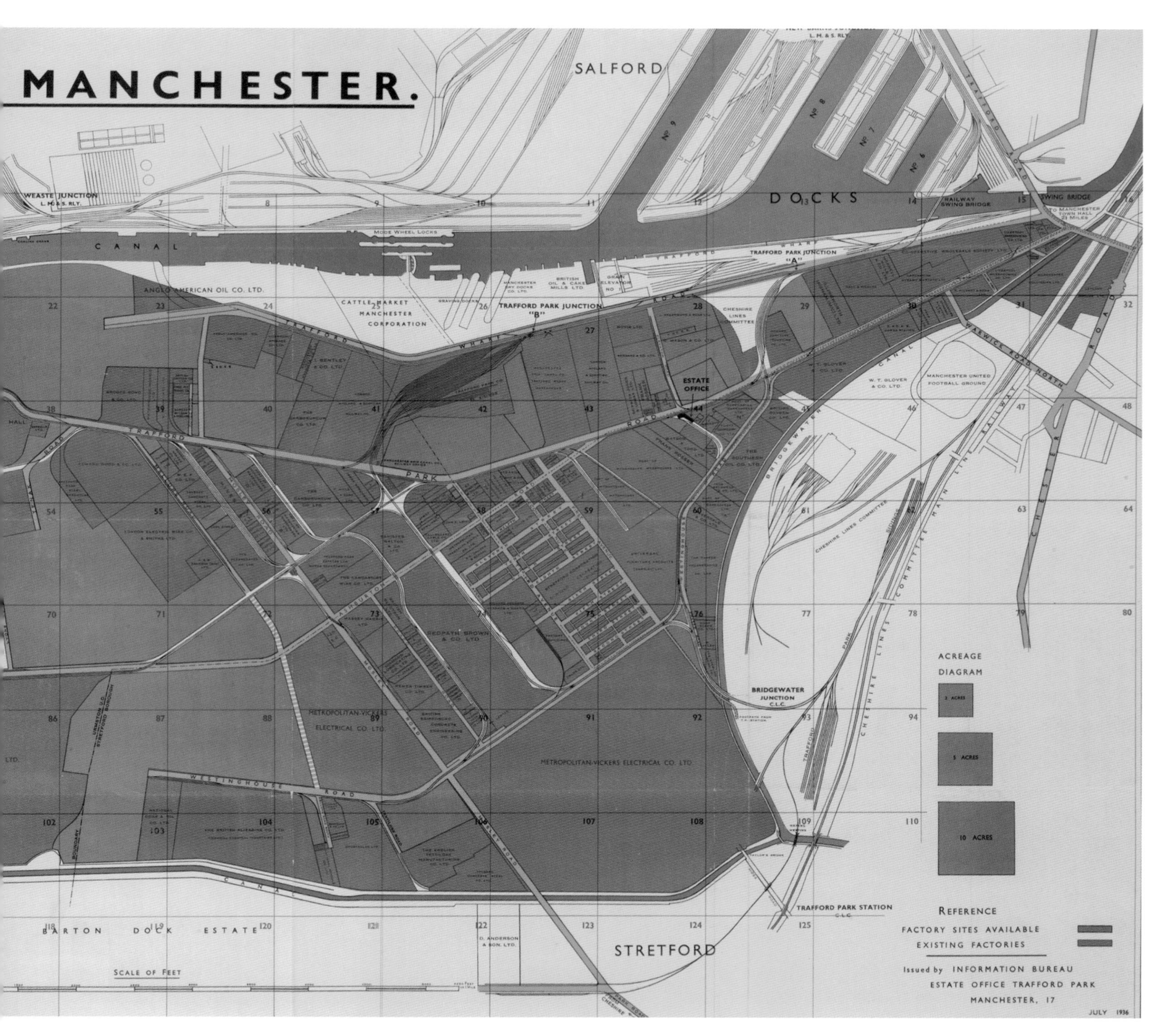
MANCHESTER.
SALFORD
DOCKS
CANAL
WEASTE JUNCTION
L. M. & S. RLY.
MODE WHEEL LOCKS
ANGLO AMERICAN OIL CO. LTD.
CATTLE MARKET
MANCHESTER
CORPORATION
TRAFFORD PARK JUNCTION
"B"
TRAFFORD PARK JUNCTION
"A"
TRAFFORD WHARF ROAD
RAILWAY
SWING BRIDGE
SWING BRIDGE
TO MANCHESTER
TOWN HALL
2½ MILES
TRAFFORD ROAD
CHESHIRE
LINES
COMMITTEE
ESTATE
OFFICE
TRAFFORD PARK ROAD
W. T. GLOVER
& CO. LTD.
MANCHESTER UNITED
FOOTBALL GROUND
WARWICK ROAD NORTH
CHESTER ROAD
BRIDGEWATER CANAL
CHESHIRE LINES COMMITTEE
ASHBURTON ROAD
REDPATH BROWN
& CO. LTD.
METROPOLITAN-VICKERS
ELECTRICAL CO. LTD.
METROPOLITAN-VICKERS ELECTRICAL CO. LTD.
WESTINGHOUSE ROAD
BRIDGEWATER
JUNCTION
C.L.C.
TRAFFORD PARK STATION
C.L.C.
BARTON DOCK ESTATE
D. ANDERSON
& SON. LTD.
STRETFORD
SCALE OF FEET
ACREAGE
DIAGRAM
5 ACRES
10 ACRES
REFERENCE
FACTORY SITES AVAILABLE
EXISTING FACTORIES
Issued by INFORMATION BUREAU
ESTATE OFFICE TRAFFORD PARK
MANCHESTER, 17
JULY 1936

REFERENCE NOTE TO COLOURING.
BUSINESS PROPERTY
Offices, Warehouses, etc.
Works, Factories, etc.
Railways, Stations, Sidings, Canals, Docks, River Irwell, Ponds, etc.
DWELLING-HOUSE PROPERTY
Back-to-back houses
Slum property
Converted back-to-back property
Property which complied with earlier Bye-laws
Property which complies with modern Bye-laws
Suburban houses with gardens
UNBUILT AREAS
Public Parks, Recreation Grounds, & Cemeteries
Other unbuilt on ground is left white
Scale of Half a Mile
BROUGHTON
CHEETHAM
Cheetwood
Stocks
COLLYHURST
MILES PLATTING
NEWTOWN
STRANGEWAYS
VICTORIA STATION
EXCHANGE STATION
PEEL PARK
ANCOATS
HOLT TOWN
BESWICK
LONDON ROAD STATION
CENTRAL STATION
KNOTT MILL STA.
OXFORD RD. STA.
ARDWICK
ARDWICK STA.
Ardwick Cemetery
GREAT CENTRAL RAILWAY
MIDLAND RAILWAY
LANCASHIRE & YORKSHIRE RAILWAY
ROCHDALE CANAL
BOLTON CANAL
RIVER MEDLOCK
CORNBROOK
HULME
CHORLTON UPON MEDLOCK
OLD TRAFFORD
Site for Model Dwellings
Goods Yard (L.&N.W.)
WHITWORTH
Art Gallery

1904

Mapping Manchester's slums

Thomas Marr's dramatic map charts the legacy of a century of house building in the city. It traces out roughly concentric rings: at the centre, a large commercial core; next, a collar consisting of slums, back-to-backs and 'converted' back-to-backs, interspersed with industry; further out is a ring of earlier and later bye-law housing; and finally, suburban houses with gardens. It largely reflects the age of housing, but this is complicated by an overlay of sectors largely determined by the network of railways cutting their way towards the centre of the city. Clearly, the coloured areas were not homogenous, but the categories depict the predominant types of housing within them. The map was compiled by a local surveyor, J.R. Corbett, and it illustrated Marr's 1904 book on housing conditions in Manchester and Salford. The report drew on the work of members of the Citizens' Association for the Improvement of the Unwholesome Dwellings and Surroundings of the People, a pressure group established by Thomas Horsfall, founder of the Art Museum in Ancoats and advocate of the ideas of German municipal socialism. Marr was its secretary.

The legacy of much of the city's nineteenth-century housing was clearly appalling. Jerry-built back-to-backs, cellar dwellings and common lodging houses provided accommodation that was cold, dark and dank, and this was reflected in the city's high death rates. Manchester's housing featured prominently in the Victorian parliamentary commissions into urban problems as well as in the investigations of bodies such as the Manchester Statistical Society, which conducted numerous surveys of living conditions in the city and elsewhere. A survey in 1834 of over 4,000 families in the St Michael's and New Cross areas found that three-quarters lived in houses, about one-fifth in cellars and the remainder in rooms. Only one-third of the dwellings were considered 'comfortable'. Follow-up surveys across Manchester found no fewer than 3,600 cellar dwellings with an average of over four persons per cellar and

---

OPPOSITE. T. Marr, *Map showing housing conditions in Manchester & Salford* (1904) [NLS]

estimated that 11 per cent of the town's population lived in cellars. Words failed Friedrich Engels when he attempted to describe the living conditions of the poor in the centre of Manchester, and Elizabeth Gaskell deliberately set a poverty-wracked family in a cellar dwelling at the moral centre of her novel *Mary Barton: A tale of Manchester life*.

Pressure groups such as the Manchester and Salford Sanitary Association continued to campaign for social and housing reform. The reformers took pains to spell out the awfulness of the slums, arguing that the town's comfortable citizens must have been unaware of the circumstances in which their poor fellow citizens lived otherwise they would not have tolerated the continuation of such housing. The 'unwholesomeness' of the houses and environments was a compound of factors: the absence of effective building regulations and the lower rates of return on working-class housing, which encouraged the use of inferior materials and the cramming of as many houses as possible into whatever plots of land were available.

Nonetheless, Manchester was a pioneer in tackling such conditions through local legislation. A local Police Act in 1844 led to the banning of new back-to-backs. In 1853, the Police Commissioners sought to prevent living in cellars, and the number of inhabited cellars slowly began to decline. John Leigh, the city's first Medical Officer of Health, appointed in 1868, gave priority to closing the remaining cellars. In 1867, the Manchester Waterworks and Improvement Act gave the Corporation power to declare individual properties unfit and enforce improvement. This gave Leigh power to convert back-to-backs by encouraging landlords to drive through internal walls to convert them into 'through' houses, or to demolish pairs of back-to-backs built in sets of four, or to demolish one of a pair of dwellings in order to give the remaining house a backyard and access to air.

Progress, however, proved slow. Marr's map shows the extent of the 'slum' areas that still remained at the turn of the century. By 1904, few unconverted back-to-backs remained, reflecting the work done by the city. However, the map does show that there were still many in Angel Meadow and a spattering in other areas across the city. Marr looked at seven sample areas in some detail, and two examples provide a flavour of what remained to be tackled. One area, in Angel Meadow, included 273 houses with a mix of lodging-house tenements and back-to-backs. Only 20 had water closets and the remainder had pail privies used by numerous families. In one street, 40 houses shared access to a single water tap. The OS map shows the details of the area.

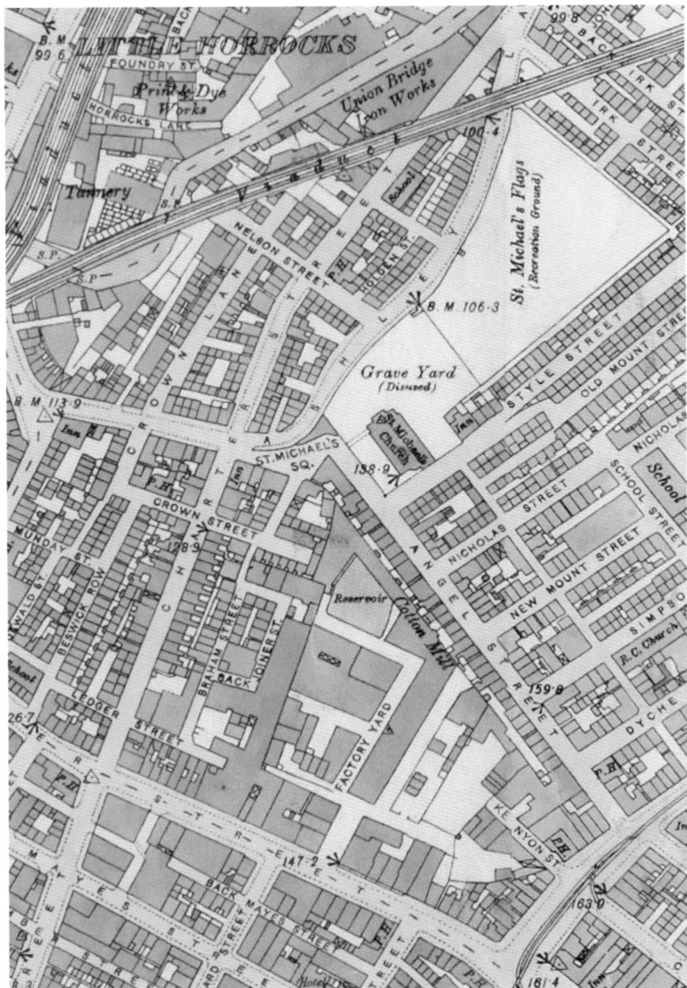

Ordnance Survey 25-inch (1892) [UML] The ill-named Angel Meadow with numerous back-to-back houses in the inner parts away from the main roads.

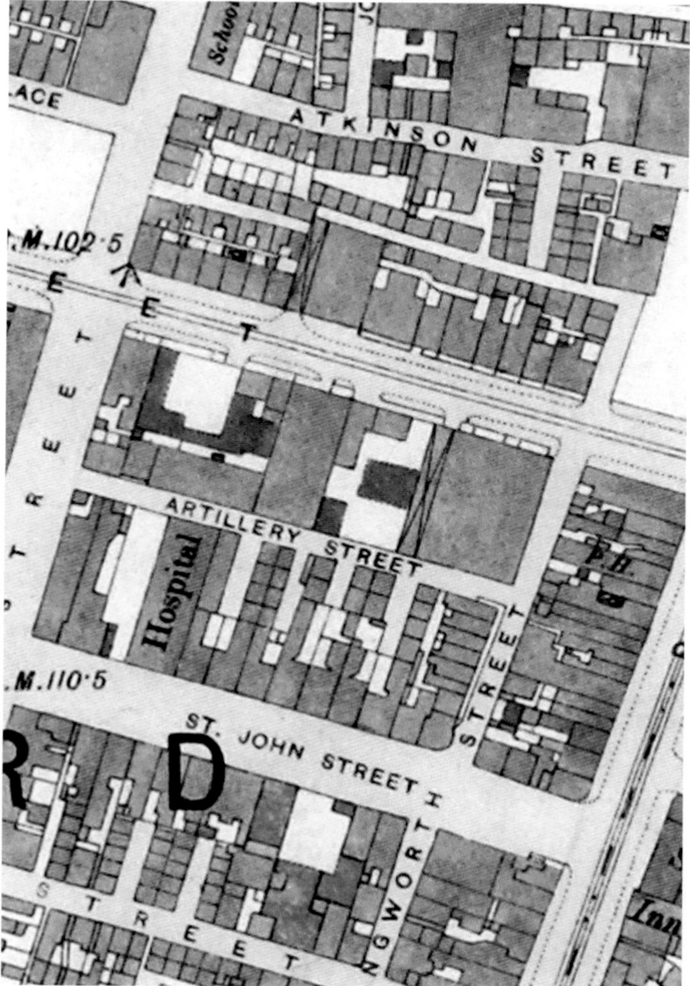

The contrast between large houses on St John Street and the narrow courts and back-to-back houses lying behind.

A second example illustrates cases where builders squeezed back-to-backs into small spaces behind more substantial houses. In the area behind St John Street, in which large handsome Georgian houses were then (as now) largely used as chambers for professional businesses, were five courts, three of which were extraordinarily narrow cul-de-sacs, one a mere 10 feet wide. In total, they contained 25 back-to-backs, a lodging house and a shop, with one pail closet for every three dwellings. The juxtaposition of grand houses and back-to-backs is shown on the OS plan. It has strong overtones of Engels's view of the ecology of the city that the 'money aristocracy can take the shortest route to their places of business, without ever seeing that they are in the midst of the grimy misery that lurks to the right and left'.

The Council's first major foray into the provision of new accommodation was the large-scale clearance of some 36 acres off Oldham Road in Ancoats where a third of houses were back-to-backs. The Council declared it an 'Unhealthy Area' in 1889, demolished 239 houses and displaced 1,250 people. The first municipal building was erected on the site. This was Victoria Buildings, completed in 1894 and modelled on London's Peabody buildings as a five-storey quadrangle with 285 tenements. The ground-floor frontage had shops, and each tenement had a pantry, coal store and gas meter. A water closet and sink were provided for every two tenements. Refuse chutes were incorporated and communal laundry facilities were provided. The development stands out boldly on Marr's map with its yellow colour, showing it as an oasis of 'property which complies with modern bye-laws'. The Council also developed further tenements on the west of the site. The pertinently named Sanitary Street (later renamed Anita Street) built in 1897, comprised two model terraces of two-storey tenements along a 36-foot-wide road, with pairs of ground-floor and first-floor tenements sharing a common entrance. Each flat had a sink and WC. Even more ambitious was George Leigh Street where five-room cottages were built. Their novelty was that they included a third bedroom in an attic.

Marr was commissioned by the Council to lead the process of converting back-to-backs, and by the outbreak of the First World War most of Manchester's back-to-backs and courts had been cleared or renovated. He left Manchester in 1919 to do similar work in Yorkshire, but his influence was already evident in a new generation of housing reformers, among whom Ernest Simon was to have a significant impact on housing policy, both local and national.

PADIHAM
ROSEGROVE
BLACKBURN
BURNLEY
HEBDEN
PRESTON
ACCRINGTON
BAMBERBRIDGE
FARINGTON
LEYLAND
HODDLESDEN
DARWEN
RAWTENSTALL
BACUP
WATERFOOT
TODMORDEN
HASLINGDEN
FACIT
CHORLEY
HOLCOMBE BROOK
RAMSBOTTOM
ASTLEY BRIDGE
ROCHDALE
HORWICH
BURY
ROYTON
MILNROW
SHAW
RADCLIFFE
WIGAN
BOLTON
MIDDLETON
ATHERTON
MOORSIDE & WARDLEY
ASHTON
PENDLETON
MANCHESTER
WARRINGTON
STOCKPORT

1906

Victoria Station's railway wall map

Having ushered in the Railway Age, Manchester, unsurprisingly, was to become a major railway centre. By the end of the nineteenth century, the scramble by railway companies to establish a station as close as possible to the city centre had resulted in the building of four terminal stations – London Road, Victoria, Exchange and Central – and there would have been another in the very heart of the business district had Edward Watkin's ambitious plans to build a double-decker station been realised. The scramble continued with the completion in 1898 of the Great Northern Railway Goods Warehouse station on Deansgate – the last of the colossal goods stations to be built in British cities. Large-scale maps of the town record the increasing presence of railways in Manchester, land being taken up by engine sheds, sidings and marshalling yards. But what they do not reveal is that all of the main lines were carried in on viaducts, creating a striking townscape of new prospects and boundaries. Railways also accelerated changes in the social landscape, most obviously by encouraging easier commuting and hence the diaspora of the middle classes to the suburbs. Once again, Manchester was able to claim to be in the forefront of this transition, the Manchester South Junction and Altrincham Railway (1849) ranking as one of the world's earliest commuter lines.

The Manchester and Leeds Railway was one of the pioneering railway lines. When the first section opened from Manchester to Littleborough in July 1839, its Manchester terminus was Oldham Road. By the time the line was fully opened in 1841 (following the completion of the Summit tunnel), a new station was planned, closer to but not in the city centre, on a line that would connect the Manchester and Leeds Railway to the Liverpool and Manchester Railway. Hunts Bank was far from ideal as a location, being close to the river Irk and the parish burial ground. Neither proved to be insurmountable obstacles. The Irk was eventually culverted,

---

OPPOSITE. Victoria Station wall map, *Lancashire and Yorkshire Railway* (1906) [AUTH]

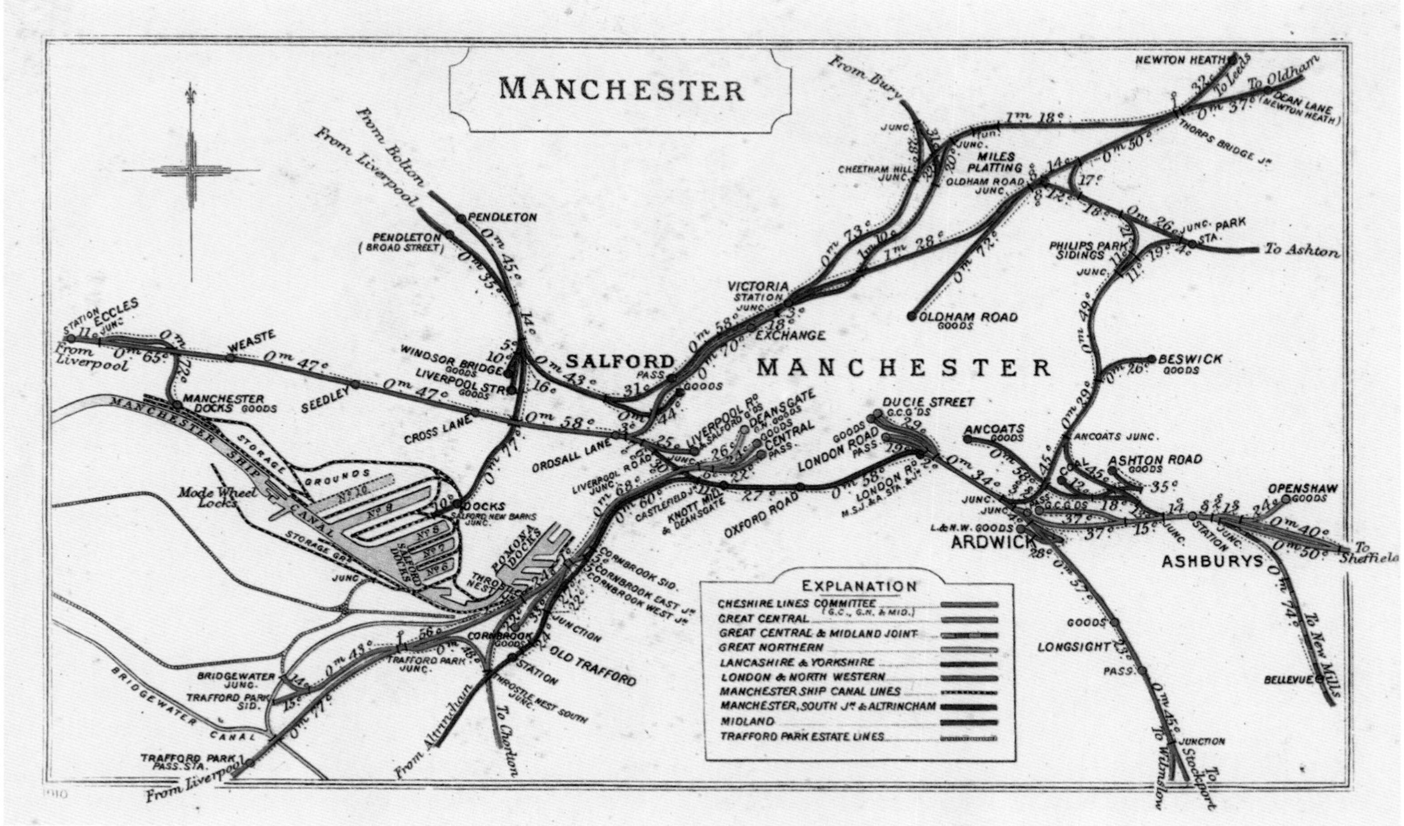

Railway Clearing House, *Manchester* (1910) [AUTH] Goods stations were a vital part of the network centred on Manchester.

while, as the sharp-eyed Friedrich Engels informed his German readers in his *Die Lage der arbeitenden Klasse in England* (1845), the pauper dead were simply dug up with 'disgusting brutality' to make way for the railway. Victoria Station was opened at Hunts Bank in January 1844, the Manchester and Leeds being the first railway company to be granted permission to name their station in honour of the Queen.

Access to the new station was via Hunts Bank, at the bottom of which was the Palatine Hotel, one of the earliest railway hotels. The station comprised a single platform and, initially, because of the gradient, a stationary steam engine was used to pull trains from the station to the junction at Miles Platting. The station building was shared with the Liverpool and Manchester Railway, which completed its line from Salford in May 1844. The new station resulted in both companies converting their earlier stations in Oldham Road and Liverpool Road into goods stations. Traffic using Victoria increased enormously, boosted by the opening of lines to the textile towns of Lancashire and Yorkshire. Extensions and improvements were frequent and within a generation the station boasted 17 platforms.

Passengers using the new railways needed to consult timetables. Different firms provided these, but it was the Manchester firm led by George Bradshaw that compiled and published what became the eponymous Victorian railway guide. Some, but by no means all, of these guides included maps. Railway companies also issued maps of their systems, though most of these were for operational purposes not public consumption. This was also true of the Manchester maps issued by the Railway Clearing House, which by the Edwardian period identified the lines of ten railway companies, two of which served the recently opened Ship Canal Docks and Trafford Park industrial estate. Independent publishers recognised that there was a market for national railway maps, Francis Whishaw, Thomas Surplice and Charles Cheffins producing some of the earliest examples. But railway companies were generally indifferent to displaying maps in their stations, presumably expecting passengers to obtain such information at the booking office. This attitude was evident at Euston station, where in the 1860s the idea of installing a public map was turned down on the grounds that it was not in keeping with the decor of the station.

In the early twentieth century, companies began to recognise the informational and promotional value of displaying maps of their network in their stations. This appears to have been the case with the Lancashire and Yorkshire Railway, which decided to include a wall map as part of the extensions being made to Victoria Station. This decision may have been influenced by the North Eastern Railway, which, in 1903, began to display tiled wall maps of its system in its larger stations, employing the Shropshire-based firm of Craven Dunnill to produce the high-quality tiles. However, the map at Victoria Station, although much larger than the North Eastern Railway station maps, was far inferior in design. Located on a wall by the booking office, the routes and names of the company's stations were merely painted directly onto the white tiled bricks, the major stations being distinguished by larger lettering. It was completed in 1906. A similar map was also provided at the company's Liverpool station.

Eye-catching because of its size, it was nevertheless a cheap and feeble feature, contrasting sharply with the exuberance of the tiling and lettering on the exterior of the new restaurant and bookstall on the expanded concourse. Of course, the glazed tiles meant that it was easier to keep clean – a necessary consideration given the smoke from steam trains – but as the

names were only painted, care was needed to make sure they were not washed away.

The Victoria Station map, however, unlike the one in Liverpool Exchange, was to survive and become a feature of the station. In part, this was due to the installation below the map of the memorial to the Lancashire and Yorkshire Railway employees who lost their lives in the First World War. The map remained unaltered during the subsequent changes in the railway's ownership and name, beginning in 1923 when the Lancashire and Yorkshire became part of the London, Midland and Scottish, followed by the nationalisation of the railways in 1948. After the Beeching rationalisation programme of the 1960s, the wall map provided passengers with a nostalgic record of once-active stations. In 2014–15, following years of neglect, Victoria Station was modernised. The map, now recognised as one of the historical features of the station, was cleaned, enabling one to appreciate the influence that the Lancashire and Yorkshire Railway once had across this part of northern England.

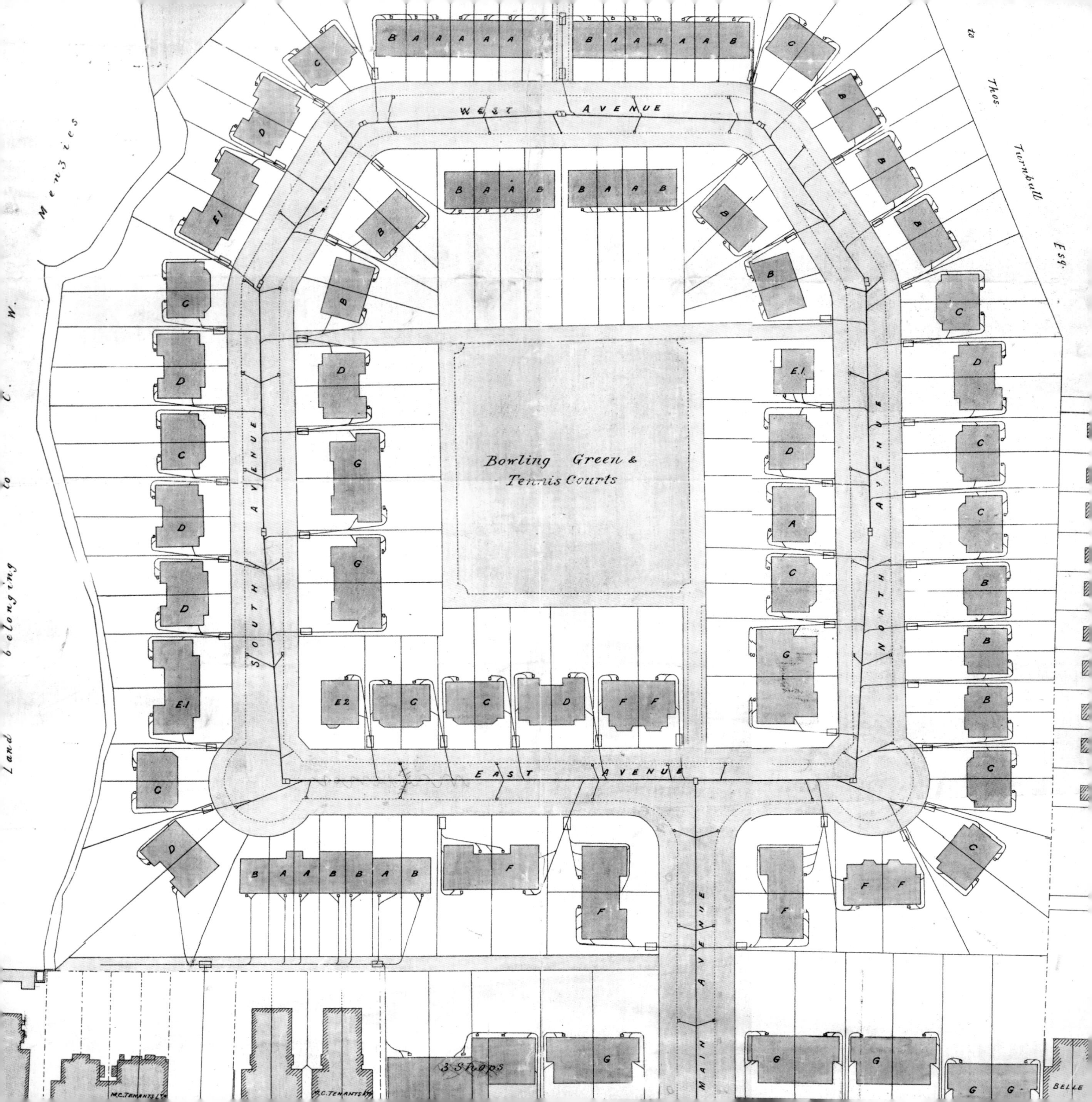
WEST AVENUE
SOUTH AVENUE
NORTH AVENUE
EAST AVENUE
MAIN AVENUE
Bowling Green & Tennis Courts
Land belonging to C. W. Menzies
to Thos. Turnbull Esq.
3 Shops
M.C. TENANTS LTD
M.C. TENANTS LTD
BELLE

1908

Garden suburbs: Burnage and Chorltonville

The end of the nineteenth century was a significant period for town planning and urban design. Ebenezer Howard's advocacy of garden cities, combined with the Arts and Crafts Movement that was inspired by the ideas of John Ruskin and William Morris, produced a powerful two-pronged response to the need to tackle the problems of urban density and squalor. Howard's aim was to address the problems linked to over-crowding and the deterioration of cities. He argued that, by creating self-contained medium-sized cities, his garden city concept would combine the best of town and country. The parallel Arts and Crafts Movement argued that good design could revolutionise living conditions. In its design of houses, it laid emphasis on maximising exposure to fresh air and sunshine, building at lower density and creating more living space, for example by the use of catslide roofs, which created deeper dwellings without raising the height of the eaves.

Manchester was quick to pick up on both of these planning ideals. Thomas Horsfall had been involved in the development of Howard's Letchworth, and he and Thomas Marr were prominent advocates of the growing town planning movement. They founded the Citizens' Association for the Improvement of the Unwholesome Dwellings and Surroundings of the People, which created the context in which the idea, not of Howard's full-blown concept of garden cities, but of the more modest but more readily achievable ideal of the garden suburb became widely accepted by the city.

Manchester's earliest garden suburb was built in Burnage. In 1906, Manchester Tenants Limited was formed (significantly at a meeting in the Co-Operative Society's rooms) as a co-partnership company to pursue the goal of creating a garden suburb on the lines of Bourneville or Port Sunlight. Marr served as chairman. It raised capital by issuing shares and stock, which

---

OPPOSITE. Manchester Tenants Ltd, *Burnage Lane Estate* (1908) [Manchester Tenants Ltd]
The proposed large houses (marked G) on the front were never built.

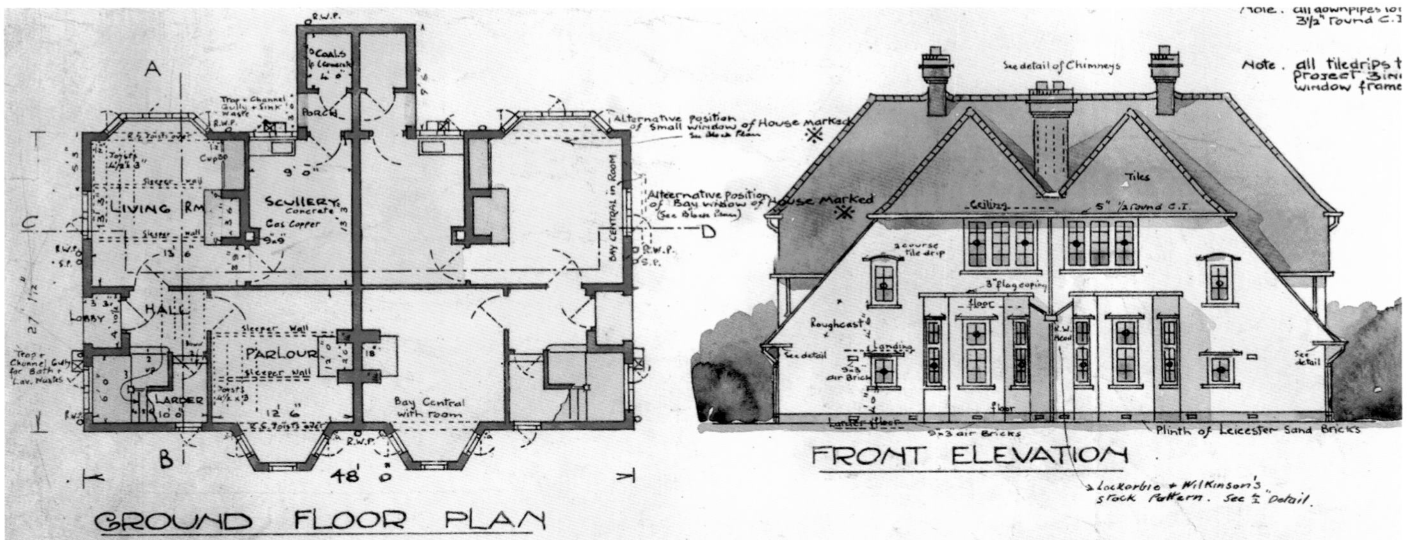

Classic Arts-and-Crafts designs with catslide roofs [Manchester Tenants Ltd]

were bought by many of the prominent Manchester families – the Simons and Reynolds, academics like Schuster, and local philanthropists like Henry Gaddum and Thomas Horsfall himself.

The outcome was the creation of Burnage Village, a development of some 136 houses situated on an 11-acre site that was bought from the Egertons some 4 miles south of the city centre. The layout that was chosen by Manchester Tenants was probably designed by a group of architects that included Raymond Unwin (although the plan is not signed). It embodied all the town planning ideals of the time: low density, at around 12 houses to the acre; characteristic grass verges, with hedging, front gardens and trees lining every street; houses oriented to catch the maximum light and air; and recreational facilities, with tennis courts and a bowling green at its centre and a recreation ground and allotments in the west. The houses ranged from two-bed cottages to four-bed semis. Their design – by the local architect C. Gustave Agate – drew on the Arts and Crafts styles developed by Barry Parker, Raymond Unwin and Charles Voysey, and the plan of the 'Village' identifies each house as being one of a limited range of designs. Each house had a bathroom, hot and cold water, and electric light – all a real advance for lower-rent houses. In order to enable lower rents, the streets were built at a width of only 18 feet, narrower than the dimensions of 24 feet, which were stipulated under the then city bye-laws. This was perfectly feasible at the outset of the Village since there were no through roads and traffic was light.

The development was clearly a social as well as a design experiment since the whole concept of a 'village' was underpinned by the creation of a village association, a drama society and the provision of a club house. As Michael Harrison showed, the inhabitants were predominantly clerks, salesmen and travellers – middle or lower-middle class households – not the casual labourer and working poor. They saw themselves (and were seen from the outside) as 'villagers', somewhat set apart and forming a community that made its own entertainment and activity.

A more dramatic example of the application of the garden suburb principles allied with Arts and Crafts design is Chorltonville, built in 1910–11 in the south of the city and driven by two local businessmen, Herbert Dawson and William Vowles, who had businesses in Hulme. Their original intent had been to provide healthy planned homes for residents from the unhealthy areas of Hulme, but, like many such schemes,

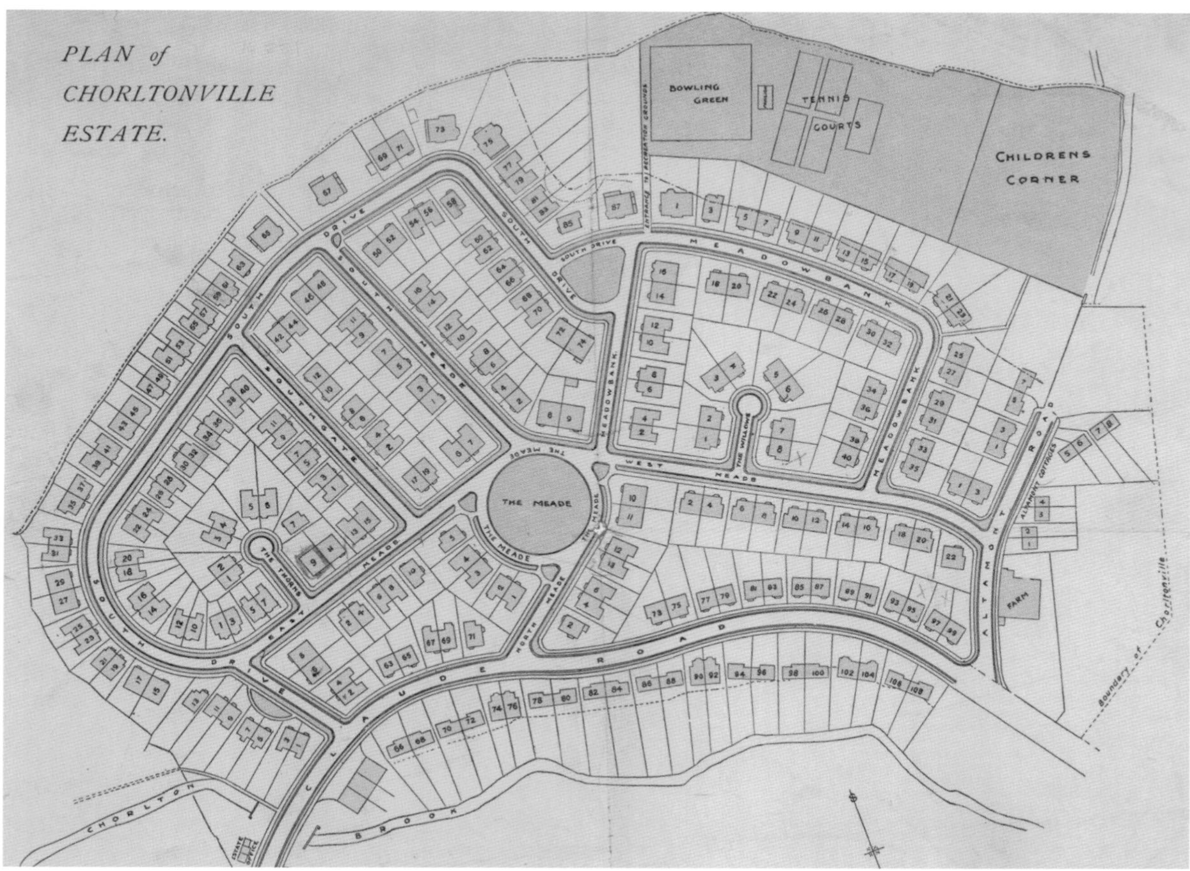

A. Cuneo, *Plan of Chorltonville Estate* (1910) [MLA]

the cost and standard of its houses saw it occupied instead by professional families.

The development was on a 36-acre site and comprised some 260 houses. Albert Cuneo, a young Manchester architect, was responsible for the design, which was based on winding streets, two 'village green' areas, numerous trees, grass verges and 15 slightly different designs for the sets of semi-detached houses, with the bay windows, catslide roofs and exposed beams, all characteristic of the Arts and Crafts Movement. The houses were built to a high standard and, unusually for the time, were supplied with both gas and electricity.

The houses were initially not for sale, but rented, and restrictive covenants were introduced. As a further protection, an association of residents was formed to maintain the roads, paths, verges, drainage gullies, street furniture and trees. The costs associated with the preservation of the estate were met by a levy paid by residents over and above the tax payable to the City Council. The combination of collective ownership, covenants and resident oversight has proved remarkably successful in maintaining the character of Chorltonville, which has now been designated as a conservation area.

Neither Burnage nor Chorltonville met the aim of widening access to good housing for the deprived households of Manchester. However, both had the effect not only of acting as models for future estate development – such as the Fairfield and Alkrington – but they also played a part in raising the profile of good house design and the significance of access to light and air for the benefit of the health of residents.

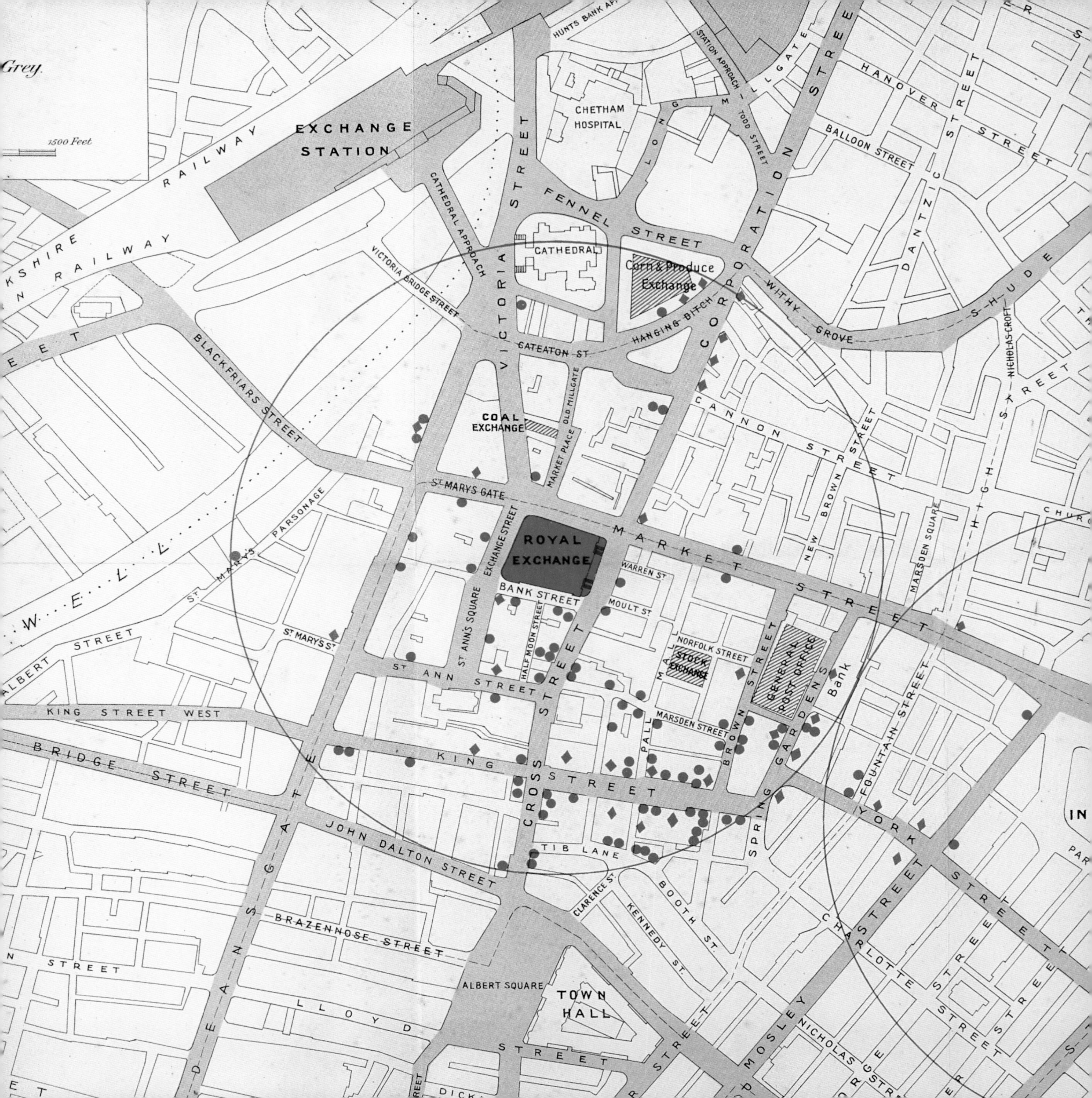

Grey.
1500 Feet
EXCHANGE STATION
RAILWAY
KSHIRE
N RAILWAY
HUNTS BANK
CHETHAM HOSPITAL
LONG
STATION APPROACH
TODD STREET
CORPORATION STREET
HANOVER
BALLOON STREET
DANTZIC STREET
FENNEL STREET
CATHEDRAL APPROACH
CATHEDRAL
Corn & Produce Exchange
VICTORIA BRIDGE STREET
VICTORIA STREET
WITHY GROVE
SHUDEHILL
HANGING DITCH
CATEATON ST
BLACKFRIARS STREET
COAL EXCHANGE
OLD MILLGATE
MARKET PLACE
CANNON STREET
NICHOLAS CROFT
HIGH STREET
BROWN STREET
NEW BROWN STREET
ST MARYS GATE
ROYAL EXCHANGE
EXCHANGE STREET
MARKET STREET
MARSDEN SQUARE
WARREN ST
BANK STREET
MOULT ST
ST MARYS PARSONAGE
WELL
ST MARYS ST
ST ANNS SQUARE
HALF MOON STREET
CROSS STREET
NORFOLK STREET
STOCK EXCHANGE
GENERAL POST OFFICE
Bank
ALBERT STREET
ST ANN STREET
PALL MALL
SPRING GARDENS
MARSDEN STREET
KING STREET WEST
FOUNTAIN STREET
BRIDGE STREET
KING STREET
YORK STREET
DEANSGATE
JOHN DALTON STREET
TIB LANE
CLARENCE ST
BOOTH ST
KENNEDY ST
CHARLOTTE STREET
BRAZENNOSE STREET
ALBERT SQUARE
TOWN HALL
LLOYD STREET
MOSLEY STREET
NICHOLAS

1912

The Royal Exchange: 'parliament of the cotton lords'

Of the buildings which came to be regarded as symbols of Victorian Manchester, none was more significant than the Royal Exchange. The 'parliament of the cotton lords', as it was dubbed by journalists, was the commercial heart of the city. To be invited to watch the merchants and manufacturers at Thursday's 'High Change' was to witness the Cottonopolis business community at full steam.

Manchester's first exchange, located close the Market Place, was opened towards the end of the 1720s at a time when the cotton trade was in its infancy. By the early nineteenth century, the astounding growth of the now increasingly factory-based industry led to the building of a new exchange on the corner of Market Street and St Ann's Square. Opened in 1809, it was designed by Thomas Harrison, architect of the Portico Library, in the fashionable neoclassical style. It was a highly visible part of the development that was creating a distinctive commercial zone in the centre of the town, a process marked by rising land prices, which encouraged residents to sell their houses and move to the suburbs. The relentless growth of the cotton industry was to result in the building of ever-larger exchanges. A new exchange was opened in the late 1840s, a building that became the Royal Exchange following Queen Victoria's first visit to the city in 1851 when she was officially received in the Exchange rather than in the Town Hall. The continuing expansion of King Cotton prompted the Royal Exchange Company to embark on further rebuilding in the 1870s, producing a larger and more architecturally impressive building that included a magnificent entrance on Cross Street. But as the membership of the Exchange continued to increase so did the demand for an even larger trading floor. A new scheme that proposed absorbing Bank Street and part of Half Moon Street was announced in 1908. Ambitious as this was, it proved

OPPOSITE. Parliamentary Papers, Manchester Royal Exchange Bill, *Manchester Royal Exchange* (1912) [MLA] A clear financial quarter evolved, focused around the Royal Exchange.

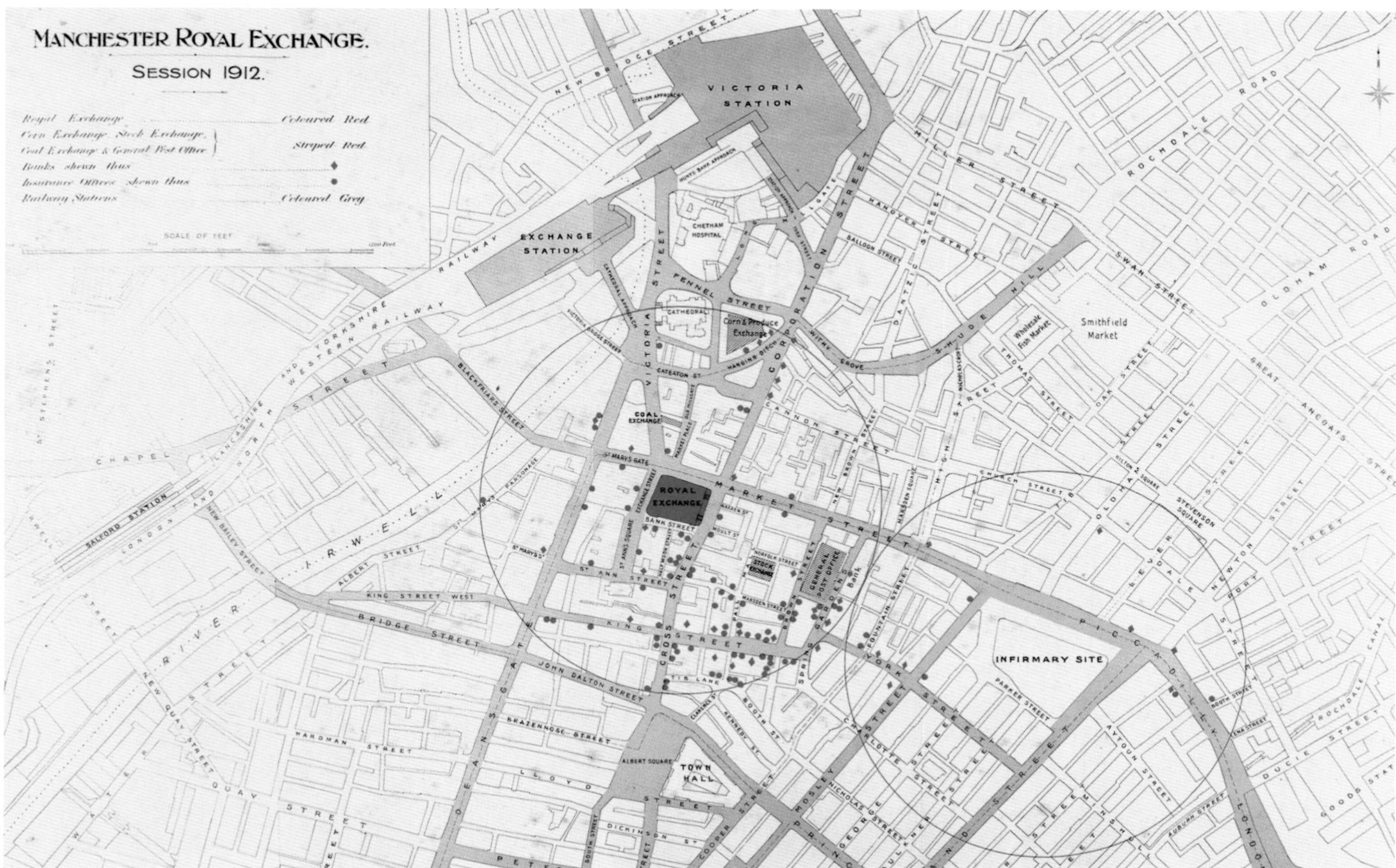

contentious as the Corporation wanted the building extended to St Ann's Street. The argument was finally settled in favour of the Royal Exchange Company, though traffic improvements on Cross Street meant the loss of its grand entrance. Bradshaw Gass & Hope were chosen as architects of the new building. Work on what was to be the final extension to the Exchange began just before the First World War.

Many of the city's leading financial businesses and institutions were located in the streets around the Exchange. By the 1840s, the existence of a commercial zone comprising banks, offices and warehouses, busy with workers during the day but with few permanent residents, had become a commonplace of contemporary observers. 'Nearly the whole city is abandoned by dwellers, and is lonely and deserted at night' noted the young Friedrich Engels. Banks providing essential financial services predominated in what became new streetscapes, notably in Upper King Street and Cross Street. Manchester's importance as a financial centre had been recognised by the establishment of a branch of the Bank of England in the city, its offices removing to a new building in Upper King Street in 1847. The street became a prestigious business address, and the larger commercial banks looked to locate in this part of the city. This process continued into the twentieth century, exemplified by the opening of Lloyds Bank in 1913, which was located on the corner of Upper King Street and Cross Street, the site of Manchester's first town hall, which had itself required the demolition of one of the largest private houses in the town.

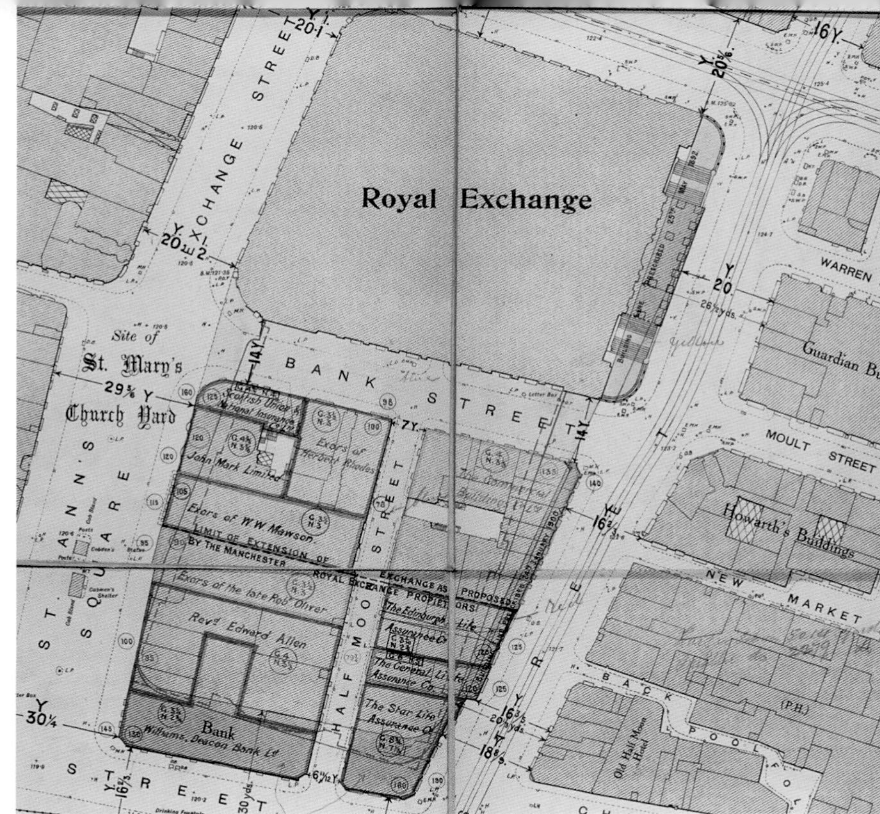

Parliamentary Papers, Manchester Royal Exchange Bill, *Proposed extension of the Royal Exchange* (1912) [MLA] The landownership plan was required for parliamentary approval of the enlargement of the Royal Exchange.

Insurance companies, whose origins were in the fire insurance business in the eighteenth century, were also to be found in the streets around the Exchange, notably Cross Street and Corporation Street. The insurance business, as in banking, became ever more professional and specialised, typified by companies such as Vulcan Boiler and National Boiler, which focused on power systems in cotton-mills and engineering works. Manchester also became the headquarters of leading life assurance companies, including the Co-operative Insurance Society and the Refuge Assurance Society, the latter having to locate its Edwardian offices in Oxford Street, outside the immediate financial district, but on a site that was large enough for Alfred and Paul Waterhouse to provide the company with the palatial headquarters that proclaimed its standing in the financial world.

Another key institution in the city's financial sector was the Manchester Stock Exchange. Founded during the railway boom of the 1830s, it was to survive and become the largest and most important of the provincial stock exchanges. This was underlined when it moved into new Baroque-style premises (also designed by Bradshaw Gass & Hope) in Brown Street, almost halfway between the Royal Exchange and the city's colossal post office in Brown Street. Other financial institutions continued to find accommodation in and around the old market place. These included the Coal Exchange, which, having extended its premises in 1908, was within sight of both the recently rebuilt Corn Exchange and the Royal Exchange.

The 1912 parliamentary inquiry into the proposed extension of the Royal Exchange provided abundant evidence of the dynamism of the financial and business centre of Manchester, many of the businesses being within easy walking distance (250-yards radius) of the Exchange. Suggestions that the Exchange should be moved to another part of the city – the empty site in Piccadilly created by the removal of the Infirmary was one possibility – were strongly opposed, it being argued with some justification that this would have 'upset the business of the city', inflicting 'enormous damage on all the property owners near the present Exchange'.

When George V officially opened the new Royal Exchange in 1921, few of the numerous merchants who filled what was asserted to be the largest trading floor of any commercial building in the world seemed aware that the industry was in fact at the beginning of its long and terminal decline. Membership numbers of the Royal Exchange fell during the following decades, providing a singularly dismal index of the performance of the wider economy of 'Textile Lancashire'. Cotton trading was finally to finish on 'Change' in December 1968. After some uncertainty, which included proposals to demolish the building, the central space was taken over by the 69 Theatre Company, and in 1976 an innovative theatre-in-the-round was opened on the former trading floor. The exotic names and prices of the last day of cotton trading remained on the indicator boards as evidence of the building's earlier history, while more curious theatregoers could crane their necks to read the inscription around the dome which reminded Victorian merchants that those 'Who seek to find eternal treasure must use no guile in weight or measure'.

COUNTY
BOROUGH
OF BOLTON
AINSWORTH
BOROUGH
OF HEYWOOD
OF
RADCLIFFE
UNSWORTH
BOROUGH
OF MIDDLETON
LITTLE
LEVER
FARNWORTH
OUTWOOD
WHITEFIELD
KEARSLEY
LITTLE
HULTON
PRESTWICH
CLIFTON
THERTON
TYLDESLEY
SWINTON
CITY
OF
MOSTON LANE
ROCHDALE ROAD
MIDDLETON ROAD
WATERLOO ROAD
HILL ROAD
CHEETHAM
Car Shed
QUEEN'S
ROAD
OLDHAM
ROAD
UGH
WORSLEY
COUNTY
BOROUGH
OF SALFORD
EIGH
ASTLEY
BOROUGH
OF ECCLES
BRADFORD RD
ST. ANCOATS ST
ASHTON NEW ROAD
LONDON RD
OXFORD ST
ASHTON OLD
CHESTER RD
CITY ROAD
STRETFORD ROAD
BRUNSWICK ST
Car Shed
CLOWES ST
HYDE ROAD
STOCKPORT RD
PLYMOUTH GROVE
HIGH STREET
DAVYHULME
STRETFORD
ALEXANDRA ROAD
DICKENSON ROAD
SLADE LANE
Car Shed
MANCHESTER
URMSTON
WILBRAHAM ROAD
BARLOW MOOR ROAD
PALATINE ROAD
ASHTON
UPON
MERSEY
SALE
Boundary in Centre of Road
COUNTY
BOROUGH O
STOCKPORT
NORTHENDEN
DUNHAM
REFERENCE
Tramways worked by the Manchester Corporation { Existing / Proposed
Tramways worked by other Municipal Authorities { Existing / Proposed
Tramways worked by Various Companies - - { Existing / Proposed

1916

The handy penny tram

The year 1824, unlike 1830, does not immediately spring to mind as a pivotal date in the history of transport, but it deserves to be better remembered. It was in that year that John Greenwood established a horse-drawn passenger omnibus service in Manchester and Salford, connecting its affluent and growing suburbs to the city centre (initially only from Pendleton). Greenwood's service ran several times a day, the initial fare being sixpence. It pre-dated the much better-known omnibus service set up in London by George Shillibeer, which is generally regarded as the first of its kind in the world. Before the coming of horse-drawn passenger coach services, almost everyone wanting to reach a destination in another part of the town, whether for work or pleasure, had little option but to walk. Ownership of a private carriage was only for the very wealthy. Greenwood's business heralded what was to be a profound change whereby ever more efficient transport systems transformed city life. It was the fact that middle-class households had already begun to settle in the suburbs of Ardwick and Chorlton-on-Medlock that encouraged the development of such omnibus services. No longer having to live within walking distance of their place of work opened up the possibility of developing new residential districts ever further from the city centre.

Greenwood's success led to the establishment of rival companies, although his company's omnibuses were to remain the leading firm in Manchester and Salford. Greenwood's 'Exhibition Omnibuses' carried 1,239,820 passengers from the city centre to the 1857 Art Treasures Exhibition at Old Trafford, easily exceeding those who travelled to the event by train. Passenger numbers increased, especially as fares were

---

OPPOSITE. Manchester Corporation, *Existing and proposed tramways in Manchester and neighbouring boroughs and districts* (1916) [Museum of Transport Greater Manchester] The tram revolution connected the city to the wider region.

reduced, and by the middle of the nineteenth century firms were providing services on the principal commuter routes as well as for those travelling shorter distances within the town. In 1861, Greenwoods pioneered one of the country's first street tramways, their foundation route once again being between Pendleton and Manchester. In order to eliminate wasteful competition, negotiations between the local coach companies resulted in the establishment of the Manchester Carriage Company in 1865. Demands to install tram tracks on existing roads brought local government more directly into this area of city life and, under the terms of the 1870 Tramways Act, they were given the powers to construct tramways but not to operate them.

Competition was fierce among independent operators, and the year 1888 was notable for the introduction by the Manchester Carriage Company of penny fares on its trams. The business continued to expand and, by the mid 1890s, the Manchester Carriage & Tramways Company owned 5,244 horses, 515 tramcars, 26 horse-drawn omnibuses and 71 cabs. These were housed in 20 depots, premises which necessarily included extensive stabling facilities. At All Saints, Oxford Road, for example, the company had a large depot with tramcars on the ground floor and stables on three upper floors.

The development of tramways combined with lower fares benefited both businesses and consumers of all classes. By the end of nineteenth century, the development of suburban tram lines offering cheap fares allowed skilled and semi-skilled workers to consider moving out of the city centre. Cheap trams – Richard Hoggart's 'gondolas of the people'– also meant that women were able to take the tram for a shopping trip to the city centre, while the operation of late evening services was helping to transform other leisure activities. Many of the vast crowds of supporters attending sporting events, notably football and cricket matches, travelled to and from games by tram. People seeking more genteel pleasures such as a Sunday band concert in a municipal park also made use of the tram service. When Manchester City Council purchased the vast Heaton Park estate in 1902, it was pointed out that although the new park was located in what some ratepayers regarded as the remote north-west of the city, the tram made it accessible to all Mancunians.

By the early twentieth century, the age of the private horse-drawn bus company was coming to an end. Manchester and Salford councils were to take over control of the tram network. From the 1890s, the electrification of the system was underway. Under municipal control, the network was further extended. Fares remained competitive and passenger numbers continued to increase. The growth in the system was breath-taking. By 1913, the main tram routes out of the city (Stockport Road, Palatine Road, Oxford Road/Wilmslow Road, Rochdale Road, Oldham Road, Cheetham Hill Road, Hyde Road, Ashton Old Road) were among the most intensively used in the country. One of the benefits of municipal ownership was a closer cooperation between different districts, allowing passengers to travel more easily and cheaply outside their own town.

Cheap municipal transport was to be a key force in making the twentieth-century city. No city was capable of operating without an infrastructure of trams and buses. Once more, Manchester was often at the forefront of new developments, though not all of its transport problems found a solution. In a city of terminal railway stations, the ambitious but entirely intelligible project of building an underground line between London Road and Victoria stations, first seriously advanced in 1901, was never realised.

Electric tramways were to be challenged and eventually eclipsed by the internal-combustion engine. Manchester's last electric tram ran in January 1949. However, the tram was to return in a new guise when, some 50 years later, the Metrolink light railway service began to be constructed. Its first line, running from Bury to Altrincham, opened in 1992. As in the past, the carrying of commuters was a priority. Further lines followed, including in 1999–2000 a line connecting Eccles to Manchester, a connection that would no doubt have been recognised by and received the approval of John Greenwood.

OPPOSITE. Manchester Corporation, *City of Manchester, Map of bus and tram routes* (1939) [Museum of Transport Greater Manchester] Showing the density of tram routes in the town centre.

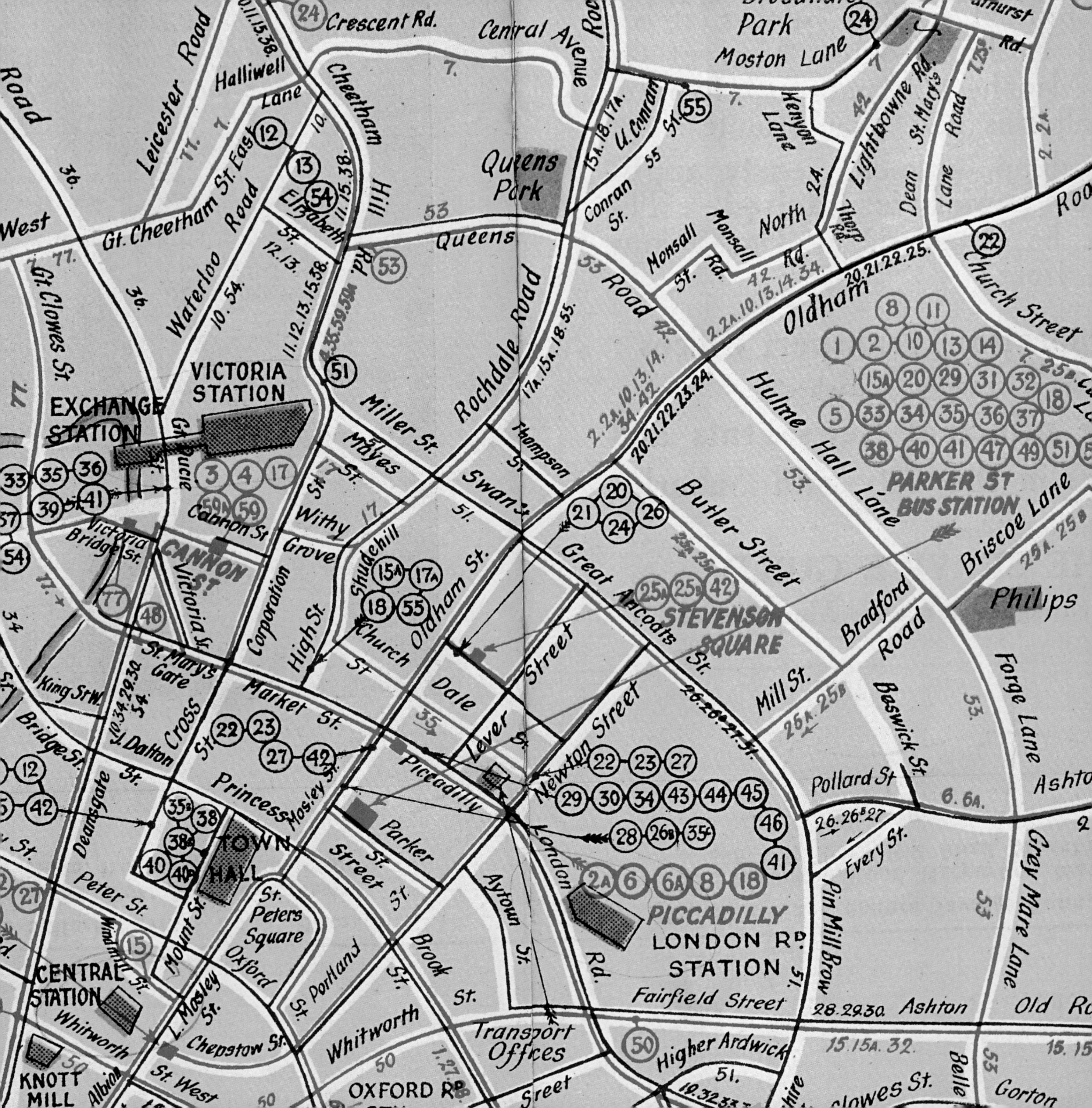

Crescent Rd.
Central Avenue
Park
Moston Lane
Leicester Road
Halliwell Lane
Cheetham Hill Rd
Queens Park
Gt. Cheetham St. East
Waterloo Road
Queens Road
Conran St.
U. Conran St.
Monsall St.
Monsall Rd.
North Rd.
Kenyon Lane
Lightbowne Rd.
St. Mary's Road
Dean Lane
Church Street
West
Gt. Clowes St.
Oldham Road
Rochdale Road
VICTORIA STATION
EXCHANGE STATION
Miller St.
Mayes St.
Swan St.
Thompson St.
Hulme Hall Lane
PARKER ST BUS STATION
Butler Street
Briscoe Lane
Philips
Withy Grove
Cannon St
CANNON ST
Victoria Bridge St.
Victoria St.
Corporation St.
Shudehill
High St.
Church St
Oldham St.
Great Ancoats St.
STEVENSON SQUARE
Bradford Road
Forge Lane
St. Mary's Gate
King St. W.
Market St.
Dale St.
Mill St.
Beswick St.
Bridge St.
J. Dalton St.
Cross St.
Lever St.
Newton Street
Pollard St
Ashton
Deansgate
Princess St.
Mosley St.
Piccadilly
Parker St.
TOWN HALL
Peter St.
Mount St.
St. Peters Square
Oxford St.
Portland St.
Aytoun St.
London Rd.
PICCADILLY LONDON RD. STATION
Every St.
Pin Mill Brow
Grey Mare Lane
CENTRAL STATION
Windmill St.
L. Mosley St.
Whitworth St.
Brook St.
Fairfield Street
Ashton Old Rd
KNOTT MILL
Chepstow St.
St. West
Transport Offices
Higher Ardwick
OXFORD RD.
Clowes St.
Belle
Gorton

PROPOSED NEW GROUND FOR THE MANCHESTER UNITED F. C. LTD.

GENERAL ARRANGEMENT.

EXIT GATES 5 PAYBOXES
EXIT GATES 2 PAYBOXES
EXIT GATES 2 PAYBOXES
EXIT GATES 2 PAYBOXES
EXIT GATES 2 PAYBOXES
EXIT GATES 2 PAYBOXES

COVERED TERRACE

CENTRES

300' 0"

509' 6" CENTRES OF OUTER COLUMNS

629' 6" CENTRES OF OUTER COLUMNS

FIELD ENTRANCE

GRAND STAND

336' 0"

UPPER TERRACING HALF PLAN

HALF PLAN LOWER TERRACING

EXIT GATES TO GRAND STAND

TO STAND, 10 PAYBOXES

SCALE 30 FEET TO 1 INCH

1923

Moving the goalposts: Old Trafford and Maine Road

By 1900 professional football had established itself as the people's game. The rising numbers of paying spectators encouraged the leading clubs to provide purpose-built stadiums. The popularity of football was nowhere more evident than in Manchester where both of the city's leading clubs moved into new stadiums in the first quarter of the twentieth century.

Manchester United began life in 1878 as the Newton Heath Lancashire and Yorkshire Railway Football Club, having been formed by the Dining Room Committee of the Carriage and Wagon Works of Lancashire and Yorkshire Railways. Its first ground was at North Road, Newton Heath, next to the railway works. In 1893, the year after it had joined the Football League, the club moved to a larger ground in Bank Street, Clayton. The Edwardian years saw dramatic changes in the fortunes of the club, which had been renamed Manchester United in 1902. John Henry Davies, a local brewer, took a prominent role in the running of the club and his business acumen and willingness to invest in the club placed it on a more secure financial footing, enabling it to buy star players and to improve the facilities at Bank Street. Success followed; the club won the first division league title in 1908 and the FA Cup in 1909. Rising attendances, dissatisfaction with the Clayton ground and the club's ambitions led to the bold decision to relocate and build a new stadium in Old Trafford. A 16-acre plot between Chester Road and the Bridgewater Canal was purchased.

Locating the new ground at Old Trafford, a district close to the city boundary, was an unexpected choice since it was over 3 miles from the working-class districts in the east of the city from which it drew many of its supporters. By the Edwardian period, Old Trafford, once a residential suburb for servant-keeping families was experiencing rapid change. In the wake of the building of the Ship Canal docks and the opening

OPPOSITE. A. Leitch, *Proposed new ground for the Manchester United F.C. Ltd* (1923) [Manchester United F.C. Museum]

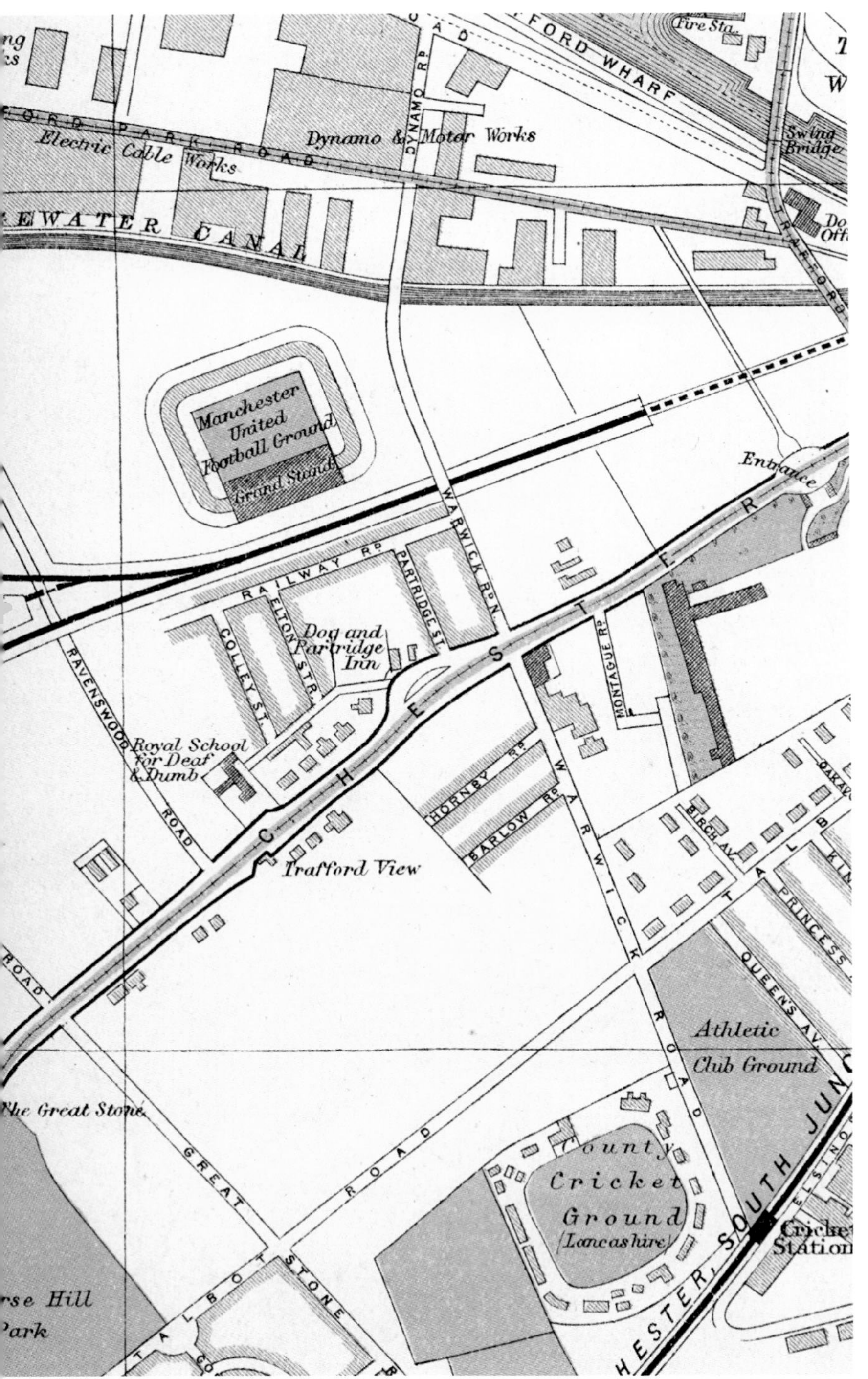

Bartholomew, *Manchester* (*c.*1920) [AUTH] As well as the football ground, the map shows the county cricket ground a short distance to the south.

of the Trafford Park industrial estate, Old Trafford was being brought within the orbit of the 'great machine shop' of Manchester. This meant that the new stadium was within easy walking distance of a booming industrial district employing thousands of skilled and unskilled workers, the majority of whom stopped work at noon on Saturdays. The opening in 1907 of the White City pleasure park on the site of the Botanical Gardens provided another local point of development, a commercialised leisure complex that was in direct competition with Belle Vue as a place of mass entertainment. The district had good tramway links to the city as well as a railway station, the latter having been built for the Art Treasures Exhibition of 1857, but was now better known as the 'cricket stop', being adjacent to the ground of the Lancashire County Cricket Club.

The Old Trafford stadium was designed by Archibald Leitch, a Scottish industrial engineer who had become the most prolific designer of football stadiums, which was an unexpected achievement given that the collapse of one of his earliest stands in Glasgow had resulted in numerous fatalities. Leitch's design was widely praised when the stadium opened in 1910. It was said to be capable of accommodating 80,000 spectators. The vast majority of supporters stood on open terraces, separated by crush barriers, while the more affluent could sit under cover in cushioned tip-up seats that were, according to the *Manchester Guardian*, 'for all the world like theatre stalls'. The stadium was estimated to have cost £60,000, an enormous sum at the time. The holding of the 1911 FA Cup Final replay at Old Trafford confirmed it to be one of the country's first-rate grounds, but for most league games the capacity far exceeded demand. Attendances did increase – from an annual average of 18,185 (1905–09) to 22,219 (1910–14) – but this still left the club struggling to pay off debts.

The other major team, Manchester City, did not move to a new purpose-built stadium until the 1920s. Its origins were also in east Manchester, traceable back to West Gorton (St Marks), a church club that was founded in 1880. The club was to develop away from the church, playing on pitches that were easier to criticise than praise. It eventually found a more permanent home on land close to Hyde Road, belonging to the Manchester, Sheffield and Lincolnshire Railway. Being in Ardwick rather than Gorton led the club to change its name to Ardwick FC. Support continued to increase, and in 1892 it was admitted into the second division of the newly established Football League. Promotion to the first division came in 1899. These early years in the league were a struggle, but it had the good fortune to sign Billy Meredith who helped the team win the FA Cup in 1904. Local derby matches underlined the popularity of football in Edwardian Manchester. On the occasion in December 1906 when City played United, one football reporter who described the immense crowds making their way towards Hyde Road, noted that 'five in a hansom was no uncommon sight'. City continued to attract support, and its annual average attendance of 23,581 between 1910–14 placed it among the best-supported clubs in the country.

Although the club enjoyed more support than United, it continued to follow a policy of piecemeal improvements to the Hyde Road ground rather than building a new stadium on its own land. This was to change in the immediate post-war years when the search began for a suitable location. A serious fire in December 1920 gave further impetus to the idea of moving, as did a run of good results that saw the club end the season as runners-up. Discussions had been taking place with Belle Vue where land was available for the club to build a stadium. This was in many ways an attractive site, close to the existing ground and with a transport infrastructure capable of handling tens of thousands of people. But suitable terms could not be agreed and the search for a new location ended when the club purchased a 16.25-acre site in Moss Side, a still growing and largely working-class suburb, which had finally been incorporated into the city in 1904. The new stadium, Maine Road, was designed by the Manchester architect, Charles Swain, and was a fitting 'symbol of size and power' of the club. It was said to hold at least 80,000 spectators, but as in the case of Old Trafford, the numbers attending in the interwar years, with the exception of cup ties and derbies, rarely approached that number.

Hence, by 1923 Manchester's two major clubs were housed in large modern stadiums, confirming it as one of the country's leading football cities. Both stadiums became an important part of the social identity of supporters who regarded their home ground as a special place, invested with the memories of past matches and glories.

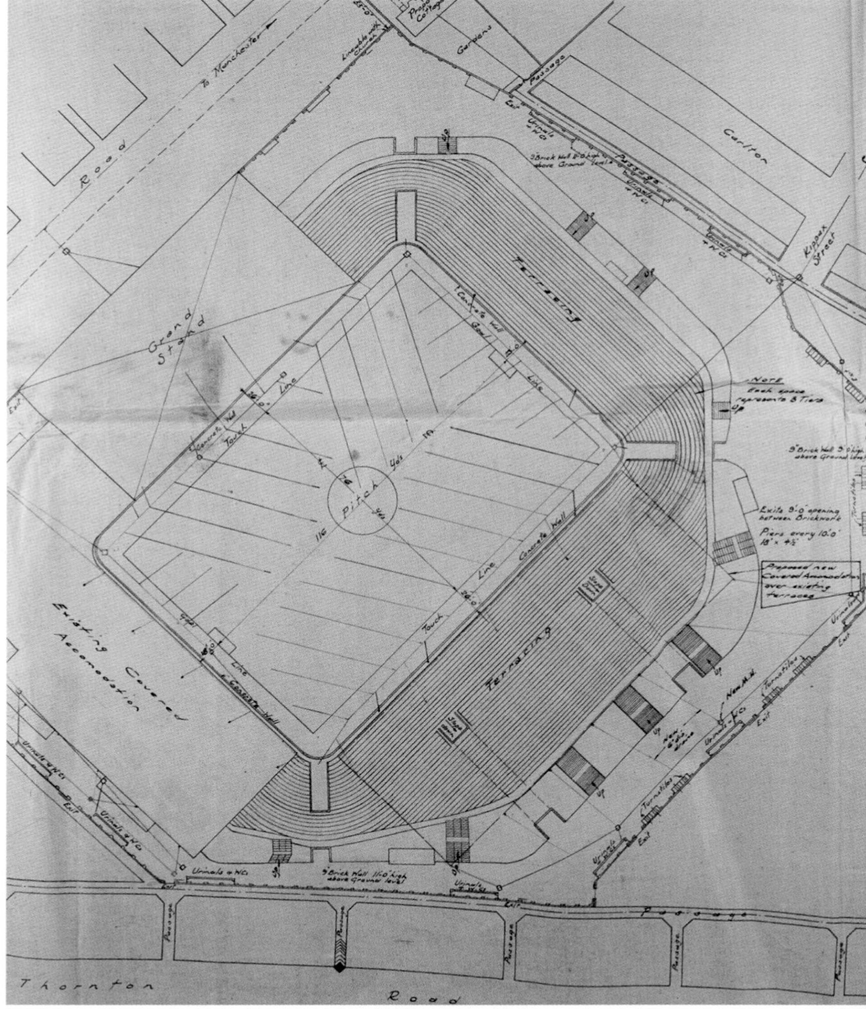

C. Swain, *Manchester City F.C. new ground, Maine Road* (1922) [MLA]

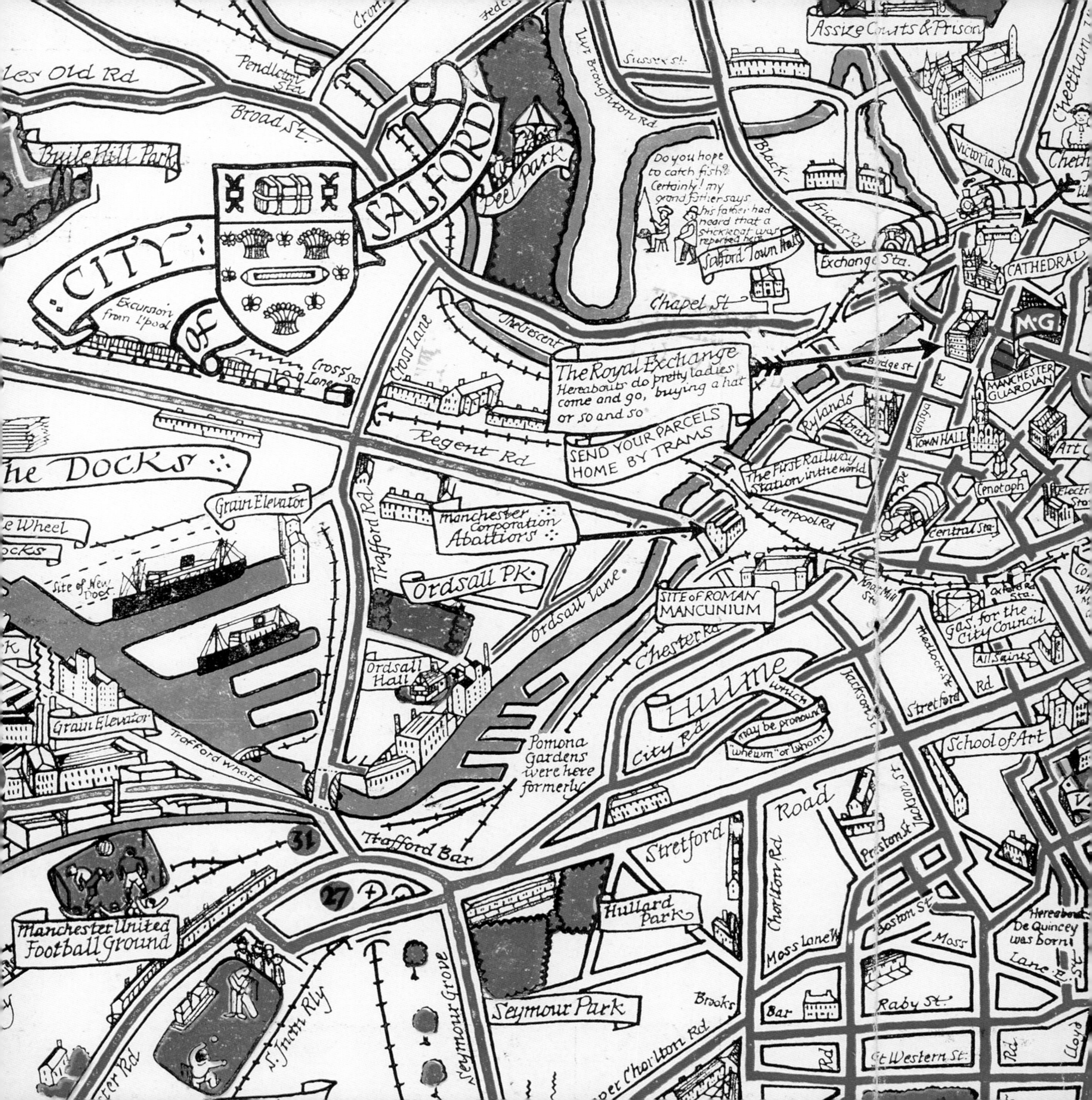

Assize Courts & Prison
Broad St.
Buile Hill Park
City of Salford
Peel Park
Do you hope to catch fish? Certainly! my grandfather says his father had heard that a stickleback was reported here
Salford Town Hall
Victoria Sta.
Exchange Sta.
Cathedral
Chapel St
The Crescent
Excursion from L'pool
Cross Lane Sta
Cross Lane
The Royal Exchange Hereabouts do pretty ladies come and go, buying a hat or so and so
M·G
Manchester Guardian
Rylands Library
Town Hall
Regent Rd
Send your parcels home by trams
The Docks
Grain Elevator
The First Railway Station in the world
Liverpool Rd
Cenotaph
Central Sta.
Manchester Corporation Abattoirs
Trafford Rd
Ordsall Pk.
Site of New Dock
Ordsall Lane
Site of Roman Mancunium
Gas, for the City Council
All Saints
Chester Rd
Ordsall Hall
Hulme which may be pronounced "whewm" or "whom"
City Rd
Stretford Rd
School of Art
Grain Elevator
Trafford Wharf
Pomona Gardens were here formerly
Trafford Bar
Stretford Road
Preston St
Jackson St
31
27
Manchester United Football Ground
Hullard Park
Chorlton Rd
Moss Lane W
Boston St
Moss Lane E
Hereabouts De Quincey was born
Seymour Grove
Seymour Park
S. Jnctn Rly
Brooks Bar
Raby St.
Gt Western St.

1926

Civic Week: a cartoon cartography of the city

Civic Week in 1926 was a seven-day celebration of Manchester packed with large-scale public events – parades, marches, exhibitions, historical pageants and musical concerts. It was aimed to foster a sense of greater local patriotism, but coupled with overt commercial boosterism and municipal tub-thumping. Part of the pressure to hold a Civic Week in Manchester arose from the enmity with Liverpool, which had held a successful Civic Week celebration at the British Empire Exhibition in 1924.

The Civic Week map offers an amusing vision of the central area of Manchester in the mid 1920s. It is full of quirky details, puns and cartoon characters expounding on various interesting things to see and do during the celebration. It is both forward looking in its stress on dynamic transport services and commercial activity across the city, but also nostalgic with its presentation of aspects of Manchester's past that were being lost in the onward match of technological progress. The map-maker's conscious use of cursive script for the title and text labels, along with the replica manuscript scrolls, banners and other mock-medieval motifs, combined with old-fashioned cartographic design, essentially look backwards. The old-fashioned scale bar features human characters in conversation with the reader, one of the quirky features of the map, and more generally of the genre of cartoon maps which deliberately break the sense of objectivity between creator and consumer that characterises scientific cartography.

Importantly, the map does not mask the fact that Manchester was an industrial city: there are plenty of smoking factory chimney and chuffing steam trains. However, it is also shown as a surprisingly clean and green city. The presence of parks is emphasised, as is the 'bright blue water' of the Irwell and the Ship Canal – although there is a brief sarcastic comment tucked

---

OPPOSITE. Manchester Guardian, *A map of the City of Manchester in its first Civic Week . . .* (1926) [AUTH]

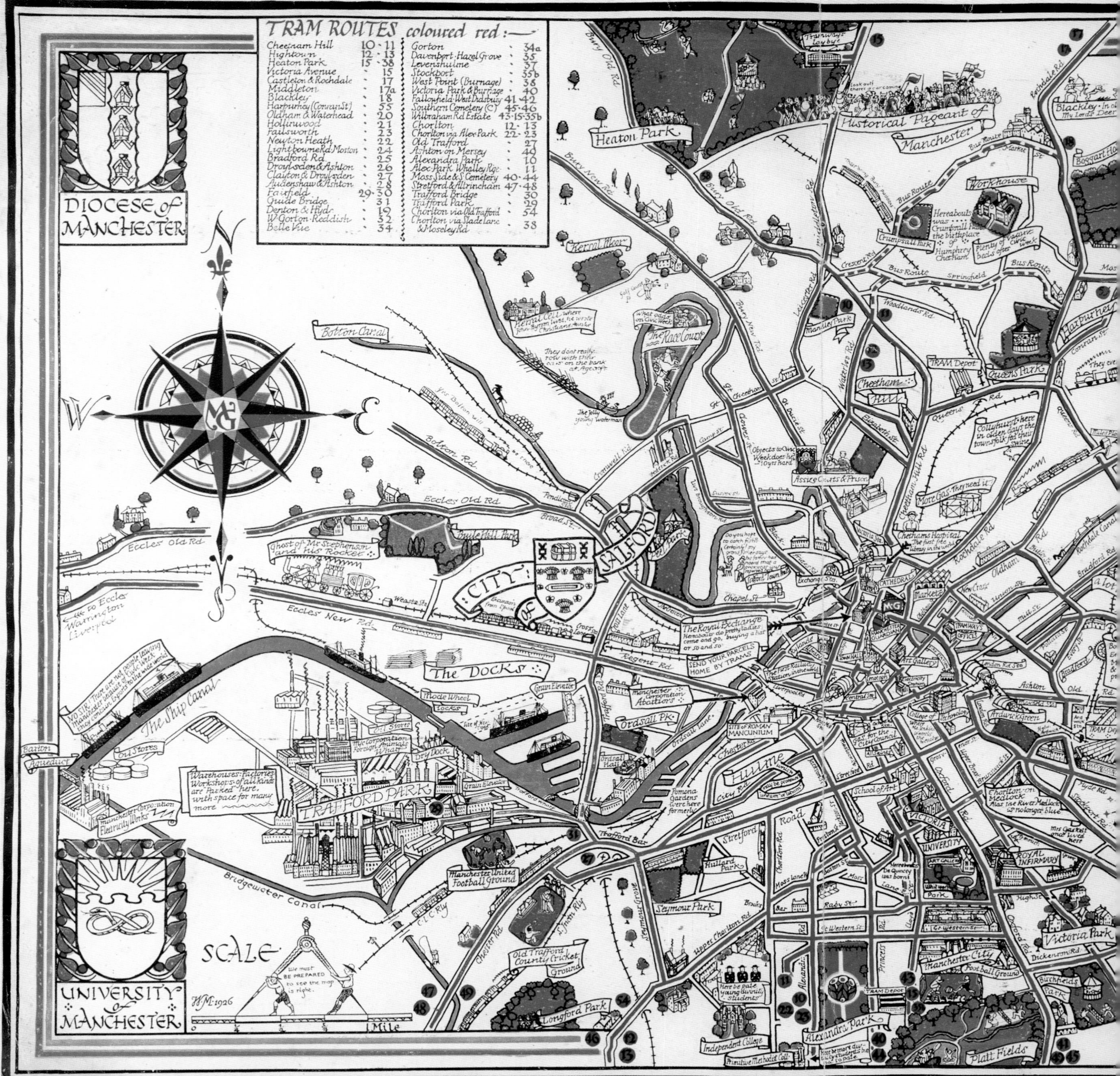
TRAM ROUTES coloured red :—
Cheetham Hill 10 · 11
Hightown 12 · 13
Heaton Park 15 · 38
Victoria Avenue 15
Castleton & Rochdale 17
Middleton 17a
Blackley 18
Harpurhey (Conran St) 55
Oldham & Waterhead 20
Hollinwood 21
Failsworth 23
Newton Heath 22
Lightbowne Rd Moston 24
Bradford Rd 25
Droylsden & Ashton 26
Clayton & Droylsden 27
Audenshaw & Ashton 28
Fairfield 29 · 30
Guide Bridge 31
Denton & Hyde 19
W. Gorton · Reddish 32
Belle Vue 34
Gorton 34a
Davenport · Hazel Grove 35
Levenshulme 37
Stockport 35b
West Point (Burnage) 36
Victoria Park & Burnage 40
Fallowfield · West Didsbury 41 · 42
Southern Cemetery (C) 45 · 46
Wilbraham Rd Estate 43 · 15 · 35b
Chorlton 12 · 13
Chorlton via Alex Park 22 · 23
Old Trafford 27
Ashton on Mersey 49
Alexandra Park 10
Alex Park Whalley Rge 11
Moss Side & S. Cemetery 40 · 44
Stretford & Altrincham 47 · 48
Trafford Bridge 30
Trafford Park 29
Chorlton via Old Trafford 54
Chorlton via Slade Lane & Moseley Rd 38
DIOCESE of MANCHESTER
UNIVERSITY of MANCHESTER
SCALE
We must BE PREPARED to see the map is right.
1 Mile
H.M. 1926
Heaton Park
Historical Pageant of Manchester
Trainways lay bye
Kersal Moor
Kersal Cell where John Byrom lived, he wrote Christians Awake
The Race Course
Bolton Canal
Bolton Rd.
Eccles Old Rd
Eccles New Rd.
Ghost of Mr Stephenson and his Rocket
To Eccles Warrington Liverpool
CITY OF SALFORD
Peel Park
The Docks
The Ship Canal
Barton Aqueduct
Oil Stores
Manchester Corporation Electricity Works
Warehouses · Factories · Workshops of all kinds are packed here, with space for many more
TRAFFORD PARK
Mode Wheel Locks
Grain Elevator
Dry Dock
Ordsall Pk
Manchester Corporation Abattoirs
Manchester United Football Ground
Bridgewater Canal
Old Trafford County Cricket Ground
Longford Park
Seymour Park
Hullard Park
Hulme
Pomona Gardens were here formerly
Trafford Bar
Stretford Road
Site of Roman Mancunium
The Royal Exchange
SEND YOUR PARCELS HOME BY TRAMS
Regent Rd
Liverpool Rd
Exchange Sta.
Cathedral
Chethams Hospital
Assize Courts & Prison
Cheetham Hill
Manchester Guardian
Town Hall
Art Gallery
Central Sta
Tramways Office
London Rd Sta
College of Technology
School of Art
Victoria
Ardwick Green
Chorlton on Medlock
University
Royal Infirmary
Whitworth Park
Victoria Park
Manchester City Football Ground
Birchfields Park
Platt Fields
Alexandra Park
Independent College
Primitive Methodist Coll.
Workhouse
Crumpsall Park
Queens Park
Harpurhey
TRAM Depot
More Gas - they need it
Rochdale Rd
Oldham
Market
Blackley
Boggart Hole

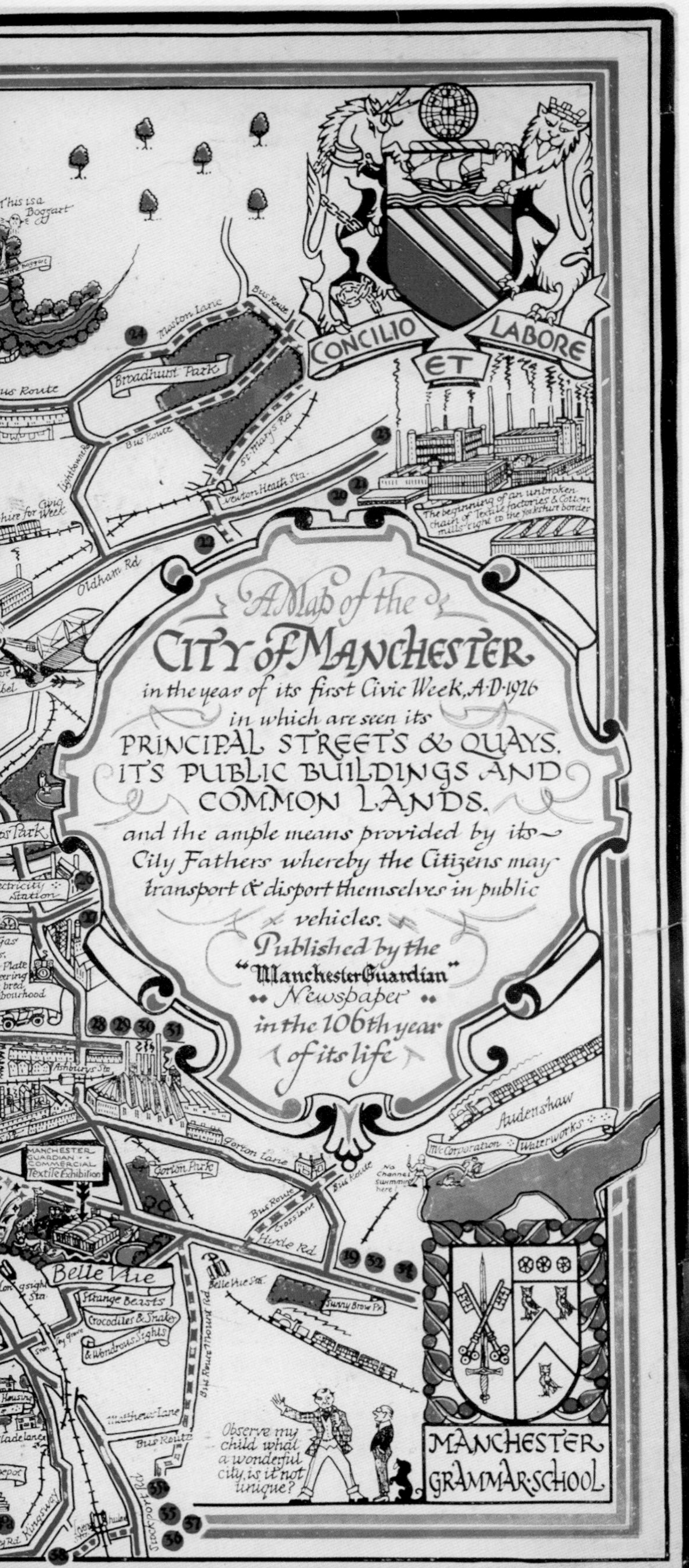

away about the Medlock not being so blue. A primary practical objective of the map was as a guide to public transport for visitors, hence the bright red tram routes that are visually dominant, resembling the veins of the city and suggesting a comprehensive system able to meet all people's travel needs.

Interestingly, utilities, services and urban infrastructures that would not normally appear on visitor guides are featured on the map. These unglamorous features were a central component in the celebration of Civic Week, which was about pride in the work of the municipality. So the gas works, power station and water supply facilities hosted tours for visitors, reflecting the fact that the early decades of the century were the high point of power for Manchester's local authority. This was an age of municipal paternalism writ large. The Corporation controlled the buses and trams, the public markets, education, health and most essential services, including sewage, water and gas supplies.

The map was specially commissioned and prepared for the event, and therefore quite a costly endeavour. Unsurprisingly, since it was probably made for – and paid for by – the *Manchester Guardian,* the newspaper is fully acknowledged. As well as being named on the title scroll, its logo is in the middle of the compass rose and its city-centre building is identified in red. Indeed, it enjoys greater visual prominence than the nearby Town Hall. This, of course, was a time when the paper was a highly influential institution deeply engrained in Manchester's cultural life. Sponsorship probably also came from the Co-operative Wholesale Society which had full-page adverts on the verso.

A good deal of thought and effort went into the content and design of the map. The map-maker was certainly talented, but sadly its author is unknown even though the work is signed with the initials 'w.m.' beside the scale bar. The same illustrator was also commissioned to produce a sister pictorial map for the 'Port of Manchester' that was published as an insert in the Civic Week Handbook. However, that map is much less interesting in overall design and devoid of the whimsical humour of the overall map.

The aesthetics of the Civic Week map were influenced by

cartographic design for commercial and civic promotions at that time. The most significant of these was MacDonald Gill's *Wonderground Map of London Town* published by London Transport in 1914. In the interwar period, many other pictorial maps appeared in a similar vein for other cities and to promote commerce and colonial trade. However, there is a longer tradition of pictorial maps that playfully tweak the accepted bounds of 'scientific' cartography and often use cartoons to satirise those in power or to convey propaganda messages. The best known are the caricature maps and politically targeted cartographies by James Gillray and his compatriots in the eighteenth century, which poke fun at rivalries between newly emergent European nation-states.

With hindsight, Civic Week has an air of 'bread and circuses', an attempt to mask the very evident economic problems of the mid 1920s. Many parts of Lancashire were suffering great hardships as the cotton industry was in marked decline. The years of economic stagnation were punctuated by the General Strike in May 1926. Civic Week's free events and the floodlighting of the grand buildings hid the continued poverty and the widening social divisions within the city.

The Civic Week map is a fascinating example of pictorial cartography that uses cartoon-style graphics and whimsical humour to engage, entertain and inform. It was drawn in an age of onrushing technological modernity and the beginnings of mass consumption, but it seeks to evoke times past. This may have reflected a feeling that Manchester itself was past its peak as a commercial powerhouse and was looking backwards at the achievements of its Victorian city fathers. The more one looks at the map, the more unsettling it becomes; it exhibits a disconcerting mixture of wallowing in nostalgia and yet promoting the modernity of Manchester. This mix was characteristic of official representations of the city during the interwar years, such as in the 1938 municipal centenary celebrations. In many respects, the map fits into the ethos of Civic Week itself and the attempts to generate pride in the city and to boost commerce, but simultaneously to fight against this with its sarcasm and wit.

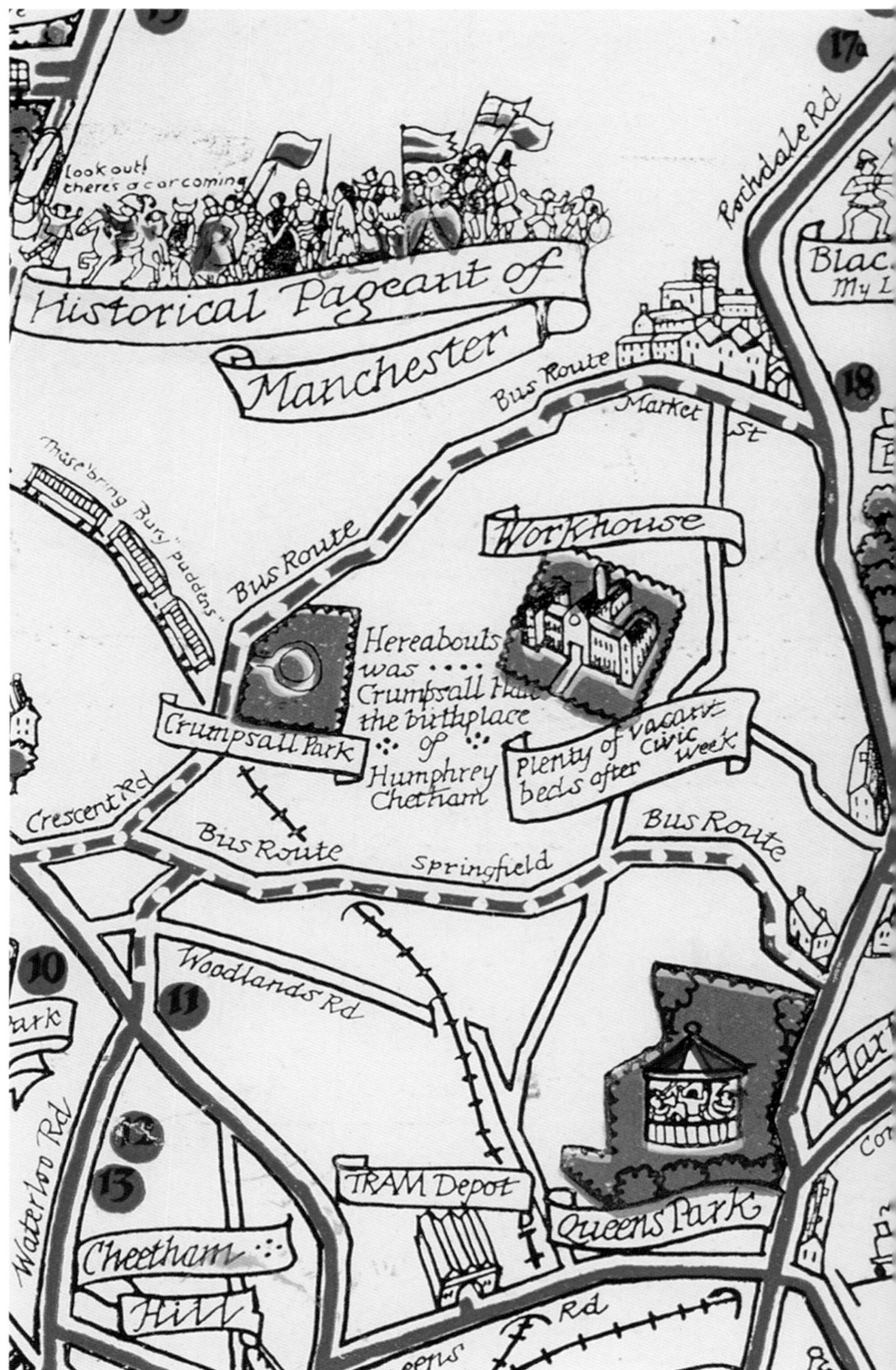

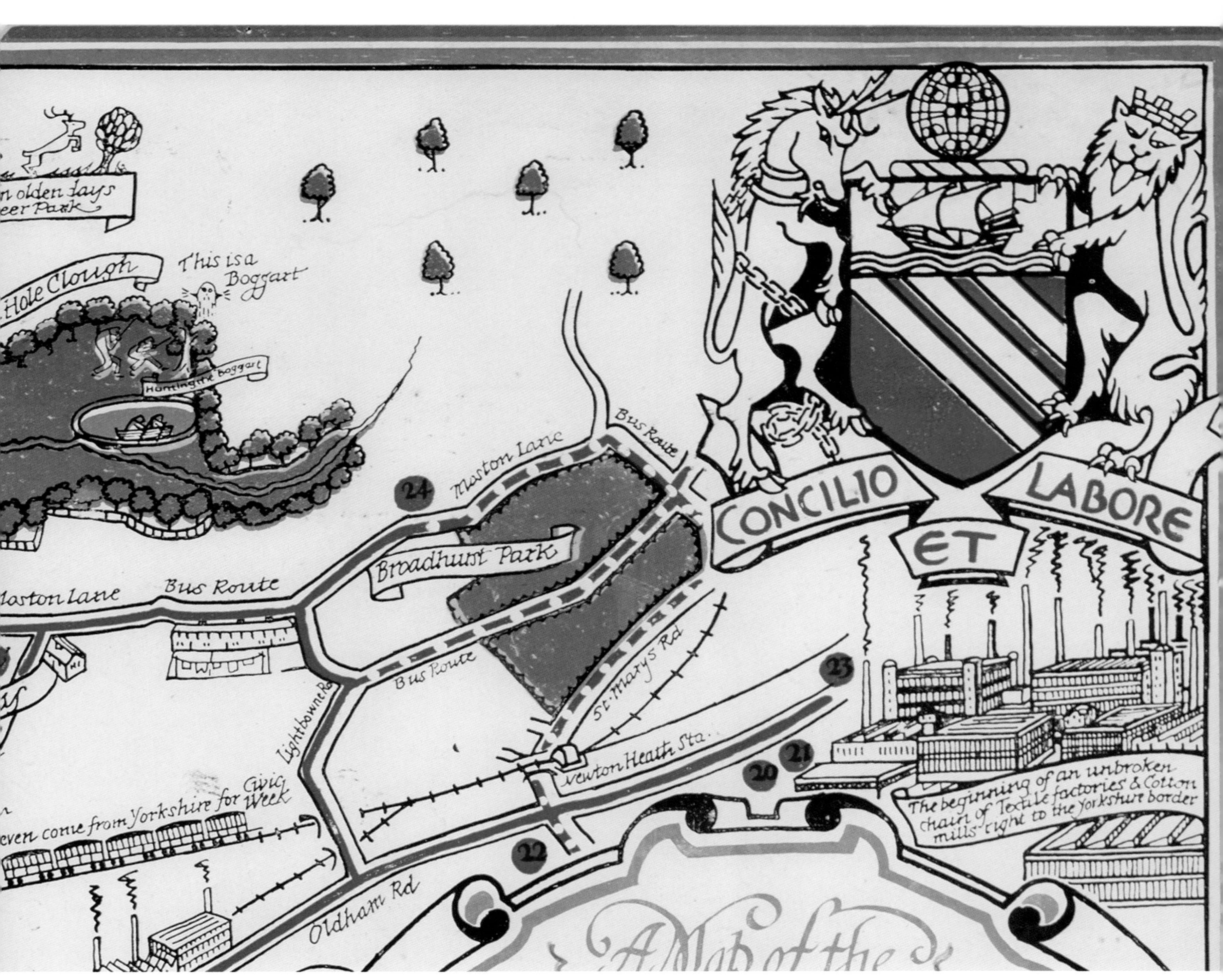

n olden days
eer Park
Hole Clough
This is a
Boggart
Hunting the Boggart
24
Moston Lane
Bus Route
Broadhurst Park
Moston Lane
Bus Route
Bus Route
St. Marys Rd
Lightbowne Rd
23
Newton Heath Sta.
20
21
Civic
Week
even come from Yorkshire for
22
Oldham Rd
CONCILIO
ET
LABORE
The beginning of an unbroken
chain of Textile factories & Cotton
mills right to the Yorkshire border
A Map of the

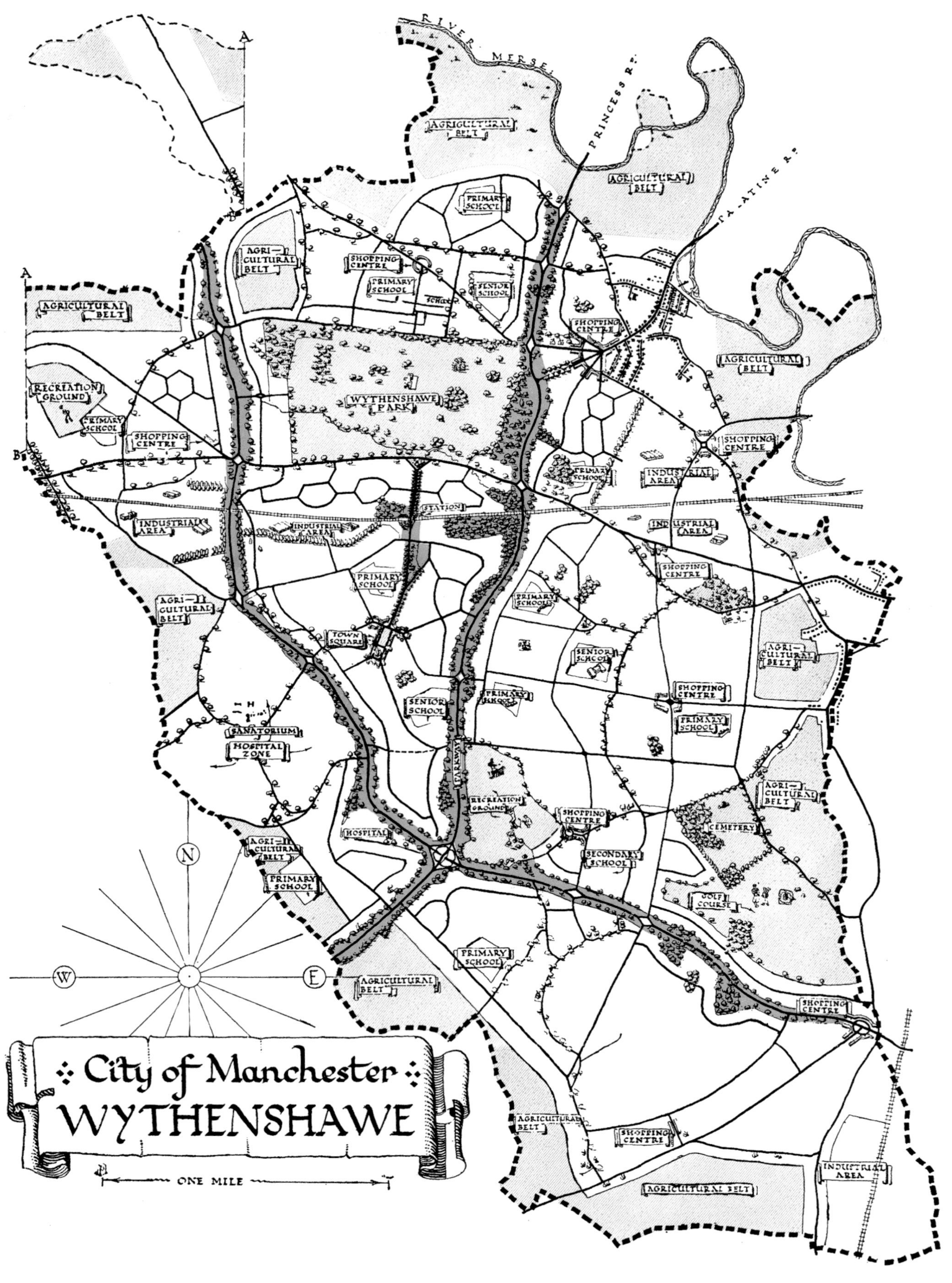

City of Manchester
WYTHENSHAWE
ONE MILE
N
W
E
A
A
B
RIVER MERSEY
PRINCESS RD.
PALATINE RD.
AGRICULTURAL BELT
AGRICULTURAL BELT
AGRI-CULTURAL BELT
PRIMARY SCHOOL
SHOPPING CENTRE
PRIMARY SCHOOL
SENIOR SCHOOL
AGRICULTURAL BELT
SHOPPING CENTRE
AGRICULTURAL BELT
RECREATION GROUND
PRIMARY SCHOOL
SHOPPING CENTRE
WYTHENSHAWE PARK
SHOPPING CENTRE
PRIMARY SCHOOL
INDUSTRIAL AREA
STATION
INDUSTRIAL AREA
INDUSTRIAL AREA
INDUSTRIAL AREA
SHOPPING CENTRE
PRIMARY SCHOOL
AGRI-CULTURAL BELT
PRIMARY SCHOOL
TOWN SQUARE
SENIOR SCHOOL
AGRI-CULTURAL BELT
PRIMARY SCHOOL
SHOPPING CENTRE
SENIOR SCHOOL
PRIMARY SCHOOL
SANATORIUM
HOSPITAL ZONE
PARKWAY
AGRI-CULTURAL BELT
RECREATION GROUND
SHOPPING CENTRE
CEMETERY
HOSPITAL
SECONDARY SCHOOL
AGRI-CULTURAL BELT
PRIMARY SCHOOL
GOLF COURSE
PRIMARY SCHOOL
AGRICULTURAL BELT
SHOPPING CENTRE
AGRICULTURAL BELT
SHOPPING CENTRE
INDUSTRIAL AREA
AGRICULTURAL BELT

1928

Wythenshawe: satellite town or garden city?

From the end of the nineteenth century, local reformers had begun to press for the city's housing problem to be tackled. When this was added to the national clamour for 'homes for heroes' at the end of the First World War, the conditions were created for a government response. It came with the Addison Act of 1919, which ushered in a new era of 'council houses' made possible by government money for municipal housing over and above that raised locally by a penny rate. Manchester was said to have an urgent need for 17,000 houses to begin to tackle the problem of its slums. What it lacked was sufficient land within its tight boundaries on which to build the houses that the situation demanded.

The battle for a site on which housing could be built was championed by William Jackson, chair of the Public Health Committee, and by Ernest Simon, chair of the Housing Committee and later an MP. Jackson persuaded the Council to commission a report from the eminent planner Patrick Abercrombie who recommended building houses in Wythenshawe, which he saw as 'virgin land, capable of being moulded to take whatever shape may be decreed'. Wythenshawe, however, lay in Cheshire and therein lay future battles. In 1926, the city bought 2,500 acres of the 4,500-acre Wythenshawe estate of the Tatton family, and Simon himself bought Wythenshawe Hall and its Park and donated it to the city. Thereafter a long battle ensued in which Cheshire and the rural district of Bucklow resisted encroachment by Manchester. This was only resolved in 1931 when a parliamentary bill extended the city boundaries. Over time, the Council gradually bought more land until it owned most of the 5,000 acres of the whole area. Simon saw this as critical since administrative control of the area only gave the city statutory powers, whereas land ownership allowed the city to do whatever was not prohibited.

Barry Parker – co-designer with his brother-in-law Raymond Unwin of Letchworth Garden City – was brought

OPPOSITE. D. MacFadyen, *Wythenshawe* (1928) [AUTH]

Ordnance Survey 2.5-inch (1951) [NLS] The slow development is evident with houses built largely only in Baguley and Sharston.

in to mastermind the design. Parker's plan outlined the framework for a huge satellite town. It embodied the low-density and neighbourhood-unit principles of Howard's garden city and drew on the American ideas of parkways and a 'Radburn' layout with widespread use of cul-de-sacs and Arts-and-Crafts-styled houses. Neighbourhoods would be served by local shopping centres and schools. Three large industrial areas were incorporated, and a hospital zone was based on the existing Baguley Sanatorium, which had been built in 1902 in the south-west of the area. The most striking feature of the plan was its two 'parkways', Princess Parkway and Western Parkway, whose aim was both to resist ribbon development and to preserve open space by building wide handsome tree-lined boulevards.

The economic depression of the 1930s and cutbacks in housing finance inevitably slowed the realisation of Parker's dreams. The first houses were completed by 1932, but relatively few were built before the war. As the OS map shows, only parts of Royal Oak, Benchill and Sharston had been built by 1939. The rest was only developed from the late 1940s as wartime building restrictions were relaxed. Princess Parkway was built as far south as Altrincham Road by 1933, and its broad tree-lined open space was a handsome testament to the parkway principle. A few of Parker's neighbourhood shops were built, but the war prevented the start of his major shopping centre and for many years most residents had to travel to shops in Northenden or rely on the vans and meat and fish carts that would arrive unpredictably as suppliers. Set against the delight at the provision of high-quality housing, the lack of amenities was one of the recurring complaints of residents over many subsequent years – not only shops, but social and recreational facilities.

Parker continued as consultant until 1941, but the post-war development was subsequently guided by the Nicholas plan of 1945. It embodied many of Parker's ideas, but made two major changes. Most dramatic was its proposal to convert the parkways into what were essentially high-speed motorways. The western parkway was never built, but Princess Parkway was extended to the airport at Ringway (which had been opened in 1938), and ultimately in the 1970s it became the M56 motorway. Parker's original parkway proposal was intended as an access road to the estate rather than a high-speed link between the airport and the city. The new high-speed road reduced east–west links and effectively cut Wythenshawe in two. The second proposed change was to move the main shopping centre into what was then a relatively isolated undeveloped area in the south of Wythenshawe so that the 'Civic Centre' was begun as the town's largest shopping centre only rather fitfully in the 1960s. By 1971, it was joined by the Forum with a library, theatre, swimming pool and restaurant. The provision of adequate civic facilities was therefore only achieved some 40 years after the start of the development as a consequence not just of war, but of successive governments' insistence on the need to prioritise house building.

After the war, the mix of house types changed. Corrugated steel-clad dwellings (popularly known as 'Tin Town') were built in Newall Green. Prefabricated houses were built in Royal Oak. Two- and three-storey walk-up flats became increasingly common as the rate of house building increased, and by the 1960s the first nine-storey flats were built in Woodhouse Park. The days of Arts-and-Craft cottages with mansard roofs were largely over.

A more positive development was the increasing provision of jobs. Initially, the only significant employment was in a Timpson's Shoe Factory on Altrincham Road. This lack of employment calls into question whether Wythenshawe should be thought of as 'garden city', 'garden suburb' or 'satellite'. Parker called it a garden city, the eminent planner Peter Hall called it the country's third garden city, but its closeness to Manchester and the absence of local jobs meant that it was not self-contained in the manner of Howard's ideal. However, employment did gradually increase. Wythenshawe Hospital was established in 1952 using the existing buildings of the Baguley Emergency Hospital, and with its expansion into a major teaching hospital it now employs more than 5,000 staff, many drawn from Wythenshawe itself. The airport has become a second major source of local employment, and the industrial estates – Sharston, Roundthorn and Moss Nook – gradually attracted a variety of light manufacturing firms. Over 40,000 people now work in Wythenshawe, and by the mid 1960s it had reached its intended population of 100,000.

Wythenshawe has long been associated with social problems. Indeed, in 2000 Benchill was classified as the most deprived ward in England. In its early years, the original residents were delighted to have new houses set in a healthy semi-rural environment. Later generations of incomers saw it less generously. Barry Parker's grand ideals were gradually whittled away over time, and Wythenshawe is now an integral and yet still a distinctive part of the wider conurbation. Its misfortune was to be sandwiched between two world wars and a succession of economic downturns, which prompted long periods of financial stringency.

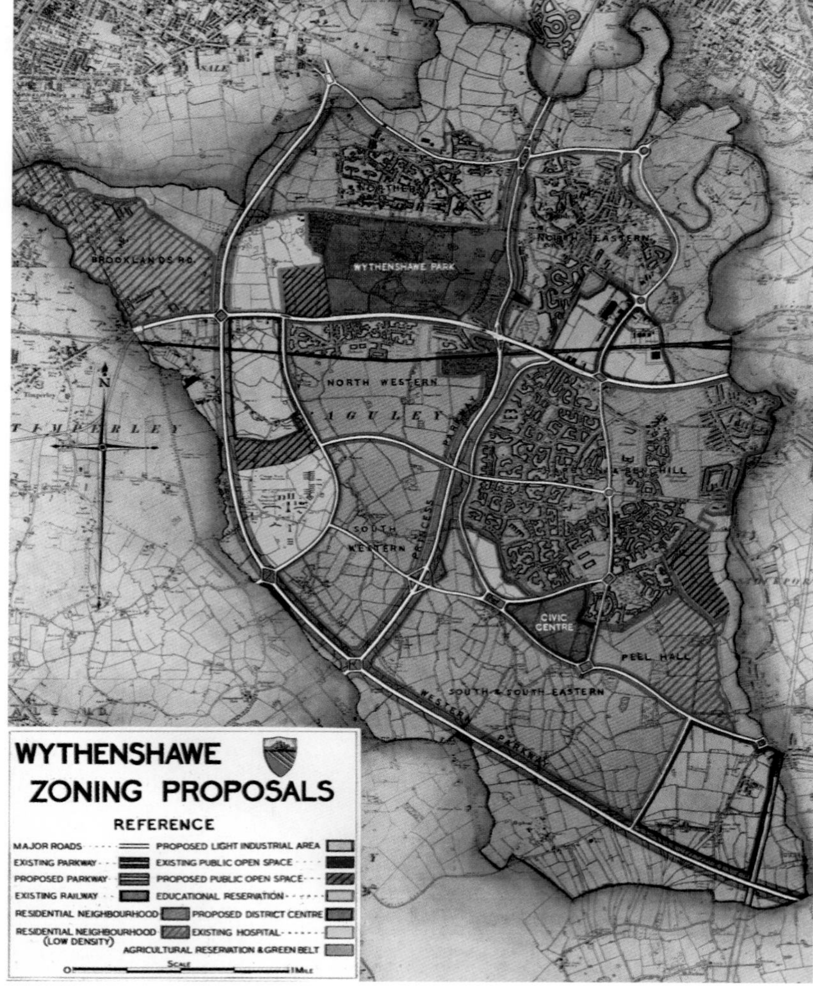

R. Nicholas, *Wythenshawe zoning proposals* (1945) [AUTH] The 1945 plan moved the proposed civic centre to an outlying location and radically reduced the number of east–west roads links across Princess Parkway.

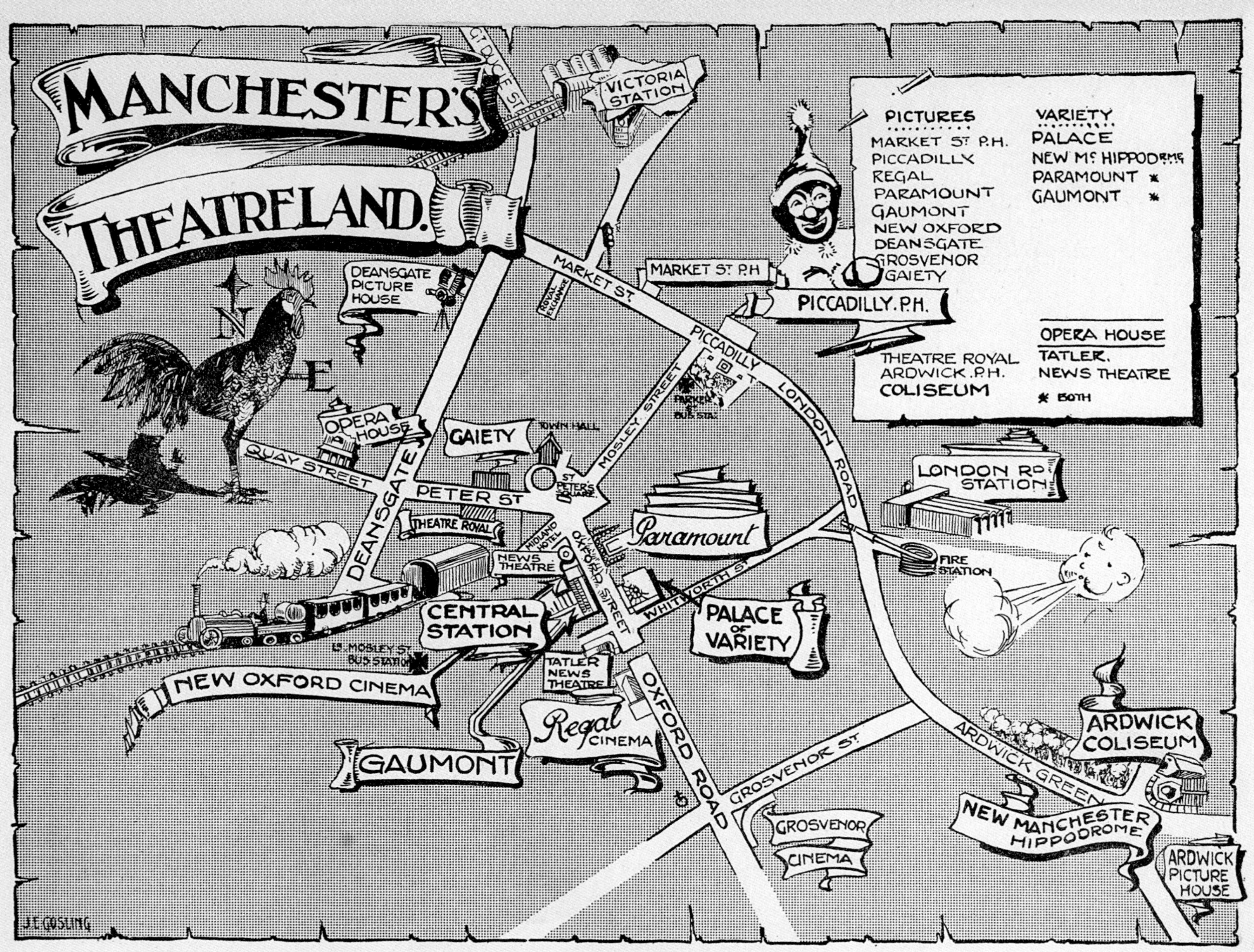
MANCHESTER'S
THEATRELAND.
VICTORIA STATION
GT. DUCIE ST
PICTURES
MARKET ST P.H.
PICCADILLY
REGAL
PARAMOUNT
GAUMONT
NEW OXFORD
DEANSGATE
GROSVENOR
GAIETY
VARIETY
PALACE
NEW MC HIPPODROME
PARAMOUNT *
GAUMONT *
OPERA HOUSE
THEATRE ROYAL
ARDWICK P.H.
COLISEUM
TATLER.
NEWS THEATRE
* BOTH
DEANSGATE PICTURE HOUSE
MARKET ST
MARKET ST P.H
PICCADILLY P.H.
PICCADILLY
N
E
OPERA HOUSE
GAIETY
TOWN HALL
MOSLEY STREET
LONDON ROAD
QUAY STREET
PETER ST
ST PETER'S SQUARE
DEANSGATE
THEATRE ROYAL
LONDON RD STATION
Paramount
FIRE STATION
NEWS THEATRE
MIDLAND HOTEL
OXFORD STREET
WHITWORTH ST
CENTRAL STATION
PALACE OF VARIETY
LR. MOSLEY ST BUS STATION
TATLER NEWS THEATRE
NEW OXFORD CINEMA
Regal CINEMA
OXFORD ROAD
GAUMONT
GROSVENOR ST
ARDWICK COLISEUM
ARDWICK GREEN
NEW MANCHESTER HIPPODROME
GROSVENOR CINEMA
ARDWICK PICTURE HOUSE
J.T. GOSLING

1937

Entertainments: stage, screen and the music scene

In the interwar years, journalists referred to the entertainment venues in Manchester city centre as 'Theatreland', borrowing the label from London. However in 1937, when the map with that title was published, it was a misnomer as only a small number of commercial theatres were left in the city centre. In what was part of a wider revolution in commercialised leisure, most of Manchester's major theatres had been devoured by the Moloch of the cinema. On closer inspection, the plan of Theatreland was evidence of this, indeed it might have been more accurate to have renamed it 'Cinemaland'.

Theatre was a prominent feature of the city's culture in the Edwardian period, with major theatres such as the Theatre Royal and the Prince's being counted among the leading venues. The Gaiety Theatre in Peter Street also occupied a special place in Manchester's theatre world as a consequence of the plays staged by Annie Horniman. Plays by Stanley Houghton, Harold Brighouse and Allan Monkhouse generated exuberant talk of a Manchester School of Dramatists, and of the Gaiety having reversed the usual perspective in which Manchester theatre looked towards London for its ideas and productions. This was not to last.

In the years after the war, Manchester theatres found themselves challenged. The Edwardian 'kinema' had continued to expand and, as more money was invested in providing larger and more comfortable venues, long-established theatres were taken over. If the staging of 'Hindle Wakes' at the Gaiety in 1912 had been Manchester theatre's *annus mirabilis*, then 1921 which saw the Theatre Royal and the Gaiety taken over and converted into cinemas was its *annus horribilis*. The Free Trade Hall would have followed had the Council not been persuaded to take over what was the city's most historically important public hall. Other city-centre theatres were also challenged. The Prince's, which had opened in 1864 in Oxford Street, limped on through crises (including one in 1937 which may

OPPOSITE. J. Gosling, *Manchester's Theatreland* (1937) [CHETS]

explain its absence from the Theatreland map) before finally succumbing. It was eventually purchased in 1940 with the intention of being demolished and replaced by a cinema, a plan which was thwarted by the war. Although the keynote of the theatre in the city centre was one of closure – Manchester, unlike London, did not build any new theatres in the interwar years – this did not mean that theatre disappeared from the city. Many theatregoers were enjoying theatre in the suburbs where there was a strong repertory movement. Local theatre companies followed a punishing schedule of performing this week's murder mystery while rehearsing next week's comedy. There was also an *avant-garde* theatre performed by groups such as the Unnamed Society, which had been formed during the war 'not to remember Manchester but to forget it', while for the politically committed there were left-wing theatre groups such as the Red Megaphones and Theatre of Action.

The popularity of theatre was also evident in the numerous amateur dramatic societies to be found across the city. Staff employed in many of the larger businesses and shops in the city centre spent their leisure hours in the company's drama group. These might have been identified on a different map, not least for the fact that in some businesses, for example for those working in the Refuge Assurance's palatial office building in Oxford Street, staff were able to put on productions in their own theatre.

Variety theatres, their origins in the Victorian music hall, were also a prominent part of the entertainment landscape. Manchester had been part of the late Victorian boom that saw businessmen, often from London, invest in modern venues. On Oxford Street the Palace Theatre, first opened in 1891, was almost entirely rebuilt in 1913 with splendid facilities, but it was overshadowed by Frank Matcham's Manchester Hippodrome, the most spectacular of all the entertainment buildings in the city centre. Every popular entertainer (from Little Tich to George Formby) appeared on its stage, as did Lloyd George when presented with the freedom of his native city. The Hippodrome's shows included water spectacles, circuses and ballets, and in 1933 as other theatres closed it responded by mounting a Shakespeare season.

However, Manchester's leisure entrepreneurs were clearly quick to recognise the potential of the 'kinema'. Even the 'Hip' could not stop the cinema juggernaut, and in 1935 it was sold and pulled down to make way for another super cinema. It found a new home in another Matcham building, the Ardwick Empire, which was renamed the New Manchester Hippodrome. However, by 1914 the city had more licensed cinemas per head of the population than any other British city. Existing entertainment buildings were taken over and converted into cinemas – which was the fate of St James's Theatre in Oxford Street – at the same time as new cinemas in architect-designed buildings were opened. Manchester's first purpose-built cinema was the Oxford Picture House in Oxford Street, opened in 1911, followed by the Grosvenor Picture House in All Saints in 1915. This was the beginning of Oxford Street becoming a faience-clad cinema strip: in 1930, as the 'talkies' replaced silent films, the Paramount (later the Odeon) opened as did the Regal Picture House which was Britain's first twin-screen cinema. The Gaumont opened in 1936, offering a level of luxury and facilities that few suburban cinemas – even the Scala at Withington – could hope to match. In the same year, Oxford Street's position as the hub of the city's popular entertainment was to be further strengthened by the opening of two smaller news cinemas – The Tatler and News Theatre.

As well as the theatres, a more detailed map of Manchester's leisure venues would have identified sporting venues, dance halls, concert halls, music halls and, above all, the cinemas, which in so short a period of time had come to dominate the city's leisure economy. That map would have changed over the decades, particularly in the final quarter of the twentieth century, which saw the city's internationally famous Hallé Orchestra move from the Free Trade Hall into a new concert hall. Manchester's popular music scene also came to have a global rather than local appeal due to groups such as New Order, Stone Roses and Happy Mondays. Venues such as the Haçienda became synonymous with 'Madchester'. In the narratives created by journalists, fans and academics, the Sex Pistols' thinly attended gig at the Lesser Free Trade Hall in 1976 was identified as the formative event in a series of musical revolts.

Salford City Council, *Salford music map* (2008) [AUTH] Present-day cultural tourism involves exploring the city's recent musical history.

Following the example of Liverpool, Manchester's status as a world music city also saw maps produced and walking tours organised for fans. Salford too saw a blossoming of popular music, as reflected in the celebratory map of performers and venues. Of the many places of pilgrimage, none was more photogenic than the Edwardian Salford Lads' Club in Ordsall (shown as '8' on the map), which, although not mentioned in The Smiths' much analysed lyrics, featured as an illustration on their 1986 album, *The Queen is Dead*.

Trinity Station
MARKET STREET
KING STREET
DEANSGATE
PETER STREET
MOSLEY STREET
1
2
3
4
5
6
7
8
10
11
12
13
15
16
17
18
20
20
20
20

1945

Post-war visionary planning

In 1945, two important plans were published under the guidance of Rowland Nicholas, Manchester's powerful City Surveyor and Engineer. One focused on the city itself, and the second covered the wider Manchester region. Both were based on detailed analyses begun before the war.

It was a time of high hopes and grand visions. Town planners saw scope for implementing the planning principles that had generally come to be accepted within the profession, if not more widely. Instead of a patch-and-mend approach to tackling the problems faced by cities they visualised large-scale comprehensive intervention. Manchester was very much alive to these ideas. The Manchester and District Joint Town Planning Advisory Committee held an exhibition and conference in 1922 at which many of the leading figures of the nascent planning profession spoke: Patrick Abercrombie on regional planning, Ebenezer Howard on garden cities, George Pepler on town planning and Thomas Adams on planning for industry. The planning of Wythenshawe was in the hands of Barry Parker from the 1930s. The city also hosted the National Housing and Town Planning Conferences in 1928 and 1936. Manchester was clearly very much on board with the contemporary planning ideals.

War damage also offered new potential for redrawing urban townscapes. While Manchester had suffered bomb damage during the war, it was never on the scale of towns such as Coventry, Plymouth or Hull. Indeed, in his report Nicholas noted, somewhat resentfully, that it was:

> . . . increasingly certain that the end of the war would see the city little altered. For this we have cause to be grateful, but from a planning point of view it has meant that our redevelopment scheme must take more careful account of the financial consequences of removing undamaged buildings. Thus our initial idealism has been tempered by a growing preoccupation with present realities.

OPPOSITE. R. Nicholas, *City of Manchester central area* (1945) [AUTH]

MANCHESTER 2045 A.D.

However, idealism there certainly was. The plan drew on the then widely accepted planning concepts: zoning land uses to remove cheek-by-jowl areas of housing, industry and commerce; neighbourhoods with an appropriate mix and number of goods and local services; a clear hierarchy of roads to distinguish major arteries from more local distributors; reduction of housing densities; and open spaces to create healthier environments. The redeveloped city shown in Nicholas's plan contained all of these elements. Had its broad outlines been implemented, Manchester would have been a dramatically different place. The endpaper in the report shows the scale of the proposed transformation to the city centre. It would have created a somewhat Soviet-style townscape with few of the then existing buildings surviving. Even Waterhouse's Town Hall would have disappeared, to be replaced by a smaller civic centre with a prominent central tower and a striking perspective down a broad boulevard formed by the demolition of Brazennose and Queen Streets and closed off to the north by a new complex of Law Courts. It was not quite a *tabula rasa*, but not far short.

The zoning proposals also stamped a reconfigured geometry of new neighbourhoods across the city. Residential neighbourhoods were to be redeveloped so that areas with housing densities of more than 24 houses per acre would be rebuilt at lower densities. Interestingly, two 'low-density' areas would have been preserved – one in Rusholme and a second in Didsbury. This aimed to provide desirable residential environments to attract back to the city some of Manchester's 'leaders in commerce and business who have been accustomed to seek homes outside its boundaries'. The reduction in housing densities meant that half of the houses in the wider district would need to be demolished, and this inevitably raised questions of 'overspill'. Three of the local authorities – Manchester, Salford and Stretford – would generate substantial overspill (and their populations would fall dramatically – Manchester from 731,800 to 475,000, Salford from 190,000 to 89,400 and Stretford from 59,700 to 51,400). Yet across the district, few areas had sufficient land to accommodate their own housing needs, let alone accept overspill. Some 42,500

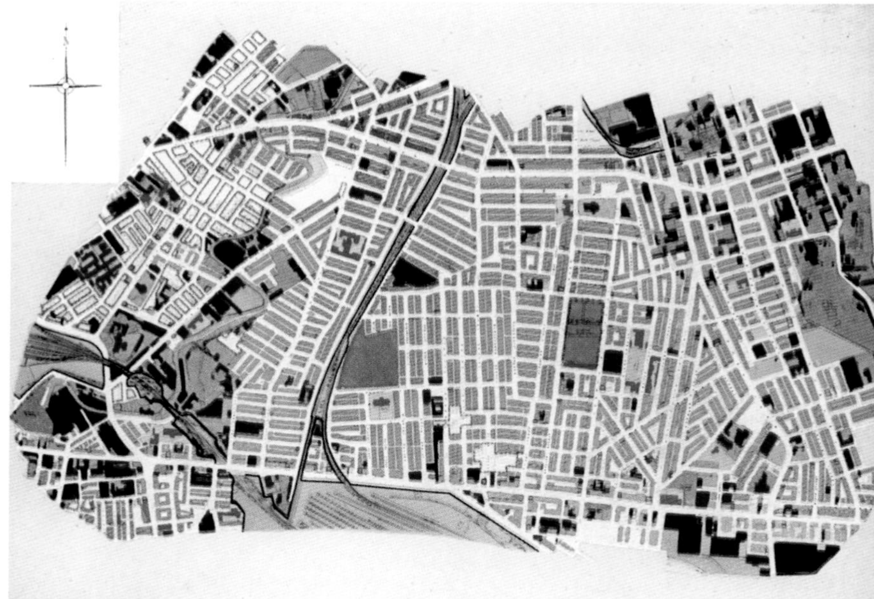

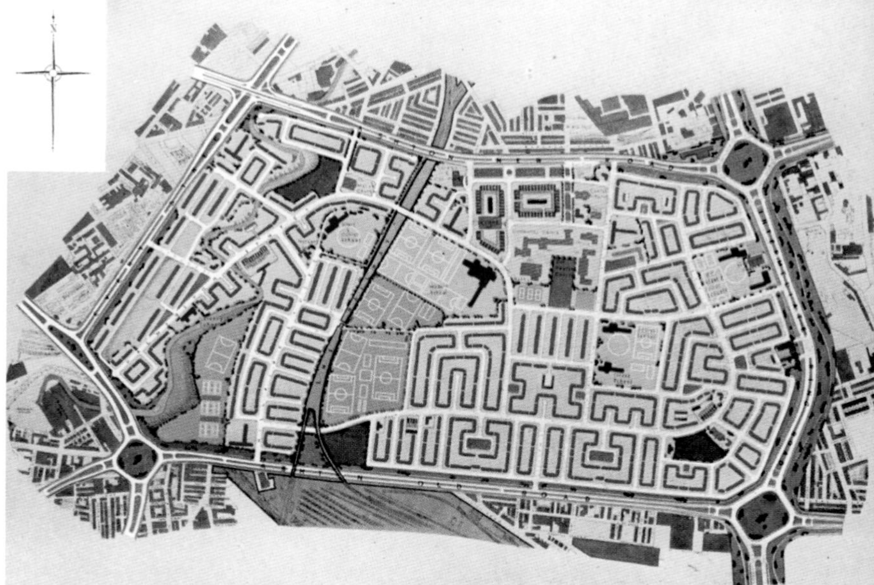

ABOVE. R. Nicholas, *Beswick*. Virtually nothing would have survived from the old Beswick.

OPPOSITE. R. Nicholas, *Manchester 2045 A.D.* The enormous proposed Trinity Station is shown at the top.

houses would therefore be needed outside Manchester's boundaries, and this raised issues about building garden cities or satellite towns or adding houses to existing small towns outside the region.

Industry was also to be zoned into categories: 'special' noxious industry, 'general' industry, light industry and 'domestic', such as bakeries or repair depots, which would simply be absorbed into local neighbourhoods. The proposals for the design of both the housing and industrial areas were specified in detail and again would have required major comprehensive renewal. The example of Beswick shows the degree of change entailed.

Roads and transport formed an important part of the plan. The map in the District report shows a hierarchy of four ring roads: a city circle surrounding the centre itself and three main rings – inner, intermediate and outer – designed to relieve pressure on the inadequate radial roads into the centre of the city. The road network was also deployed in part to demarcate the various neighbourhoods into which the city was to be divided. The plan also shows the focus on the designation of green belts both to provide lungs for the city and to maintain the physical separateness of the constituent towns within the conurbation.

The proposals for railways were in many ways even more drastic than those for roads. A major new station, Trinity, was proposed on the border of Salford and Manchester. This would have combined the functions of Victoria, Exchange and Salford stations, effectively closing them down. In due course, new rail links would also have made Central Station redundant so that the city would have been left with two principal stations, the new Trinity and London Road.

The report speculated about future developments in air travel, suggesting that there may be 'an extension of air liners, with air taxis stationed on flat-roofed buildings, and with folding-winged autoplanes housed in garages'. Whatever the future, however, a major airport was seen as essential to the city's prosperity. Ringway was seen as being 'as vital to the city in this century as was the building of the Ship Canal in the last'.

Three areas were extensively written up in detail. First was Wythenshawe, which clearly provided a model for many of the principles underlying the report. Second was a proposal for a 'Learning, Medicine and the Arts' precinct along Oxford Road from All Saints to the proposed Intermediate Ring Road. At its northern end would be a grand processional way leading to a new civic hall, with two theatres, a concert hall, baths and Broadcasting House. South of this would be a greatly enlarged university with new departments on both sides of Oxford Road. Third, the hospital complex would be greatly expanded. The whole precinct would be terminated by a substantially extended Whitworth Park. Prefiguring the later Wilson and Womersley proposals, Oxford Road would be closed to through traffic and Cambridge Street and Upper Brook Street would become the main approaches to the city from the south.

Cynics could well dismiss the report's proposals as wildly unrealistic at a time when the country was cash strapped. Despite Nicholas's introductory disclaimer about reining back the initial idealism, the report makes few concessions to the political and financial circumstances of the time. Yet many of its suggestions are ones which, in whole or in part, came into being in the fullness of time or which remain as issues yet to be addressed. Low-density neighbourhoods, satellite towns beyond the boundaries of the city, major ring roads, expansion of the airport, the creation of a university precinct are some of the suggestions that eventually materialised. Others, such as the reconfiguration of the railway network to facilitate better interconnectivity, are now being implemented. And one of the most far-sighted arguments in both reports was their emphasis on collaboration between the local authorities across the sub-region.

OPPOSITE. Manchester and District Planning Committee, *Manchester and district . . . proposed road system* (1945) [AUTH]

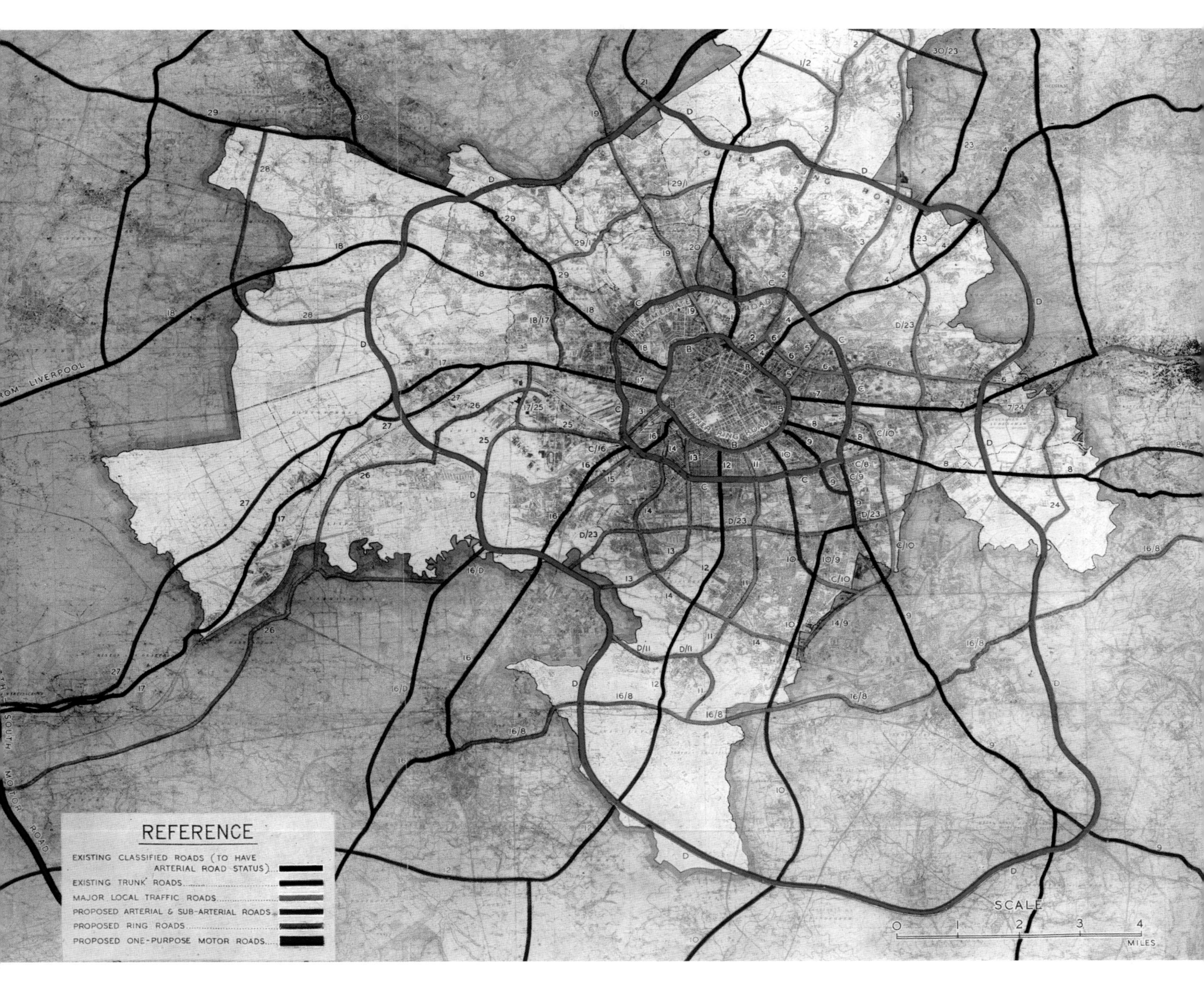

OUTER RING ROAD
INNER RING ROAD
LIVERPOOL
SOUTH MOTOR ROAD
REFERENCE
EXISTING CLASSIFIED ROADS (TO HAVE ARTERIAL ROAD STATUS)
EXISTING TRUNK ROADS
MAJOR LOCAL TRAFFIC ROADS
PROPOSED ARTERIAL & SUB-ARTERIAL ROADS
PROPOSED RING ROADS
PROPOSED ONE-PURPOSE MOTOR ROADS
SCALE
0 1 2 3 4
MILES

CITY OF MANCHESTER
M723/82
APPROACH TO LANDING STRIP 'B'
N
VEHICLES
LUGGAGE
LANDING STRIP 'A'
LANDING STRIP 'B'
CONTROL
HELICOPTER WAITING SPACE
PASSENGERS
VICTORIA STATION
APPROACH TO LANDING STRIP 'A'
EXCHANGE STATION
IRWELL
LONG MILLGATE
HELIPORT
SCALE

1956

Fantasy transport: unrealised plans above and below ground

Like most busy cities, Manchester has suffered severe traffic congestion for many years. Key roads were often clogged and slow moving particularly as more people needed to travel into the city centre. Many solutions were proposed down the years, including tunnels, viaducts, elevated highways and large ring roads. Their feasibility, in terms of the engineering and economics of going under, above or around congested central streets was often questionable, but promoters would seek to make their transport proposal plausible by plotting the suggested routes on maps. Most of the schemes, particularly the more speculative ones, remained unrealised, but it is a fascinating 'what-if' exercise to re-examine the old maps and plans for transport infrastructure that might have been.

In the post-war decades, radical re-planning of Manchester was called for, and innovative forms of transport were key futurist symbols. One of these was the helicopter, which by the early 1950s promised to revolutionise travel. But there were challenges in how they would actually operate in cities and crucially where to site the heliports. While airports grew larger and necessitated edge-of-city location, many asserted that heliports needed to be centrally located to exploit the advantages of vertical flight.

The 1951 Development Plan for Manchester noted that 'there are two possibilities, one to provide a ground station and the other to provide a station on the roof of a building (which may have to be strengthened for the purpose)'. The City Engineers Department considered a range of sites for a ground-level heliport in the mid 1950s utilising vacant and semi-derelict plots within about a mile of the city centre.

There was also a more dramatic and speculative proposal for a rooftop heliport on Victoria railway station. It had landing strips at right angles to each other, along with five circular parking spots. Service buildings and a small control tower were situated around the edge of the landing platform.

OPPOSITE. R. Nicholas, *City of Manchester heliport* (1956) [MLA]

Manchester Corporation, *Bird's-eye sketch of Victoria Station heliport* (1956) [MLA]

The heliport would have covered most of the station, and the roof would undoubtedly have needed serious strengthening.

The plan shows two approach paths into the heliport. The directions are from north-west and south-west, which would have meant that noisy helicopters would have flown in over Salford and Trafford rather than over Manchester itself. An irregularly shaped bubble of light shading also encompasses the heliport zone of operations on the plan, and this may have been a notional noise envelope indicating the areas of disturbance.

The heliport plan was accompanied by a large perspective drawing of Victoria Station and the surrounding neighbourhood, with new architectural details for the heliport added, along with some colourful, cartoon-like, helicopters busily flying around. The landing platform is literally buzzing with simultaneous helicopter activity – probably in excess of what it would have been capable of safely handling. The blue helicopter just taking off would have knocked over the passengers on the tarmac with its powerful rotor downwash.

Nothing came of the different schemes for a heliport in Manchester in the 1950s, and by the early 1960s realistic prospects faded nationally for commercially viable helicopter passenger services. As a consequence, no major purpose-built city-centre heliports were constructed in Britain, except at Battersea in London. Certainly, there were no spectacular rooftop landing decks on mainline railway stations. Helicopters failed to revolutionise urban travel. They proved to be difficult to operate efficiently, they were excessively noisy at low levels and were too costly to compete with other forms of transport.

If helicopters came to nothing, so too did the many calls for an underground link across Manchester's city centre as an attempt to solve the problem of passengers connecting between the separate mainline train stations. Beside transport utility, there has also been an issue of pride: Manchester felt it needed an underground to be taken seriously as a major metropolis.

The first scheme was probably advanced as early as 1839. Myriad proposals from Town Hall officials, elected councillors, transport engineers and private speculators can be found in the archives. They range from modest cut-and-cover tram tunnels, to boring tunnels for elaborate tube circle lines and even draining the River Irwell to make a new sunken transport corridor. Similar design proposals pop up periodically, and there is a sense of *déjà vu* looking at the maps of their routes.

Shown here is a colourful plan of a 1903 scheme by a prominent local engineer, W.A. Jackson. His idea was to build new underground stops near the mainline train stations and interlink them with 2.5 miles of tunnels at an estimated cost of £750,000. Like many similar proposals, nothing came of Jackson's vision of an underground railway for Manchester.

The closest Manchester came to achieving its desired underground was the Picc-Vic scheme in the late 1960s. A heavy rail tunnel beneath Manchester, with stations underneath the Town

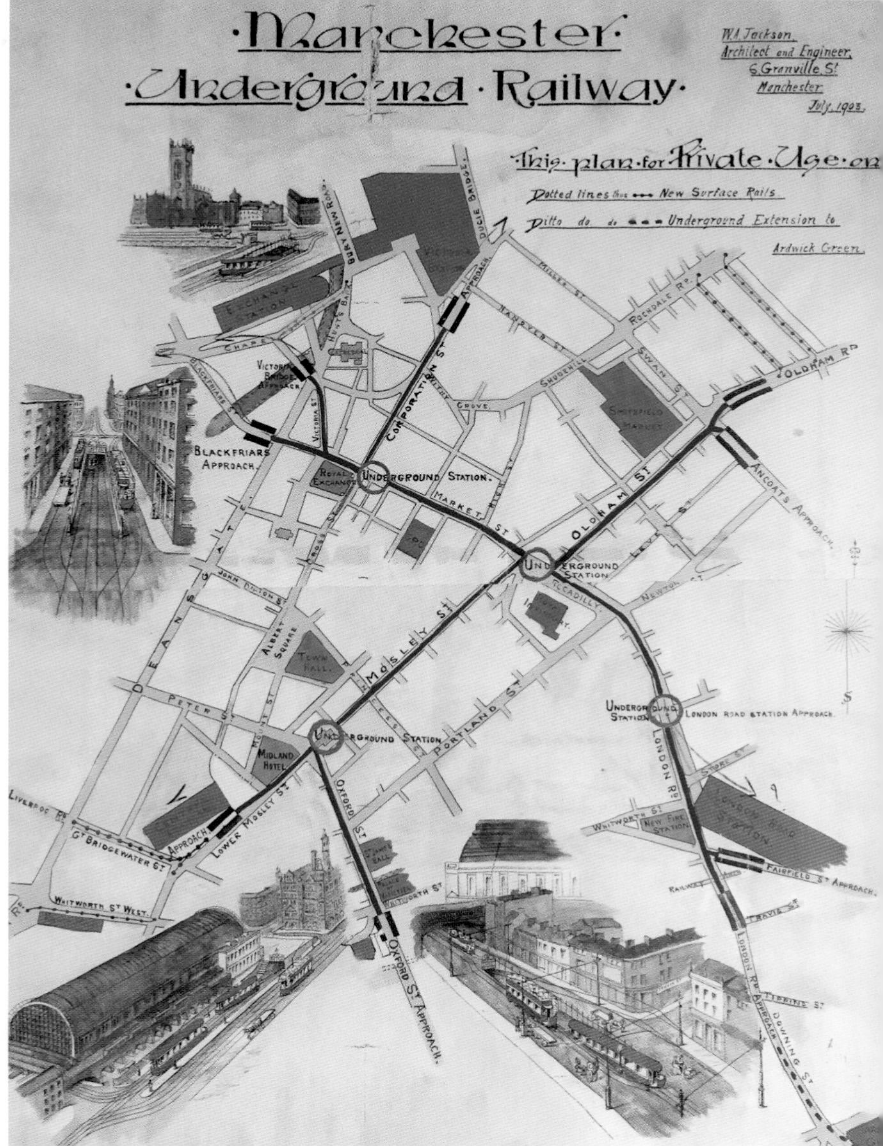

W.A. Jackson, *Manchester underground railway* (1903) [Mark Ovenden]

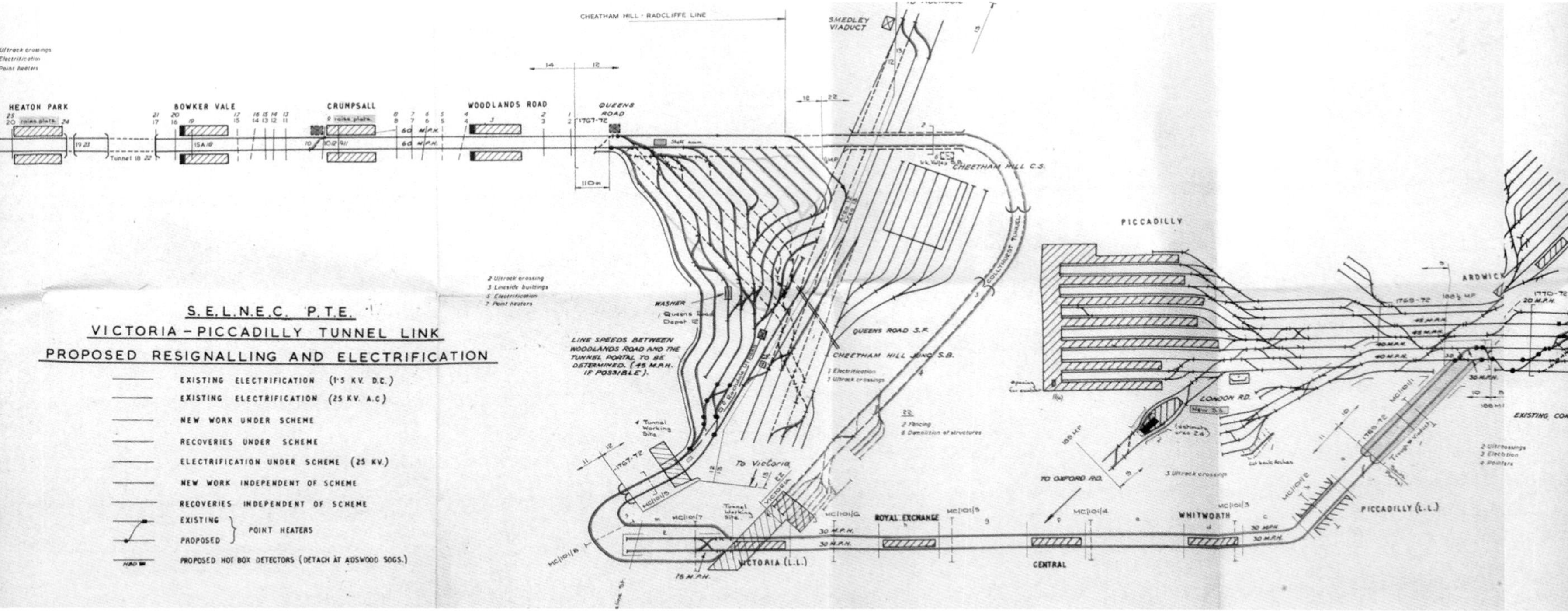

SELNEC, *Victoria-Piccadilly tunnel link* (*c*.1972) [AUTH] Proposed new portions of tracks are shown in red, including an expensive tunnel beneath the city centre with three new underground stations.

Hall and at the Royal Exchange, with access to the Arndale shopping centre, would have formed the centrepiece of a new electrified railway line running from north to south across the city region. The goal was high-frequency service with a train every two-and-a-half minutes at peak times through the city-centre section.

The Picc-Vic scheme received parliamentary powers in 1972, and detailed engineering plans were drawn up along with architectural designs for the underground stations. Yet it was going to be costly piece of civil engineering requiring over 2 miles of twin bored tunnels some 5.5 metres in diameter. It nearly got built, but was defeated in large part by the national economic recession in the early 1970s and deep central government budget cuts.

All the schemes down the years for an underground in Manchester would have been expensive and, crucially, uneconomic given the shape of the city, the relatively small size of the central zone and the scale of daily passenger movements. Yet Manchester remained envious of its near competitors like Liverpool, Newcastle and Glasgow who succeeded in building an underground transport infrastructure, however modest.

After the failure of the Picc-Vic scheme, planners and politicians looked eventually to the cheaper alternative of surface trams to solve the disconnection across the city centre. Metrolink, as Manchester's new light rail system was called, opened to passengers on the Bury to Victoria Station segment in 1992. The scale of the infrastructure is rather modest, and the station stops in the city centre are strictly functional. Yet the return of trams to Manchester streets, after a gap of 43 years, was generally welcomed and the system has gradually been expanded. Metrolink now has over 90 stops and carried nearly 38 million passengers in 2016/17.

OPPOSITE. GMC, *The Picc-Vic project* (*c*.1970) [Richard Brook]

Northbound
First train
Second train
BOLTON
OLDHAM
Exchange
« way out
Bolton
Royal Excha
CLINT
EASTWOOD
THUNDERBOLT
& LIGHTFOOT
The Picc-Vic Project
GMC

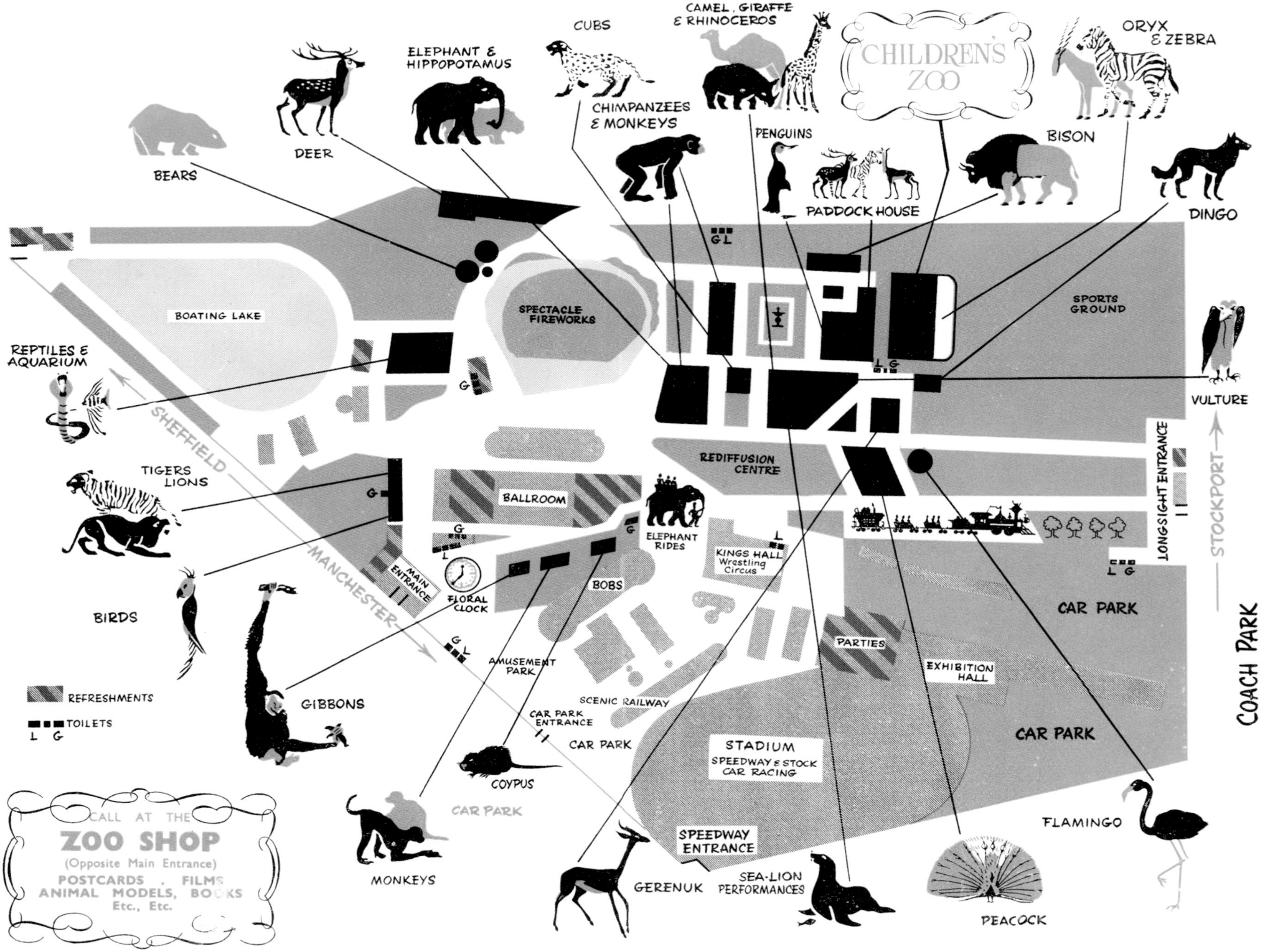

CHILDREN'S ZOO

CAMEL, GIRAFFE & RHINOCEROS

CUBS

ELEPHANT & HIPPOPOTAMUS

ORYX & ZEBRA

CHIMPANZEES & MONKEYS

DEER

BEARS

PENGUINS

BISON

PADDOCK HOUSE

DINGO

G L

BOATING LAKE

SPECTACLE FIREWORKS

SPORTS GROUND

REPTILES & AQUARIUM

G

L G

VULTURE

SHEFFIELD

REDIFFUSION CENTRE

TIGERS LIONS

G

BALLROOM

ELEPHANT RIDES

LONGSIGHT ENTRANCE

STOCKPORT

G

G

L

KINGS HALL Wrestling Circus

L G

MAIN ENTRANCE

FLORAL CLOCK

BOBS

MANCHESTER

CAR PARK

BIRDS

COACH PARK

G L

PARTIES

AMUSEMENT PARK

EXHIBITION HALL

REFRESHMENTS

GIBBONS

SCENIC RAILWAY

TOILETS
L G

CAR PARK ENTRANCE

CAR PARK

CAR PARK

STADIUM
SPEEDWAY & STOCK CAR RACING

COYPUS

CAR PARK

FLAMINGO

SPEEDWAY ENTRANCE

CALL AT THE
ZOO SHOP
(Opposite Main Entrance)
POSTCARDS . FILMS
ANIMAL MODELS, BOOKS
Etc., Etc.

MONKEYS

GERENUK

SEA-LION PERFORMANCES

PEACOCK

1958

Belle Vue: bread and circuses

As one of the Britain's leading commercial and industrial cities, Victorian Manchester produced many outstanding businessmen. Few of these were as innovative and successful as John Jennison. His business was manufacturing fun. Jennison established what became the Belle Vue Zoological Gardens in 1836, located on a 35-acre site some 2 miles to the east of the city centre, at the junction of Kirkmanshulme Lane and Hyde Road. He built Belle Vue into a business that became synonymous with popular entertainment at a time when more people had the free time and spare cash to enjoy themselves. Its customers were eventually to be drawn widely from across the north of England, the Midlands and Wales. Indeed, by the end of the nineteenth century, such was the reputation of the gardens that in the mental maps of some people Manchester was no longer 'Cottonopolis'; rather, it was a place near Belle Vue.

Jennison, who had previously run a small pub with a garden in Stockport, had clear ideas about the world of commercialised leisure. One of his early key decisions was to establish a zoo, a novel attraction that remained a central feature of his leisure park. Over the years, he invested in animals, making Belle Vue one of the largest permanent zoos in Europe. Armadillos, dingoes, monkeys, raccoons and wildcats were among the animals displayed. Maharajah, an Indian elephant purchased at the auction of a menagerie in Edinburgh and then walked (with much publicity) to Manchester, was one of the star attractions that even after death remained a must-see sight in Belle Vue's own natural history museum.

Jennison was quick to recognise the contribution that railways might make to his business, particularly in bringing in customers on cheap excursion trains. A revealing document in his business papers is a pencilled list of the dates of annual wakes holidays in Yorkshire and Lancashire towns, primitive market research, but evidence that he recognised these as places from which he might attract customers. By the early 1900s,

---

OPPOSITE. Belle Vue Company, Belle Vue Gardens (1958) [AUTH]

there were four railway stations within easy walking distance of the gardens – Longsight (opened in 1842), Ashbury (opened in 1855), Belle Vue (opened in 1875) and Hyde Road (opened in 1900). Such was the volume and profitability of the Belle Vue traffic that Longsight found it necessary to build excursion platforms. The catchment area from which excursionists were drawn increased over time: Whitsun 1897 saw excursion trains from as far away as Newcastle and Cardiff. Long-distance day trippers did not mean the neglect of more local visitors, cheap railway fares being negotiated with railway companies for those travelling from and to the city centre (London Road Station). Not everyone reached Belle Vue by train, and in the annual Belle Vue guidebooks information was provided on the location and fares of cabstands and bus stops.

Jennison recognised the importance of providing sufficient attractions to ensure that his visitors stayed in the gardens for as long as possible, preferably all day. Special events and competitions were introduced to boost attendance. Of these, Belle Vue's annual brass band competition, started in 1853, quickly established itself as one of the premier events in the banding calendar, attracting bands and their supporters from across the country. The early 1850s also saw the exhibition of monster paintings depicting historical events and military battles. Re-enactments of patriotic episodes culminating with a fireworks display soon became another popular favourite in the calendar of entertainments. Not all of these pyrodramas were historical: in 1911, the public viewed the unexpected sight of Manchester's iconic town hall being shelled by the German army, an incident based on William Le Queux's sensational bestselling novel *The Invasion of 1910*. Dancing proved to be another crowd-pleasing entertainment, and over the years money was invested in building outdoor pavilions and indoor halls, as well as in the bands that provided the music. Faced with such a range of attractions, new visitors were no doubt glad to consult plans of the gardens, which were included in the main brochure or sold separately as single sheets.

It was also recognised that new attractions were needed in order to keep people returning to the gardens. In this respect, Jennison was as much an imitator as innovator. Reinvested profits drove the business forward. A viewing tower built in 1882, if not quite the equivalent of the seaside pier, did provide a vantage point from which to survey the gardens and the city. Jennison kept a close eye on his competitors. In Manchester, it was Pomona Gardens, situated closer to the city centre on the banks of the Irwell, that was his main rival in the mid-Victorian decades, and Belle Vue was fortunate that Pomona was eventually forced to close due to the building of the Ship Canal docks. But competing as it did in the market for day trippers, Belle Vue also had to be alert to the new leisure attractions available in such Lancashire seaside towns as Blackpool.

Belle Vue was successful because it was run on the strictest business lines. Expenditure was closely monitored and wherever possible costs controlled by producing products in-house. A bakery, brewery, printing office and fireworks factory were among the service buildings. The food and drink consumed by visitors – Belle Vue claimed to be able to feed as many as 2,000 people at a single sitting – became one the main sources of its profits. Later, this expertise was not confined exclusively to the gardens, since the Jennisons extended their business to cater for events as far away as the shooting trials at Wimbledon.

Jennison set up Belle Vue in favourable times, operating in the midst of a large and increasing population, many of whom were enjoying increased disposable income and free time, and looking to find new forms of amusement. At the same time, the railways provided the means of extending the business beyond the city and its region. Belle Vue remained in the Jennison family until it was sold in the 1920s. It continued to look for new ways to entertain, opening the country's first modern greyhound racing track in 1926, followed in 1928 by a speedway stadium. Further investment followed the takeover in 1956 by London businessmen, Sir Leslie Joseph, Managing Director of Battersea Pleasure Gardens, and the caterer, Charles Forte. Some of their changes were evident when Queen Elizabeth visited the gardens in 1961. Older features had been removed, the enclosures of the more popular animals re-sited and the Children's Zoo made a more prominent attraction. The ballroom had been rebuilt and improvements made to the

Belle Vue Company, *Belle Vue* (*c.*1900) [CHETS]

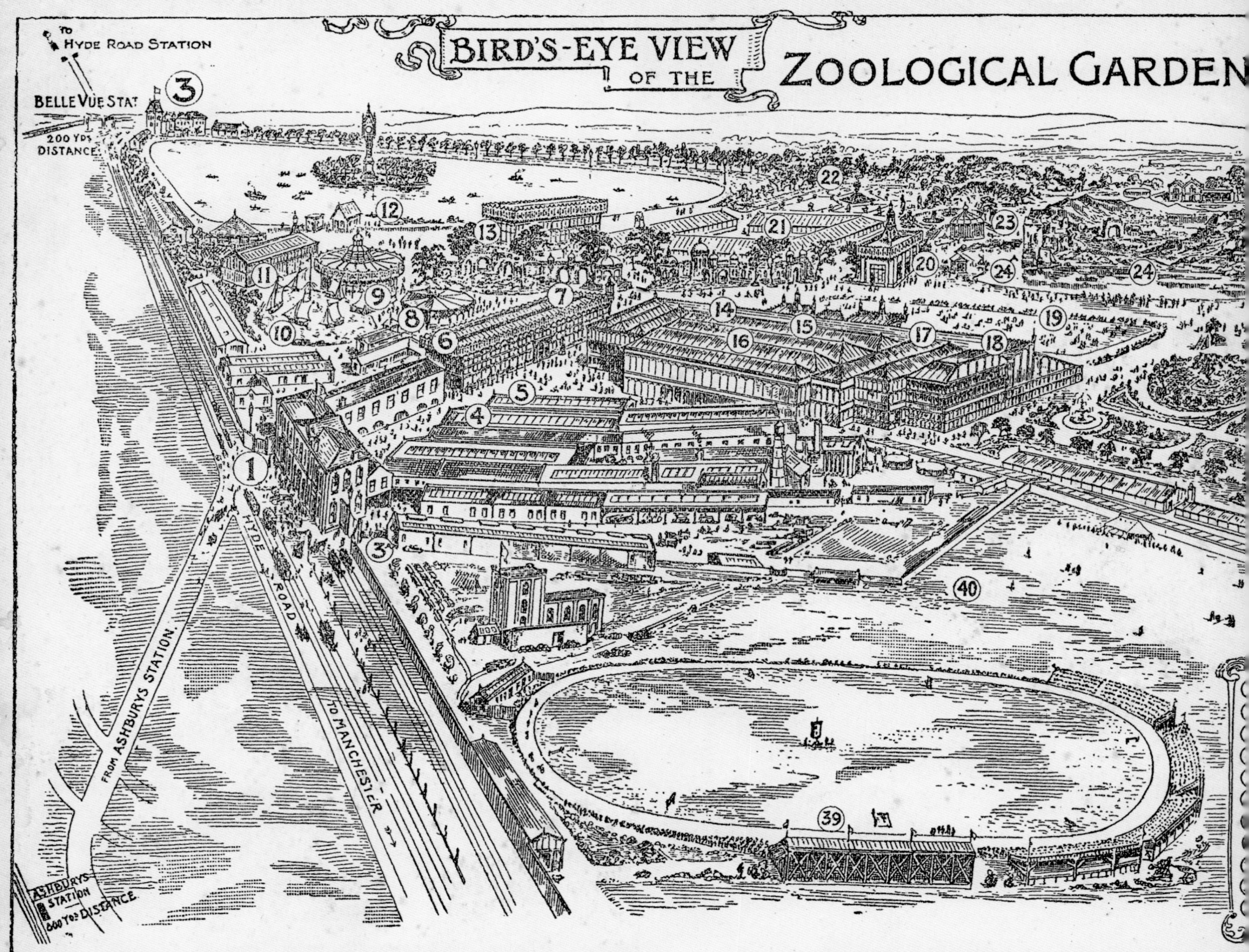

King's Hall, the venue for boxing and wrestling matches as well as the popular Christmas circus. Less eye-catching but just as important was the increasing space given over to car parking. Belle Vue continued to find ways to attract visitors – The King's Hall became a major venue for pop concerts – but it was never to achieve the visitor numbers enjoyed earlier in the century. The closure of the zoo in 1977 was the beginning of the end, and the park finally closed its gates in 1980.

G. Falkner, *Bird's-eye view of the zoological gardens, Belle Vue* (*c.*1900) [CHETS] Plans of Belle Vue were provided for visitors who were expected to spend the whole day exploring its attractions.

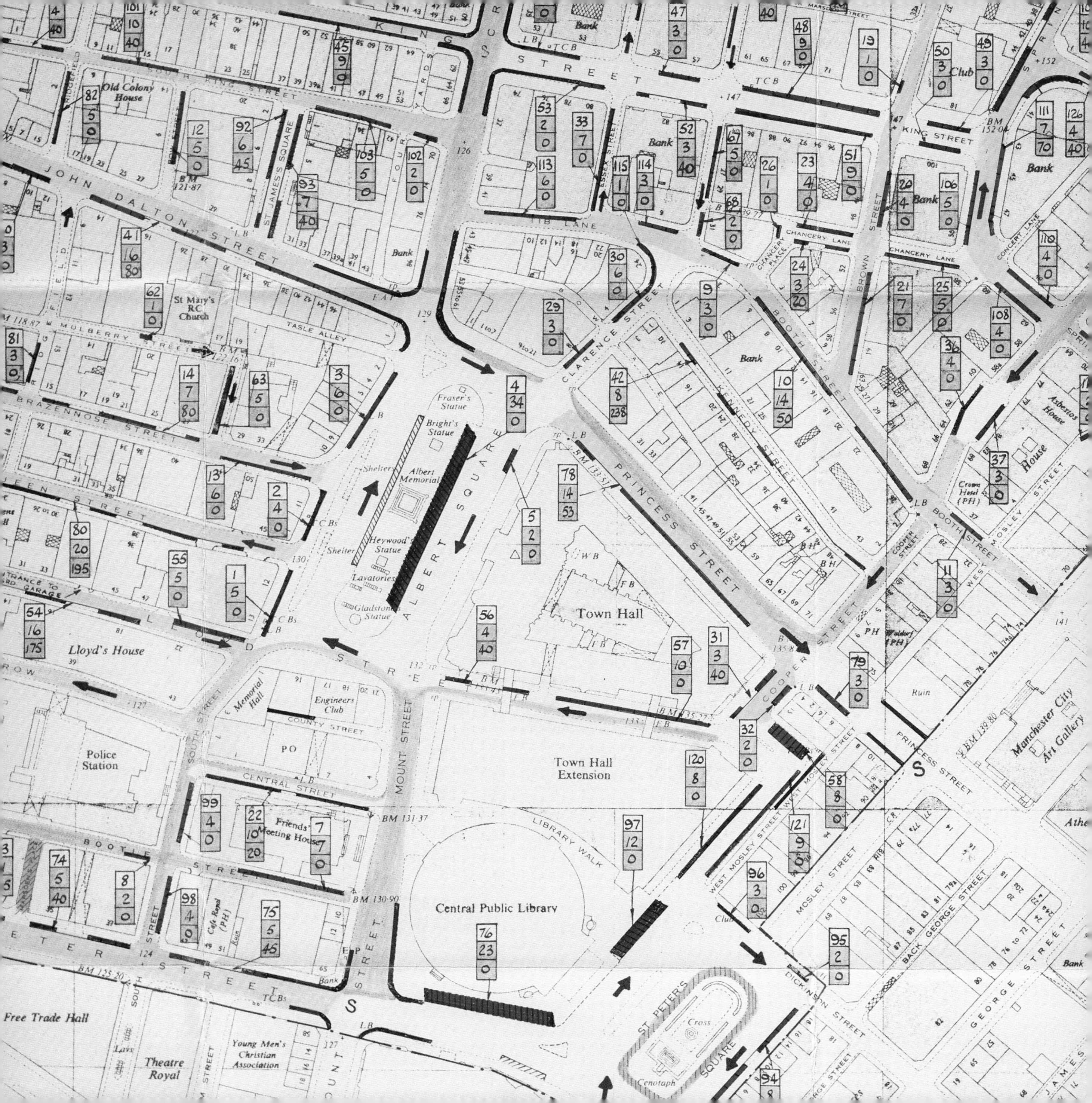

King Street
South King Street
Old Colony House
John Dalton Street
St James's Square
Tib Lane
Essex Street
Chancery Lane
Chancery Place
Brown Street
Concert Lane
Bank
Club
St Mary's RC Church
Tasle Alley
Mulberry Street
Brazennose Street
Clarence Street
Booth Street
Kennedy Street
Princess Street
Fraser's Statue
Bright's Statue
Albert Memorial
Shelters
Shelter
Heywood's Statue
Lavatories
Gladstone Statue
Albert Square
Town Hall
Asbestos House
Crown Hotel (PH)
Cooper Street
West Mosley Street
Waldorf (PH)
Ruin
Manchester City Art Gallery
Lloyd's House
Entrance to Garage
Memorial Hall
Engineers Club
County Street
Police Station
PO
Town Hall Extension
Central Street
Mount Street
South Street
Friends' Meeting House
Library Walk
Bootle Street
Central Public Library
Mosley Street
Back George Street
George Street
Peter Street
Cafe Royal (PH)
Free Trade Hall
Theatre Royal
Young Men's Christian Association
St Peter's Square
Cross
Cenotaph
Dickinson Street
James

1960

Introducing parking meters

The post-war decades saw Britain becoming a car-driving nation. The number of licensed vehicles more than doubled in the ten years from 1950, jumping from 4.4 million to 9.4 million. As a result, cars and cities came into ever-greater conflict, with problems of congestion, delays and pollution becoming chronic.

Of course, parts of Manchester city centre have always been crowded with people and busy with traffic. There are photographs of queues of horse-drawn carts and trams on Market Street in the Victorian era, and a traffic census in 1911 measured and confirmed what everyone knew about the sheer volume and slowness of the vehicles on the major feeder routes into the centre of Edwardian Manchester – London Road, Oxford Road, Cheetham Hill Road and others. By then, motor cars were also part of the jams caused by electric trams, horse-drawn carts and cabs. Interestingly, thousands of cyclists chose to ride into the city each working day, presumably in part as a response to the slowness, if not the cost, of the traffic. By the 1930s, the problem of traffic congestion in the city centre saw the Council investigating new responses. These included the possibility of constructing underground car parks in Piccadilly Gardens and Smithfield Market, but, like proposals for 'parking towers', these were to be seen as too expensive. After the Second World War, the problem of commuter traffic intensified, although the clearing of buildings after the blitz did provide a number of temporary car parks in the city centre.

By the 1950s, the scale of growth in private car ownership meant streets were being seriously clogged with parked cars as more and more people wanted to drive into the city centre for work, shopping and entertainment. By 1960, the problem was becoming so serious that radical changes were proposed. The parking map was submitted to the Ministry of Transport as the City Council sought approval to solve – or at least mitigate – the problem through the controversial introduction of

OPPOSITE. Manchester City, *Parking places in Manchester* (*c.*1960) [BL]

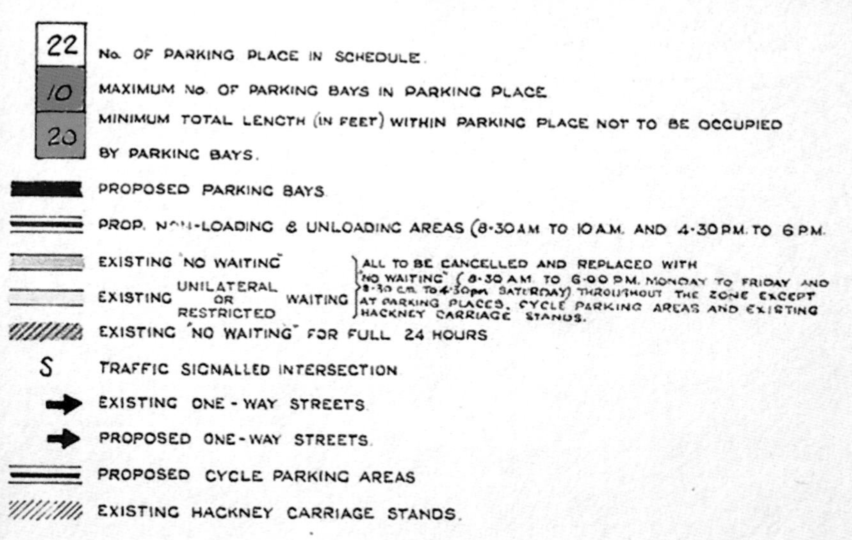

parking meters and restrictions on waiting during normal working hours.

The scheme covered street parking in the civic and commercial core, a fairly small but economically significant area between the major thoroughfares of Deansgate, Market Street and Mosley Street. The area had become the banking and office district for the city, as well as containing key trading exchanges and the main post office. Towards the Peter Street end are the major civic institutions, including the Town Hall, main library, police station and the Art Gallery. The area was densely used and heavily congested with traffic by the 1950s, and the plan was to limit drivers to 740 car parking spaces, distributed along most streets. The locations are indicated on the map by the red strips with the exact number of marked parking bays set out in the box symbols. The largest numbers of parking spaces were to be in front of the Town Hall (with space for 35 cars), beside the Central Library and around St Ann's Square.

The map also provides detail on the proposed redesign of vehicular circulation in the core, with yellow and blue shaded streets being subject to 'no-waiting' restrictions during the normal working week. Many streets had already been made one-way, but new ones were to be introduced. The goal was to keep the traffic moving as well as providing short-stay car parking, which was regarded as essential for business life to flourish.

After a public inquiry, the Ministry of Transport approved Manchester's scheme, and parking meters were installed and started operation in 1961, charging 6d per hour. The maximum stay for motorists was limited to 2 hours. This marked the end of uncontrolled and free on-street parking in the city centre with drivers subsequently having to pay ever more for the privilege of renting a small rectangle of land to park their vehicles legally. Manchester was the first city outside London to introduce parking meters.

More traffic restrictions would come in the 1960s across different parts of the city centre – including the dreaded double yellow lines – to the aggravation of many car drivers. Several large multistorey car parks and underground garages were also constructed in the city centre in the 1960s, often tied to major redevelopments such as Piccadilly Plaza and Market Place. There were also many proposals for more large-scale and disruptive interventions for Manchester's road systems, to expand capacity and speed flow significantly, including a large dual carriageway around the city centre and links to radial urban motorways. These proposals were not realised, which many would regard as a good thing, although a short stretch of elevated highway, the Mancunian Way, was built around the southern flank of the city centre as part of wider slum clearances.

By the end of the 1970s, the city centre was struggling with worn-out physical fabric and serious economic decline. Changing consumer patterns also meant that many shoppers were attracted to new retail parks and out-of-town supermarkets rather than travelling into central Manchester. Major efforts were undertaken to revive the city centre and to make it more attractive for visitors, although this also meant further restrictions on car access. After much work by planners and pressure by politicians in the Town Hall and the new Greater Manchester Council, the full pedestrianisation of some key shopping streets and public squares was achieved by the mid

Greater Manchester Council, *Paving the way for a brighter city centre* (*c.*1984) [The National Archives] Pedestrianisation efforts in the early 1980s by GMC included closing several major retail thoroughfares and public squares to cars.

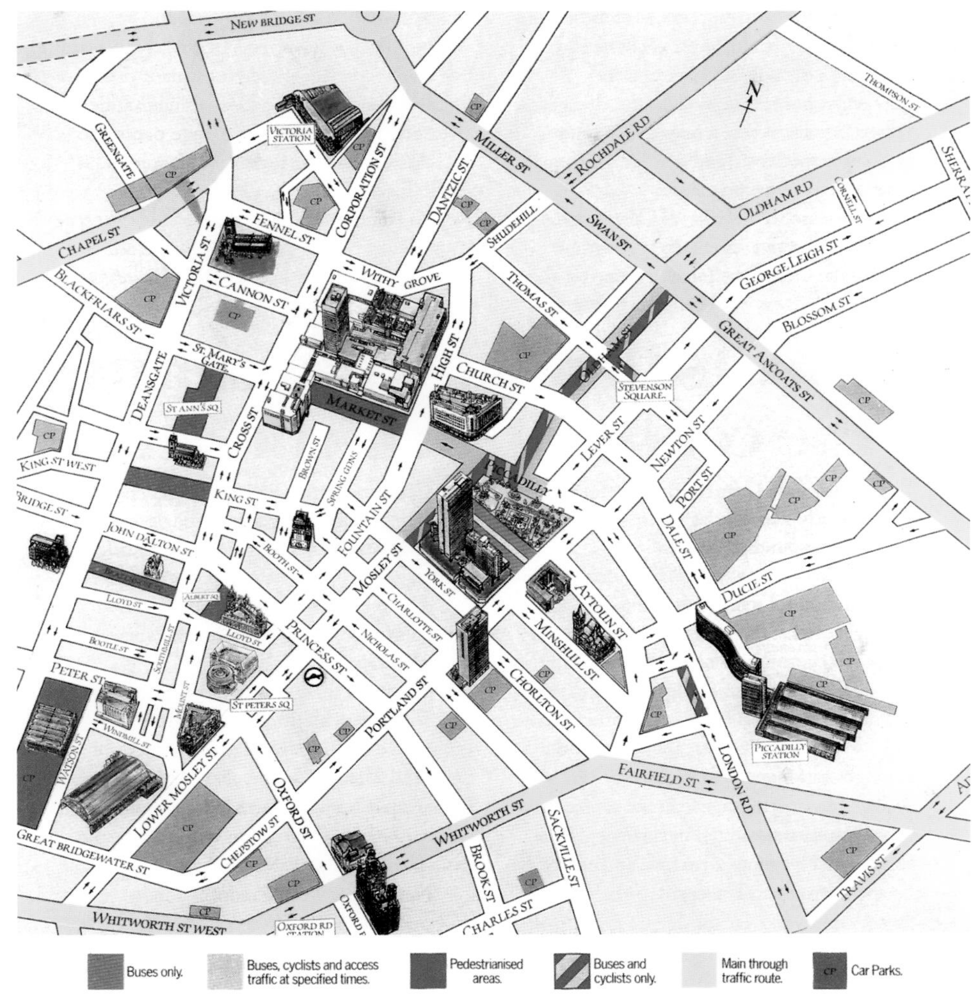

1980s. The struggle to close Market Street to traffic had been a long-fought battle, although eminently sensible to encourage people to stay and shop. The closure of Piccadilly Gardens to private vehicle traffic – one of the emblematic public spaces for Manchester – was also partly achieved. The area would be more radically reconfigured in the 1990s to accommodate the Metrolink trams.

While plans were progressing in the 1980s to make the city centre more visitor-friendly through pedestrianisation, they were also providing car parking for shoppers. The Council clearly faced a conflict of interest in this regard: on the one hand, it sought to restrict vehicles to make a more attractive and sustainable urban environment, but on the other, car parking charges offer a useful source of revenue. Despite efforts to make Manchester more pedestrian-friendly and to enhance public transport provision, there are still too many cars crowding into the city centre on a daily basis, in part because plentiful parking is available.

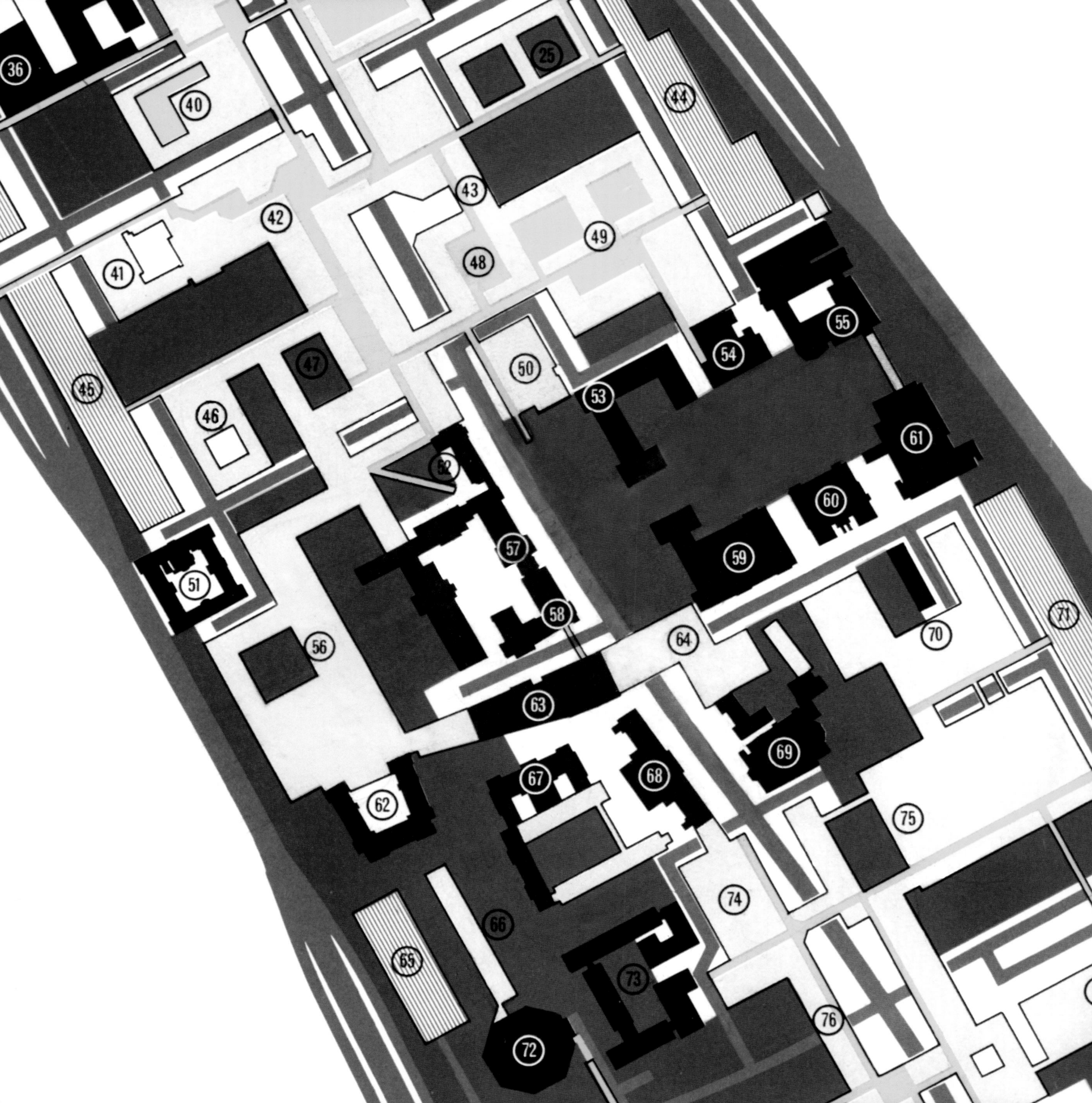
36
40
25
44
43
42
49
41
48
55
54
45
47
50
53
46
61
52
60
57
59
51
58
71
70
64
56
63
69
67
68
62
75
74
66
65
73
76
72

1967

Master-planning an educational precinct

In the makeover of the city's twentieth-century townscape, the development of the Manchester Education Precinct (MEP) was one of the most dramatic interventions. The site comprised an area of almost 300 acres less than a mile south of the city centre. It was promoted as the largest site in Europe to be tackled as a single precinct. Its redevelopment from the late 1960s involved a powerful consortium comprising the City Council, Manchester University, the University of Manchester Institute of Science and Technology (UMIST) and the United Manchester Hospitals. The city's role was twofold, both as planning authority and as local education authority. In the latter guise, it was then responsible for the Manchester College, which later became Manchester Polytechnic and then Manchester Metropolitan University. Two of its component parts – John Dalton College and the College of Art and Design – lay within the area of MEP and formed an intrinsic part of the overall proposals. Indeed, sanguinely if unrealistically, the report saw this as a 'happy relationship [between university and college] with the opportunities it affords for sharing residential, dining and Union facilities, and in particular, the sharing of a major Indoor Sports Centre . . .'

The proposals were drawn up by the architectural firm of Wilson and Womersley who had so large a hand in the 1960s' rebuilding of Manchester. They brought to the scheme the received wisdom of the time, which – for better or worse – entailed large-scale comprehensive development, a preference for concrete and a preoccupation with the impact of cars. In the latter, they were greatly influenced by the 1963 Buchanan report on *Traffic in Towns*. Buchanan's views on the car – 'a monster of great potential destructiveness, and yet we love him dearly' – reflected the growing concern that inherited townscapes could not accommodate the surge of traffic

OPPOSITE. Wilson & Womersley, *Manchester Education Precinct* (1967) [AUTH] Oxford Road runs down the centre; 36 is the College of Art now part of MMU; below are Manchester University buildings. The shopping precinct is 42.

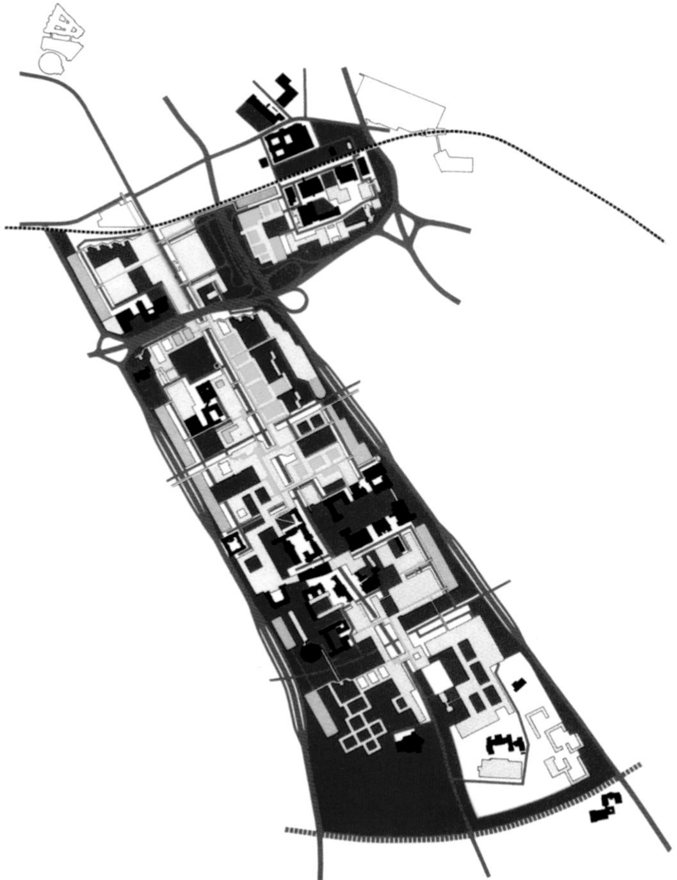

Overall MEP masterplan [AUTH] The proposed first-floor pedestrian areas and walkways are shown in yellow.

brought about by increasing car ownership. He argued that only by segregating traffic could planners create successful urban redevelopment.

These principles fed directly into the Wilson and Womersley proposals. The overall site had four nodes lying on both sides of its central spine of Oxford Road – at its centre was the University, to the south the hospital complex, to the north the Manchester College and to the north-west UMIST. Some post-war buildings (proposed in Nicholas's 1945 plan) had already been built on the university sites, but the MEP proposals covered a much larger area that would entail comprehensive demolition of most existing buildings. The report blandly asserted that 'The adjacent areas generally consist of property which has completed its useful life and is therefore available for and in need of renewal'. Some of the most significant buildings of the early redevelopment were the University's science buildings on Brunswick Street, new buildings north and south of All Saints as an extension of the Manchester College, St Mary's Maternity Hospital on Hathersage Road, the construction of the Mancunian Way and the 'transformation of sordid industrial and residential slums' south of the railway into a complex of academic buildings for UMIST, grouped around a grassed square. One of the most iconic and innovative buildings was the Maths Tower – with a 3-storey podium and 18-storey tower with contrasting façades, one with a brutalist concrete exterior and the other a modernist design with windows jutting out at varied angles.

The plan classified roads as one of four types: primary, exemplified by the high-speed Mancunian Way; district distributors to carry fast traffic to the primary network; local distributors to carry slower local traffic and link with district distributors; and service roads to take vehicles to individual buildings. Upper Brook Street and Cambridge Street, to the east and west of the Precinct spine, were to become district distributors (virtual urban motorways), and Oxford Road and Booth Street were to be mere local distributors. The conflict between traffic and pedestrians would clearly be in the local distributors, so it was here where the proposals aimed at the formal separation of the flows of cars and pedestrians. As the yellow lines on the plan show, this was to be achieved by building a network of walkways at first-floor level for pedestrians, with numerous bridges crossing Oxford Road and with ramps, stairs or escalators linking the walkways to the ground at significant points such as bus stops. A further high-level walkway was proposed to link Oxford Road Station with MEP.

The only walkways which were actually built were at the junction of Oxford Road and Booth Street where the Shopping Precinct, the jewel of the scheme, was located (reference 42 on the main map). The Shopping Precinct was a retail complex built at first-floor level on the south-west corner of the intersection of the two roads. It aimed to meet (or to create) a demand both from members of the University and local residents. The plan blithely populated it with commercial investments:

> In addition to general day-to-day shopping it will provide many specialised shops selling such things as musical instruments, artists' materials, sports equipment, surgical instruments, etc. There will also be bookshops, restaurants and banks.

Part of the retail demand would be generated by encouraging students and staff to live within the area of MEP in order to reduce commuting. This would be achieved by providing new housing between the Mancunian Way and Booth Street, and close to Whitworth Park.

Only parts of the scheme were ever realised. Most significantly, the proposal to create a new hierarchy of roads never materialised. Some east–west roads were closed so that the area became somewhat more like a precinct, but Oxford Road continued to be the principal route into the town centre, effectively slicing through MEP and bringing with it an ever-increasing volume of traffic (and, many claim, the busiest bus corridor in Europe). The two proposed district distributors – Cambridge Street and Upper Brook Street – continued to play only subsidiary roles in channelling traffic into the town. The Shopping Precinct at the heart of the scheme was left stranded by its first-floor location, literally up in the air. It initially attracted some commercial investment, including a major bookshop, bank, post office, travel agent, pub, stationers and food store, but consumers obdurately refused to behave as the architects had proposed, and by the 1980s many of the shops and businesses had closed and the Centre was eventually demolished in 2017 – along with its bridge across Oxford Road. First-floor living was clearly an architectural aspiration too far. Whether the Shopping Precinct would have been more successful had Oxford Road become a mere local distributor and the first-floor pedestrian walkway vision been extended over a larger area is a moot point. However, for many years, the first-floor walkway concept stamped a unique pattern on the Oxford Road/Booth Street area: the main walkway formed a bridge across Oxford Road, a ramp and an escalator led up to the Precinct Centre and another ramp led to a walkway which wended its way around the Maths Tower podium, dropped down to the St Peter's House Chaplaincy and climbed back up to the level of the Shopping Precinct bridge. Some of the new buildings on Booth Street had closed-up apertures waiting for bridges that were never to materialise. So, while MEP created a huge almost continuous precinct, its two most innovative features – the grading of roads, and the separation of pedestrians and traffic through a network of first-floor walkways – never materialised.

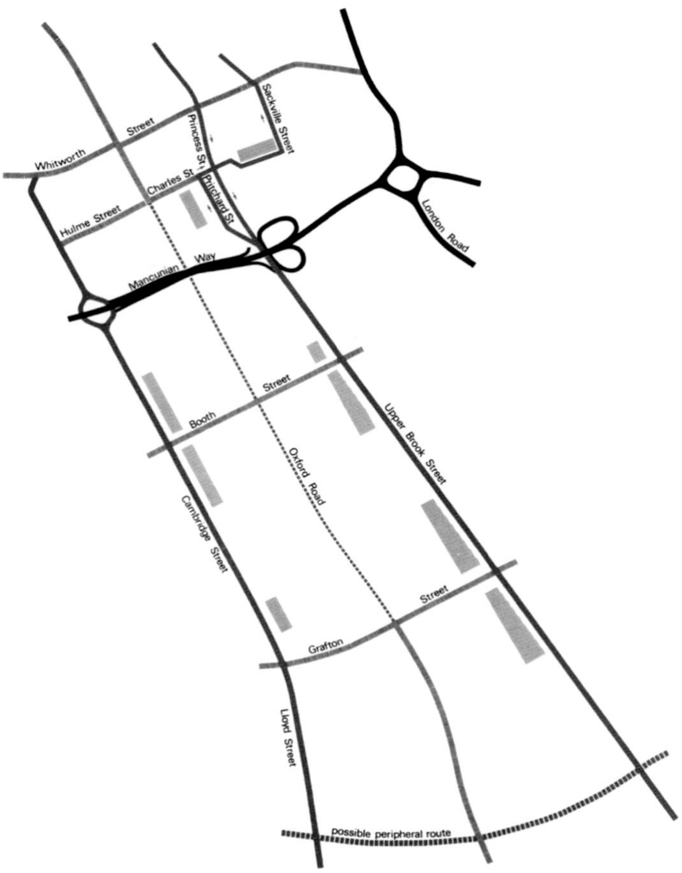

MEP road proposals [AUTH] Oxford Road was proposed to become a relatively minor service road.

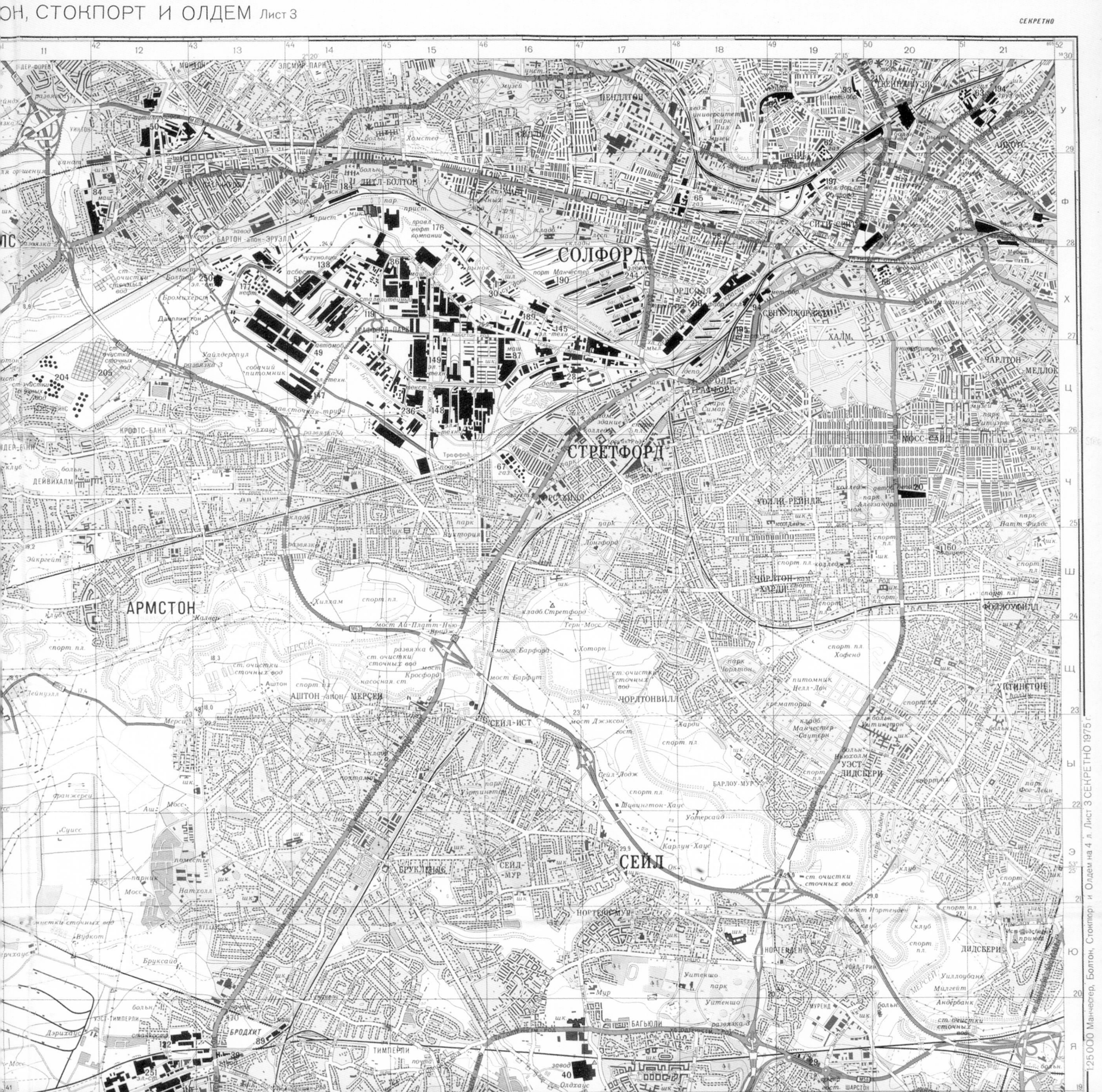

1:25 000 Манчестер, Болтон, Стокпорт и Олдем на 4 л. Лист 3 СЕКРЕТНО 1975 г.

1972a

Soviet mapping: a view from the East

Soviet military mapping during the Cold War has attracted increasing interest after the break-up of the old Soviet state, since a huge array of maps was discovered and imported in bulk into countries in the West. The mapping, done under the auspices of the Chief Administration of Geodesy and Cartography, generated coverage across the world at scales ranging from 1:1,000,000 to detailed plans at 1:10,000. John Davies has documented that they include at least 91 known plans of British towns at scales of either 1:10,000 or 1:25,000.

The map of Manchester is on four sheets. It was mapped in 1972 and published in Leningrad in 1975 at a scale of 1:25,000. It embraces Atherton and Lymm in the west; Ramsbottom and Littleborough in the north; Saddleworth and Broadbottom in the east; and Poynton and Bucklow Hill in the south. It therefore covers not only Manchester and Salford, but seven of the other eight boroughs of the current Greater Manchester. Only Wigan falls outside its coverage.

While there is some uncertainty about the sources on which the mapping draws – with the Ordnance Survey arguing that it simply copies their data – many of the maps show details not recorded on OS maps. Davies argues that the maps derive essentially from a mix of aerial photography, satellite evidence and on-the-ground data gathered (surreptitiously?) by Soviet agents and local informants. Neither is there agreement about their exact purpose. Some have seen them as 'invasion maps', but the more plausible explanation is that they were partly a general quest for information and partly to provide readily interpretable guides in the event of an occupation of the areas covered. This certainly helps explain why they include detail of civilian buildings, such as churches, post offices and administrative offices. Whatever the reasons for their production, the overall output from the huge effort that it clearly entailed was a remarkable achievement.

The detail is impressive. Like other of the town plans, the

OPPOSITE. USSR Administration of Geodesy and Cartography, *Manchester, Bolton, Stockport and Oldham, Sheet 3* (1972) [UML]

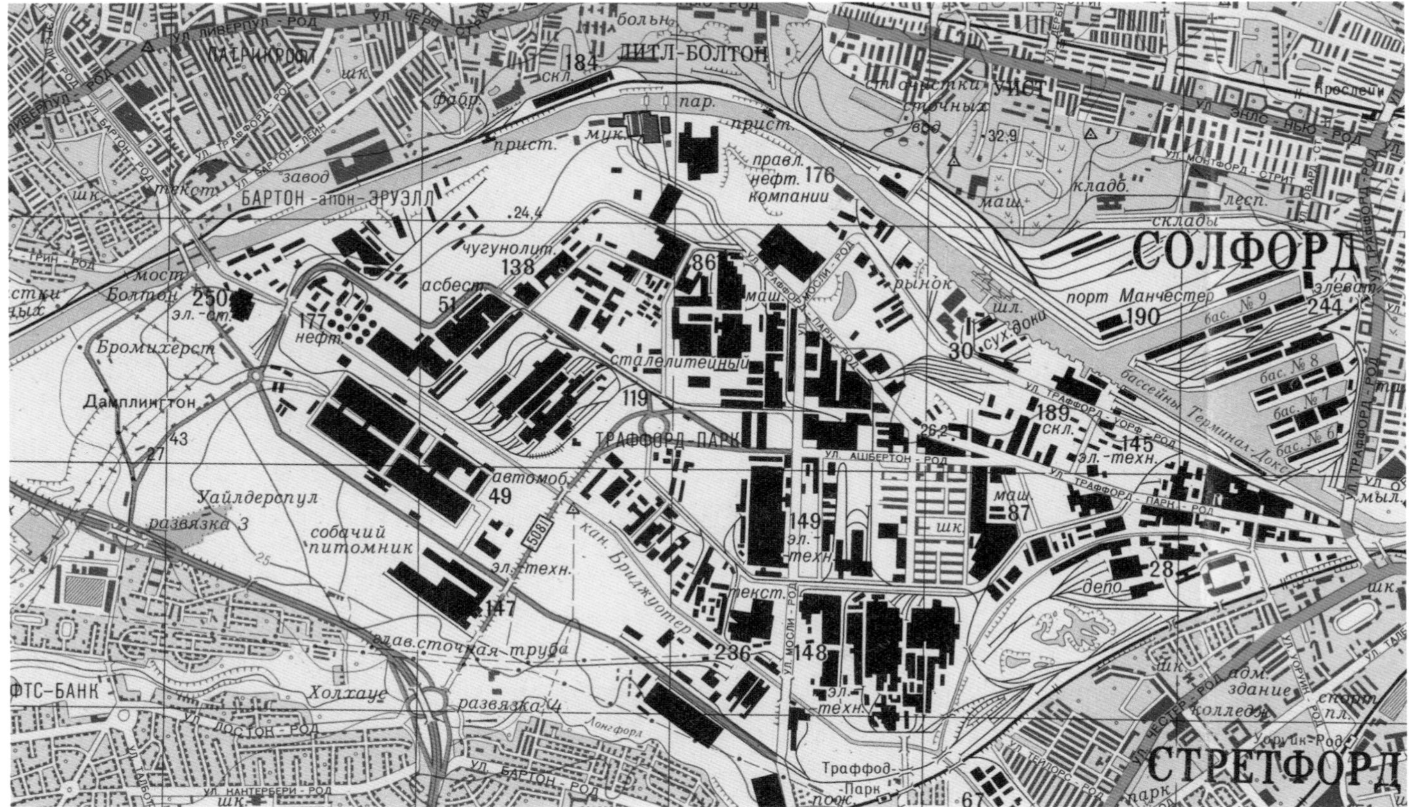

Trafford Park and Manchester Docks.

Manchester map uses a common array of colours and symbols. Significant buildings are numbered and listed, and the building types are distinguished by colour: industrial buildings black (although for some reason mills are simply shown in brown); significant administrative buildings are purple; military establishments are deep green; and other buildings – shops and offices, houses and public buildings – are brown. Main streets are generally named. The principal roads are shown in orange and given road numbers – for example the M6 is shown as both 'M6' and 'E33'; whereas A-roads are simply numbered, as '56' or '6144'. The principal roads are shown without colour wherever they cross bridges or when they enter built-up areas, a distinction that has been suggested as identifying stretches of road that could not bear heavy traffic. The symbols for both roads and railways differentiate between sections that are raised above the surrounding land and those that are sunk below. The topography is mapped in considerable detail. Contours are shown with spot heights in metres to one decimal point. Wooded areas are shown in green, built-up areas in light brown. The right-hand margin of the map has both an extensive street directory and a listing of the numbered buildings as well as a general account and description of the whole area.

The most striking aspect of the map is its use of Cyrillic script, which, to Western eyes, makes it look very 'foreign' with place names that are not readily interpretable. Moreover, the Cyrillic is shown phonetically so that Soviet users would have been able to pronounce place names correctly. For example,

Hale is shown as МЕИЛ and 'road' becomes РОД. Altrincham (with its awkward '. . . ingham' pronunciation) is helpfully shown as ОЛТИНГЕМ, but the equally awkward Beswick (locally pronounced 'Bezik') is incorrectly shown as БЕСУИК ('Bezwik'), which may have been influenced by the nearby Ardwick (which is correctly shown as АНДУИК).

This impressive cartography reinforces the view that this was a map produced essentially as a *vade mecum* for administrators after an occupation. Whatever its purpose, the map is generally accurate and comprehensive. One interesting aspect of its comprehensiveness is the detail shown for the (relatively few) military establishments in the area. For example, the camp south of Culcheth is shown in considerable detail, whereas it does not feature on OS maps or in general atlases such as A–Z street maps. The evolving motorway system is also accurately shown. In the early 1970s, the M6 was fully functional to the west of the conurbation, as was the M56 to the south, shown bending to join what then was the M63 (now the M60 ring road) as it approached Stockport. The M62 is shown coming from the north and swinging west to head for Liverpool. The M602 is shown heading towards Salford. In the east, the only motorway is again correctly shown for the time as the M67 whose construction was in progress north of Denton and running to Hattersley.

Visually, the most prominent feature of the map is the heavy black of industrial sites. In the early 1970s, Pomona Docks still had functioning industries and Trafford Park was still (just about) a relatively thriving industrial estate before the impact of successive recessions stripped it of many of its industries. Its industrial plants stand out dramatically, with almost 20 large firms numbered and identified. At the heart of the Park, the residential 'village' was still standing with its 12 numbered streets. One anomaly, however, is that the railway network is shown as though it was still completely functional, whereas it was effectively no longer in use. It seems likely that this was simply a reflection of the use of aerial photography. There is also a curious unexplained administrative area shown in purple just to the south-east of the Park near the Old Trafford cricket ground.

The map also presents other puzzles, illustrated by the section covering the central areas of Manchester. The city centre is shown without any significant administrative buildings; the Town Hall for example, is simply an unnamed public building. There are two sites shown as being of military significance, one in or close to Lever Street and a second just off Market Street. These may have been misinterpretations since there is no record of military sites in either area. The same section also shows one of the few significant administrative buildings as being the heart of Manchester University (although neither Salford University nor UMIST nor the then Manchester Polytechnic are so shown). Equally puzzling is the depiction of the core of Hulme, which is shown as virtually bereft of buildings, although with an elaborate pattern of streets. Some of the early buildings of the rebuilt Hulme are shown, but there is no sign of the four notorious deck-access Crescents, which were opened in 1972.

Despite such anomalies, the map is a striking achievement and a testament to the ingenuity of the Soviet cartographers.

The city centre. Oddities include the fact that the only 'significant administrative building' (purple) is shown as the University, and that there are two 'military establishments' (green).

KEY
garages
pedestrian routes
play areas
50'
100'
ST. WILFRED'S PRIMARY SCHOOL
CHURCH
NURSERY
CLUB
EAGLE HOTEL
WELLINGTON
WHITE HORSE
garden
SHOPS
ZION CHURCH
CHURCH
LIBRARY
service yard
SHOPPING SQUARE
SHOPS
SHOPS
LAUNDRY
POOL
CLINIC
SURGERY
service area
service road
PUB
shops under
shops under
NELSON INN
HULME 5 LAYOUT
Wilson and Lewis Womersley, Architects & Planners, Manchester

1972b

Hulme Crescents and after

The rise and fall of the Hulme Crescents has become the stuff of legend. They were part of a huge redevelopment of Hulme entailing the declaration of one of the largest clearance areas in the country and the replacement of bye-law terraces with 5,000 new homes in deck-access developments and tower blocks. The Crescents themselves were four U-shaped deck-access blocks comprising almost 1,000 maisonettes. They were designed by Hugh Wilson and Lewis Womersley who brought to the 1960s' redevelopment of the city their ideas on large-scale projects, the separation of traffic and pedestrians, and their love of concrete. Womersley had earlier been responsible for the Park Hill flats in Sheffield, and Wilson had overseen the development of the Scottish new town of Cumbernauld. Both were persuaded of the merits of the concept of 'streets in the sky' and applied this to the Crescents to create high-density housing combined with large areas of open space.

Hulme certainly needed redevelopment. It had long been recognised as one of the most deprived and ill-housed areas in the city. Overcrowding was rife, large families were packed into two-bedroom houses lacking bathrooms and internal lavatories. The whole area had been declared a clearance zone as early as 1934, and the plan of part of this zone shows the density of housing and, indeed, that the area still had back-to-back houses (between Abram and John Streets, Eagle and Phoenix Streets, and Violet and Primrose Streets). However, the only part that was redeveloped before the Second World War was Bentley House, which consisted of three-storey walk-up flats in the north of the area. Large-scale clearance of Hulme only began in the 1960s with many displaced families being rehoused in estates such as Wythenshawe and new out-of-town estates such as Langley near Middleton.

Given the scale of national housing need in the post-war years, there was widespread interest in using industrialised building methods and this was the context in which the redevel-

---

OPPOSITE. H. Wilson & L. Womersley, *Hulme 5 layout* (1972) [AUTH]

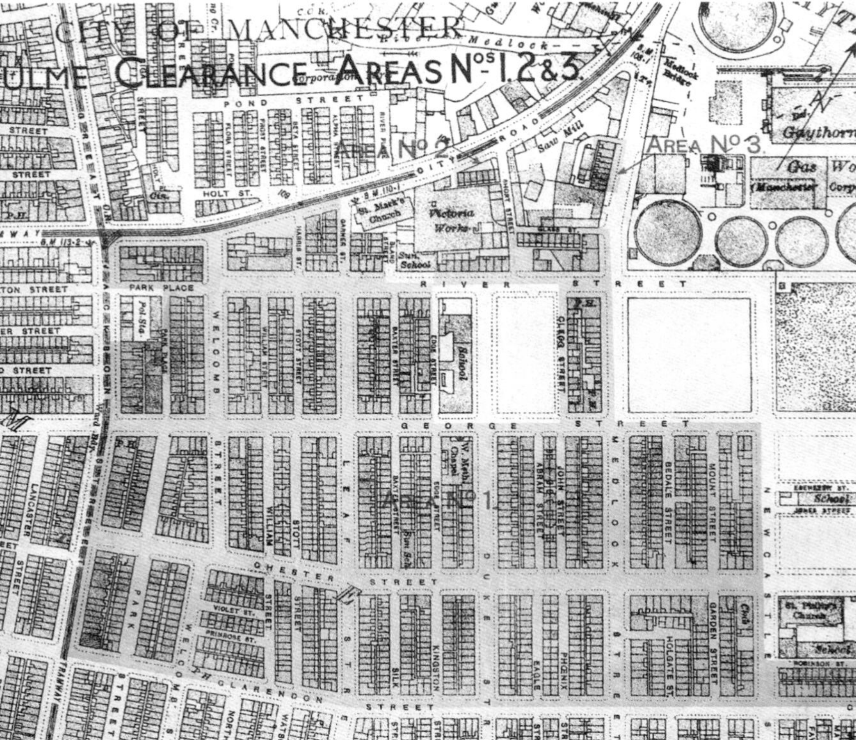

Manchester City Council, *Hulme Clearance Areas* (1932) [MLA]

opment of Hulme took place. Four main areas of largely deck-access homes were built: around Jackson Crescent; east and west of the newly built Princess Road; and, most dramatically, the Crescents. Each Crescent was named after an outstanding Enlightenment architect – Robert Adam, Charles Barry, William Kent and John Nash, together with Hawksmoor Close, a small block attached to Charles Barry Crescent. Ironically, the design was greeted with widespread enthusiasm. *The Guardian* praised its 'Georgian elegance' and the *Manchester Evening News* lauded it as 'A touch of Bloomsbury'.

The Crescents were formally opened in 1972, but almost from the start began to present horrendous problems as a result of their poor design and sloppy construction. Metal ties holding concrete panels were often missing; concrete panels fractured as reinforcing rods rusted; the 1970s' oil crisis made the underfloor heating of the flats too expensive for many residents; the heating ducts provided ready access for infestation by cockroaches and vermin; internal rubbish chutes became blocked; the wide walkways encouraged noisy 'street' play; and, since they were not classified as streets, the walkways were not policed, and crime and vandalism went largely unchecked. Moreover, the area became an isolated island trapped between the elevated Mancunian Way to the north and the new fast dual carriageway of Princess Road to the east. Critically, Stretford Road, which had been the area's principal shopping street, was closed to traffic as a result of the layout of the Crescents, and this both exacerbated the isolation of the area and prompted the closure of local businesses. In any case, businesses had suffered from the dramatic drop in Hulme's population, which fell from 130,000 in the 1930s to a figure of some 12,000. Within two years of their opening, the Council deemed the flats unfit for families and began to move residents out.

Campaigning groups rapidly began to form. The Hulme Tenants' Association and Hulme People's Rights Group were established by the mid 1970s. The articulate residents proved formidable campaigners. There was a long stand-off between the residents, the Council and the (then) Department of Environment. By 1984, unable to finance either improvements or demolition in the face of recession and financial cuts, the Council stopped charging rents and the maisonettes rapidly became home to a motley variety of bohemian groups, squatters and students. This produced a fascinating mix with a third of residents having no qualifications and a quarter with degrees or diplomas.

The impasse was eventually broken by the government's City Challenge programme, which introduced a significant change to the direction of regeneration policy. Whereas previously local authorities, seen as part of the urban problem, were sidelined so that much regeneration, as in Castlefield, was in the hands of government-run bodies such as Urban Development Corporations, City Challenge changed tack by recognising that local authorities could play a positive role in turning round the prospects of deprived areas. Partnership became a fundamental element of regeneration. The city successfully bid for City Challenge and in 1992 was rewarded with over £35

Hulme Regeneration Ltd (1996) [AUTH] The three phases of Hulme's building footprint: nineteenth-century bye-law housing; 1970s' comprehensive development; and 1990s' redevelopment

75-87 DWELLINGS/HA

million from government. Hulme Regeneration Ltd was established as a joint partnership between the city and the developer company AMEC. The city element comprised the Hulme Subcommittee (with representatives from the Housing Department, housing associations and tenants). The partnership structure therefore included city officials, local residents and the voluntary sector, but also – critically – the private sector, thereby giving confidence to private developers that development would take account of commercial realities and that the Council would play an ostensibly hands-off role. This was the 'Manchester model', which was also to prove so successful in later large-scale programmes such as the rebuilding of the city centre and the regeneration of east Manchester.

From 1993, the Crescents were demolished, and the redevelopment of Hulme was steered by a *Guide to Development in Hulme* published by Hulme Regeneration Ltd. It outlined a number of principles for the design of inner urban areas: traditional streets should be fundamental; buildings should front onto pavements to ensure oversight and maximise security; street design should be permeable so that streets lead somewhere and cul-de-sacs are avoided; streets should be 'legible', with key sites such as corners identified with distinctive buildings; there should be a sufficient density to support local shops and amenities and give a sense of 'urbanity'; and there should be a clear hierarchy of roads.

These ideas were translated into the master plan, much of which materialised on the ground. One of the key developments was that Stretford Road was reopened to traffic to reconnect Hulme to the rest of the city with a bridge (the Hulme Arch) crossing Princess Road and with a large public park stretching from Stretford Road towards the city centre. A new 'high street' with major retail stores was built in the north of the area. In place of the previous almost uniform council tenure, there is now a mix of tenures with two-thirds social housing and one-third private. One of the innovative elements, reflecting the young radical groups who came to dominate the Crescents, was Homes for Change set up as a mutual co-operative with help from the Guinness Trust. It developed a complex of some 75 flats together with work spaces for small businesses, studios, shops, a theatre and performance spaces. Its design was essentially led by members of the co-operative who consciously incorporated some of the design elements of the Crescents, with deck access, a central grassed space and buildings arrayed along three sides.

The two major makeovers of the nineteenth-century Hulme created dramatic contrasts between the three faces of the district. The map of the different building footprints nicely captures the successive transformations that the area has undergone.

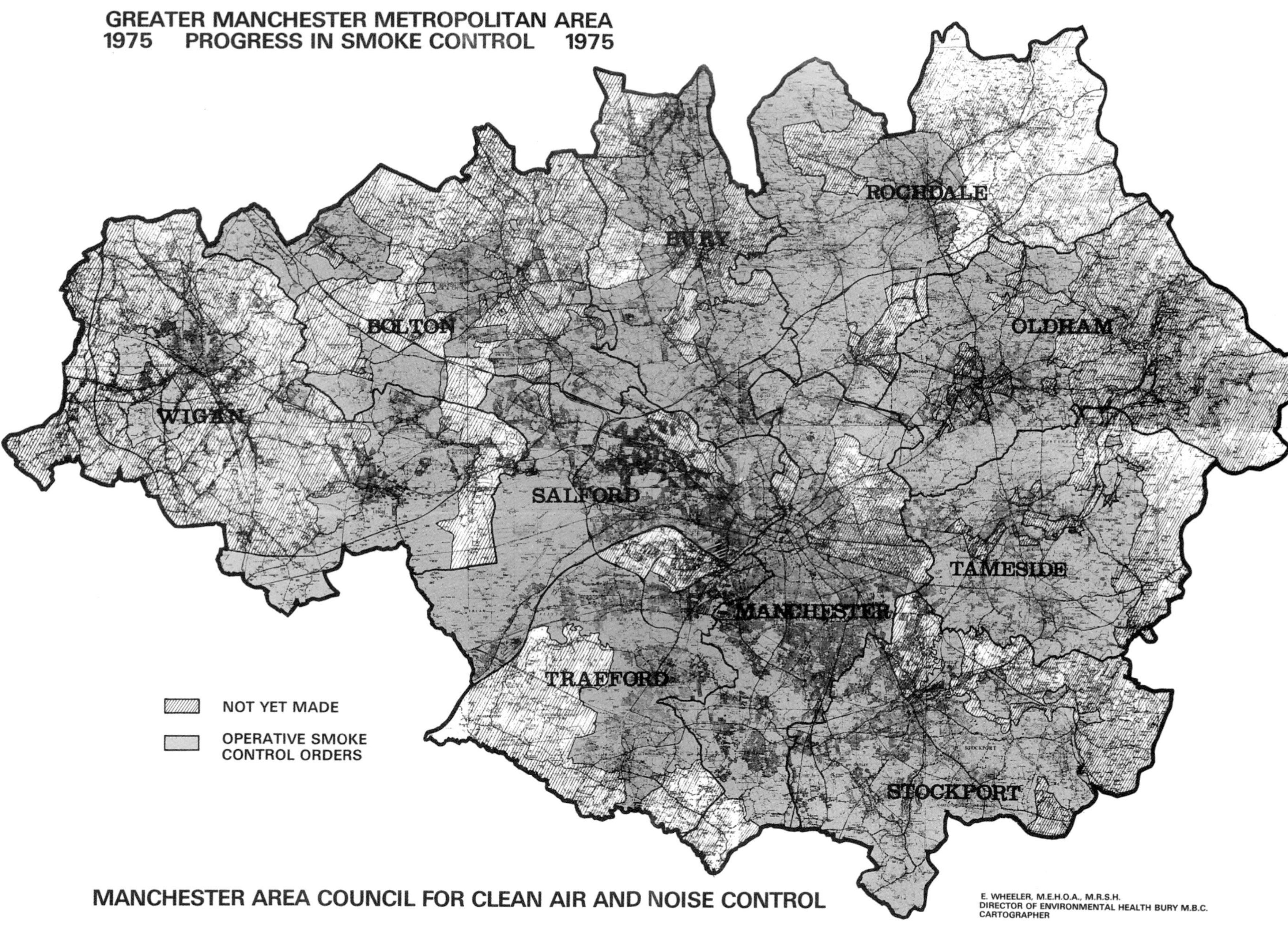
GREATER MANCHESTER METROPOLITAN AREA
1975 PROGRESS IN SMOKE CONTROL 1975
ROCHDALE
BURY
BOLTON
OLDHAM
WIGAN
SALFORD
TAMESIDE
MANCHESTER
TRAFFORD
STOCKPORT
NOT YET MADE
OPERATIVE SMOKE CONTROL ORDERS
MANCHESTER AREA COUNCIL FOR CLEAN AIR AND NOISE CONTROL
E. WHEELER, M.E.H.O.A., M.R.S.H.
DIRECTOR OF ENVIRONMENTAL HEALTH BURY M.B.C.
CARTOGRAPHER

1975

Air pollution and smoke control areas

For industrial cities in the north of England, the amount of black smoke billowing out of factory chimneys was regarded by many as an indicator of economic success. Huge quantities of bituminous coal were burnt in the steam boilers that powered the machinery that made fortunes for the mill owners in Manchester and across the textile region. It also meant that massive volumes of sooty particulates and sulphurous smoke fell on neighbouring homes and commercial buildings. Dirty, polluted air also contributed to high levels of respiratory diseases in Manchester. When smoke combined with fog, the resulting smogs significantly reduced the sunlight reaching urban residents and this was a causal factor in the incidence of rickets. The extent of poisonous industrial pollution was entwined with the low life expectancy of workers in Manchester.

Significant sanitary reform in the second half of the nineteenth century led to reductions in waterborne pollution, but attempts by campaigners, such as the Manchester and Salford Noxious Vapours Abatement Association (NVAA), to highlight the dangers of smoke and seek to limit emissions were dismissed as unrealistic or simply unnecessary. While the worst air pollution was experienced in Victorian Manchester, the problem of days of dense smog lying across the city was significant throughout the early decades of the twentieth century. This murky daytime atmosphere in Manchester was brilliantly captured by the French impressionist Adolphe Valette in his Edwardian-era paintings.

Population growth and the burning of coal in private homes through the first half of the twentieth century contributed as much to air pollution and days of smog as did the big chimneys of the mills and factories. There were some efforts to tackle the worst visible industrial air pollution in the interwar years through Nuisance Acts and Public Health legislation, however what was not confronted was the widespread domestic use of coal. Many tens of thousands of ordinary households across

---

OPPOSITE. Manchester Area Council for Clean Air and Noise Control, *Progress in smoke control, Greater Manchester area* (1975) [AUTH]

the city were daily consuming vast amounts of coal on open fires, in stoves and kitchen ranges. Lord Newton, Chairman of the Departmental Committee on Smoke Abatement could comment in 1922:

> It is no exaggeration to say that many millions of inhabitants of the north of England have never seen real sunlight in their places of residence except in the event of a bank holiday or of a coal strike, and most of them have become so inured to this deprivation that they are profoundly sceptical as to any possible remedy.

Given that Manchester was notorious for its air pollution, it was unsurprising that local campaigners and civic leaders took the lead in finding solutions in the 1930s. The National Smoke Abatement Society had its headquarters in Manchester during this period. The major innovation was to set up 'smoke control zones' that required householders and property owners to convert their heating and cooking equipment to use cleaner fuels. The idea was formalised in the mid 1930s.

The first 'smokeless zone' in Britain was legally affirmed in the Manchester Corporation Act of 1946, but it did not come into force on the ground until 1952. It covered 104 acres of the city centre and some 1,100 premises had to comply. The size of the zone was extended incrementally over the next couple of years.

The long-lasting smog of December 1952 that afflicted much of London and contributed directly to deaths of over 4,000 people spurred the government to establish an official inquiry, which resulted in the Beaver Report which recommended that formal policies should be developed to reduce smoke levels by 80 per cent in 15 years for populated areas of the country. In response, the 1956 Clean Air Act was passed, which gave powers to local authorities to designate smokeless zones.

Following the passage of the Act, Manchester and the neighbouring authorities took action on major polluters and sought to ban smoke emission from large areas. It was still laborious and time-consuming to define smoke control areas, notify occupants, mitigate local opposition and enforce compliance. However, as shown in the 1975 map, much of Manchester and surrounding towns were covered by smokeless zones within 20 years of the passing of the Clean Air Act. The process was not quite complete for Manchester – the two missing patches were Trafford Park and the industrial corridor along Ashton Old Road.

Enforcing smoke controls led to marked reductions in smoke pollution in Manchester and to a lesser extent declines in sulphur dioxide emissions. Improving air quality was also helped by the phasing out of steam trains by British Rail by 1968, and also by structural economic changes that saw the closure of heavy industry and 'cleaner' manufacturing systems being developed. In the domestic sphere, there was a shift away from coal fires especially in the 1970s with cheap natural gas from the North Sea and the change to central heating.

The dirty and dismal image that afflicted northern industrial cities like Manchester was also markedly changed by programmes from the late 1960s onwards to clean the outside of public buildings of the layers of soot that had accumulated over decades. The resulting transformation of familiar landmarks, like Manchester Town Hall, that had for living memory been stained inky black could be dramatic. Arguably, it contributed to the conservation of Victorian buildings as people could more readily appreciate their architectural merit.

The scourge of air pollution still continues. The most pressing concern now is the noxious emissions caused by the ever-growing volumes of vehicle traffic. Diesel cars, buses and HGVs are the worst polluters, and there are high concentrations of CO_2 and NO_X along commuting corridors, around major junctions and beside the motorways in Manchester. Recent concern has focused on the health risks of invisible particulate matter (PM10s) that regularly exceed legally permitted safety levels near the busiest roads. It seems likely that low emission zones will be designated to try to restrict the most polluting vehicles.

OPPOSITE. Manchester City Council, *City of Manchester central smokeless zones* (*c*.1954) [AUTH]

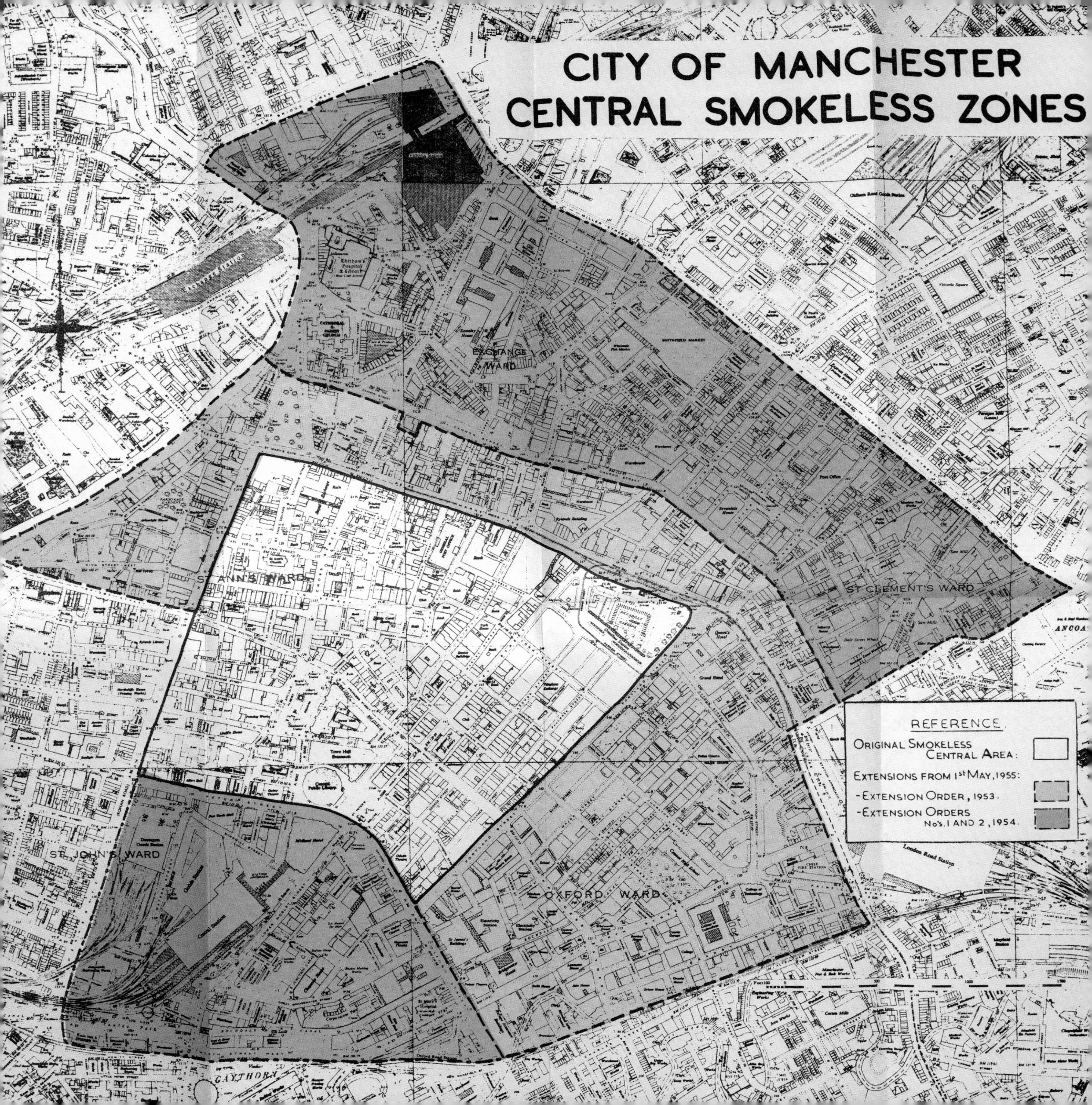
CITY OF MANCHESTER
CENTRAL SMOKELESS ZONES
EXCHANGE WARD
ST ANN'S WARD
ST CLEMENT'S WARD
ST. JOHN'S WARD
OXFORD WARD
Exchange Station
Victoria Square
Smithfield Market
Post Office
Grand Hotel
Town Hall
Central Public Library
Midland Hotel
Central Station
London Road Station
ANCOATS
GAYTHORN
REFERENCE.
ORIGINAL SMOKELESS CENTRAL AREA:
EXTENSIONS FROM 1st MAY, 1955:
-EXTENSION ORDER, 1953.
-EXTENSION ORDERS No's. 1 AND 2, 1954.

MAWDSLEY STREET
CHAMPIONS OF EUROPE
TOMMY DEAKIN'S ARCH
13. OGDENS
CORNER SHOP
11. HOWARDS
9. FAIRCLOUGHS
VIADUCT STREET
5. Vacant
ENA'S FLAT
3. BISHOPS
1. TATLOCKS
CORONATION STREET
COMMUNITY CENTRE
ROVER'S RETURN
ADVENTURE PLAYGROUND
FACTORY
ARKWRIGHT STREET
REDS
KITCHENER STREET
Dave F. Smith
10/76

1976

Mapping the most famous street in Manchester

Coronation Street was imagined by its scriptwriters and set designers in 1960, and it subsequently gained a sense of reality for many millions of fans of the soap opera. The everyday dramas of the working-class characters of 'Corrie' have, over the 60-plus years of continuous weekly broadcast, become synonymous with the city of Manchester and certain kinds of Northern-ness. The sense of authenticity throughout the show came, in large part, from its creator Tony Warren's experiences of the streets of Pendlebury in Salford where he grew up in the 1940s and 1950s. When broadcasting started in 1960, nothing like it had been seen on television.

Before airing the first episode, a decision was made to change the name to *Coronation Street* from the working title of 'Florizel Street' because it sounded too much like a disinfectant. Executives were uncertain whether it would prove popular, but it quickly became apparent that viewers had an affinity with *Coronation Street*, identifying with its archetypal characters and their daily trials and tribulations.

Coronation Street was designed in the solid brick vernacular style characteristic of northern industrial towns. It was supposedly built in 1902 and named in honour of the new king, Edward VII. It comprised a terrace of small red-brick houses, a corner shop and the local pub, backyards, alleyways and railway arches – all very resonant of Edwardian development.

While it was escapist entertainment, it was created from the stuff of real life and the struggles experienced by real people. It was, however, nostalgic in tone, and this was in part a response to significant social and economic changes happening across Britain in the 1960s. There was also comprehensive 'slum clearance' in Manchester and other northern cities that was affecting many of the viewers of the soap opera. The show clearly held up a mirror to their own experience.

OPPOSITE. D.F. Smith, *Bird's-eye sketch map of the fictional town of Weatherfield* [Coronation Street] (1976) [AUTH]

Coronation Street was renowned for the quality of the writing, its northern wit and character dialogue with distinctive Mancunian accents and phrasing. Certain key characters were part of the show for decades. It was particularly well known for having a roster of strong female characters; mentioning their first names – Elsie, Hilda, Vera – can instantly conjure into fans' minds a picture of the character as an (almost) real person.

This was coupled with evolving storylines and character relationships that kept people watching. Down the years there were deep family feuds, neighbourly disputes, notable community events and dramatic accidents, celebrations, new babies, weddings, affairs, divorces and deaths. It's been pointed out that the one ordinary thing that the characters in Coronation Street never did was to watch their own soap opera on their televisions!

The bird's-eye view of *Coronation Street*'s imagined town of Weatherfield was drawn by the illustrator David Farwell Smith and originally published in the *TV Times* in 1976. Smith was a well-known caricaturist and creator of humorous illustrations for popular books and magazines. His view of the fictional town shows rows of similar red-brick terraces and a number of pubs, shops and other businesses. The area is bisected by a canal and railway viaducts. In many respects, it is a typical late Victorian streetscape, with hundreds of near-identical houses with their front doors opening directly onto the pavement and with small backyards. There is not a single tree in the area, although allotments and a municipal park are shown in the distance. It is tidy, clean and respectable – the hallmarks of a solid working-class neighbourhood. There are some signs of post-war modernity, for example in the architectural design of the GPO sorting office and the new high-rise flats near the hospital. There is considerable pedestrian activity, a couple of buses, some commercial vehicles and a milk float on Arkwright Street. The streets are surprisingly free of the parked cars, which were quickly clogging up narrow terraced streets by the 1970s.

Coronation Street is one of a number of similar streets, cartographically only distinguished by the oversized name plate

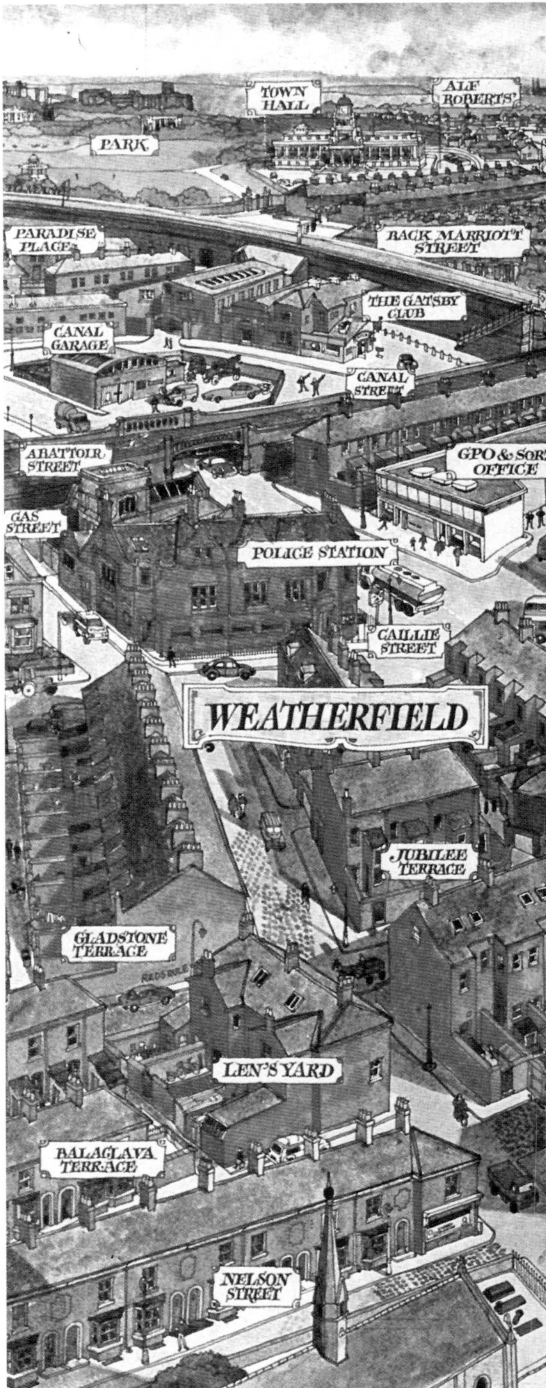

BETTY TURPIN'S
NEW FLATS
WEATHERFIELD GENERAL HOSPITAL
SHOPPING PRECINCT
CANAL
BESSIE STREET SCHOOL
BESSIE STREET
CLARENCE STREET
SYLVIA SEPARATES
RAY & DEIRDRE'S FLAT
MRS No. 19
ALLOTMENTS
ALBERT ROAD
SILK STREET
MORT STREET
INKERMAN STREET
VICTORIA STREET
JACKSON'S CHIP SHOP
NIGHTINGALE TERRACE
THE MISSION
TILE STREET
THE RED REC
THE KABIN
WEAVER'S ROW
THE FLYING HORSE
MAWDSLEY STREET
TOMMY DEAKIN'S ARCH
13. OGDENS
CORNER SHOP
9. FAIRCLOUGHS
11. HOWARDS
VIADUCT STREET
5. Vacant
3. BISHOPS
ENA'S FLAT
ROSAMUND STREET
1. TATLOCKS
CORONATION STREET
COMMUNITY CENTRE
ROVER'S RETURN
ADVENTURE PLAYGROUND
FACTORY
ARKWRIGHT STREET
MAFEKING STREET
DISRAELI ROW
KITCHENER STREET

and the labelling of the occupants of the seven houses on the north side of the road. There is a gap in the terrace, as the house at No. 7 collapsed in 1965 due to mining subsidence (it would be rebuilt in 1982). Opposite the houses on Coronation Street is a rather large modern-looking factory, some flats and a community centre.

To the general observer, it looks like a plausible rendering of the environs of The Street, but for hardcore Corrie fans, who can expertly reconstruct the cognitive geography of the area from many hundreds of hours of viewing, the view is riddled with topographic errors. These include misplacing Jackson's Chip Shop, and Arkwright Street being named on the other side of the factory when it should be Victoria Street. Len Fairclough's building yard is also shown as being behind Trafalgar Street when dedicated fans know that it was sited off Mawdsley Street.

Weatherfield was said by Tony Warren to be 'four miles in any one direction from the centre of Manchester. Emotionally, it's wherever you want it to be in your own heart.' However, the template, at least for some architectural elements, was Archie Street, a real road in the Ordsall area of Salford. In particular, the bay window frontages and front doorsteps were partly modelled on Archie Street, along with the presence of the corner shop at one end, but significantly there was not a pub at the other end. The 1960s' credit sequence showed Archie Street and a view of the steeple of St Clement's Church in Salford. Like many similar streets in Salford, it was demolished in the late 1960s.

The filming of the show was initially done inside the Quay Street studios at Granada Television, and it was only in 1968 that a replica outdoor set for *Coronation Street* was built, at three-quarter scale. During the 1980s, this became the centrepiece of Granada Studio Tours, a major tourist attraction in Manchester. Thousands of Corrie fans made the pilgrimage to stand on the short street and have their photograph taken in front of the Rovers Return.

The filming of *Coronation Street* controversially moved to a new ITV production facility on Trafford Wharf directly opposite MediaCity where a completely new – but visually identical – set for Corrie was built in 2013. The former Granada studio complex on Quay Street in Manchester, the home of *Coronation Street* for well over 50 years, was demolished to make way for a redevelopment project, the real set of the fictional terraced street being replaced by high-rise apartments and commercial towers.

Despite these changes, the enduring success of *Coronation Street* since its first broadcast on 9 December 1960 means that it remains a major part of Manchester's identity. While its themes were often nostalgic, it can now be seen as a symbol of the post-industrial reinvention of the city. In place of Cottonopolis, the late twentieth-century Manchester has become better known for popular culture, typified by television shows filmed at Granada Studios, football trophies won at Old Trafford and the dance music once played in the Haçienda nightclub.

OPPOSITE. Jenkins Design Services, *Coronation Street site location plan* [New Coronation Street set] (*c*.2016) [Jenkins Design Services]

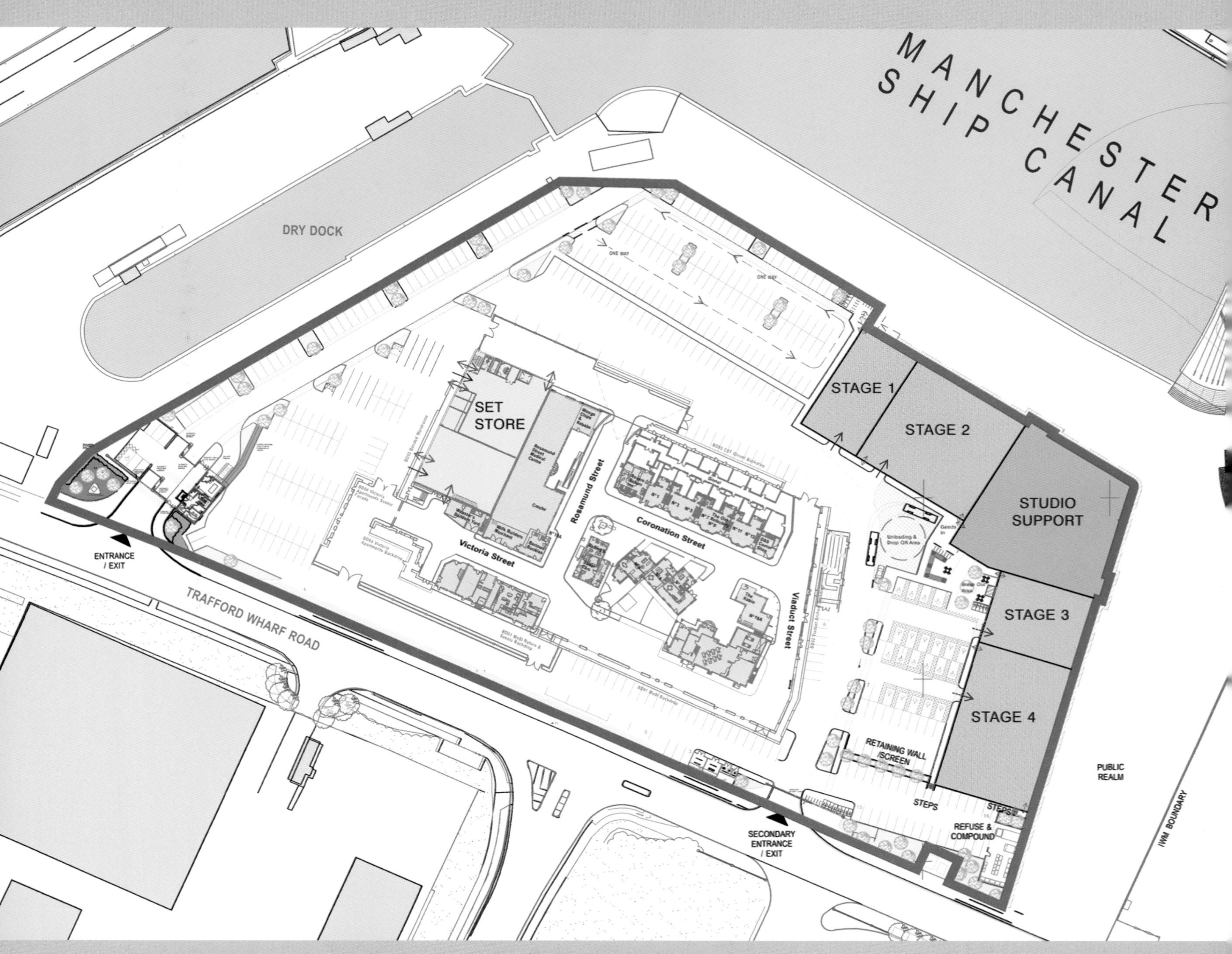

MANCHESTER SHIP CANAL
DRY DOCK
ONE WAY
SET STORE
STAGE 1
STAGE 2
STUDIO SUPPORT
STAGE 3
STAGE 4
Rosamund Street
Coronation Street
Victoria Street
Viaduct Street
Unloading & Drop Off Area
RETAINING WALL /SCREEN
STEPS
REFUSE & COMPOUND
PUBLIC REALM
IWM BOUNDARY
ENTRANCE / EXIT
SECONDARY ENTRANCE / EXIT
TRAFFORD WHARF ROAD

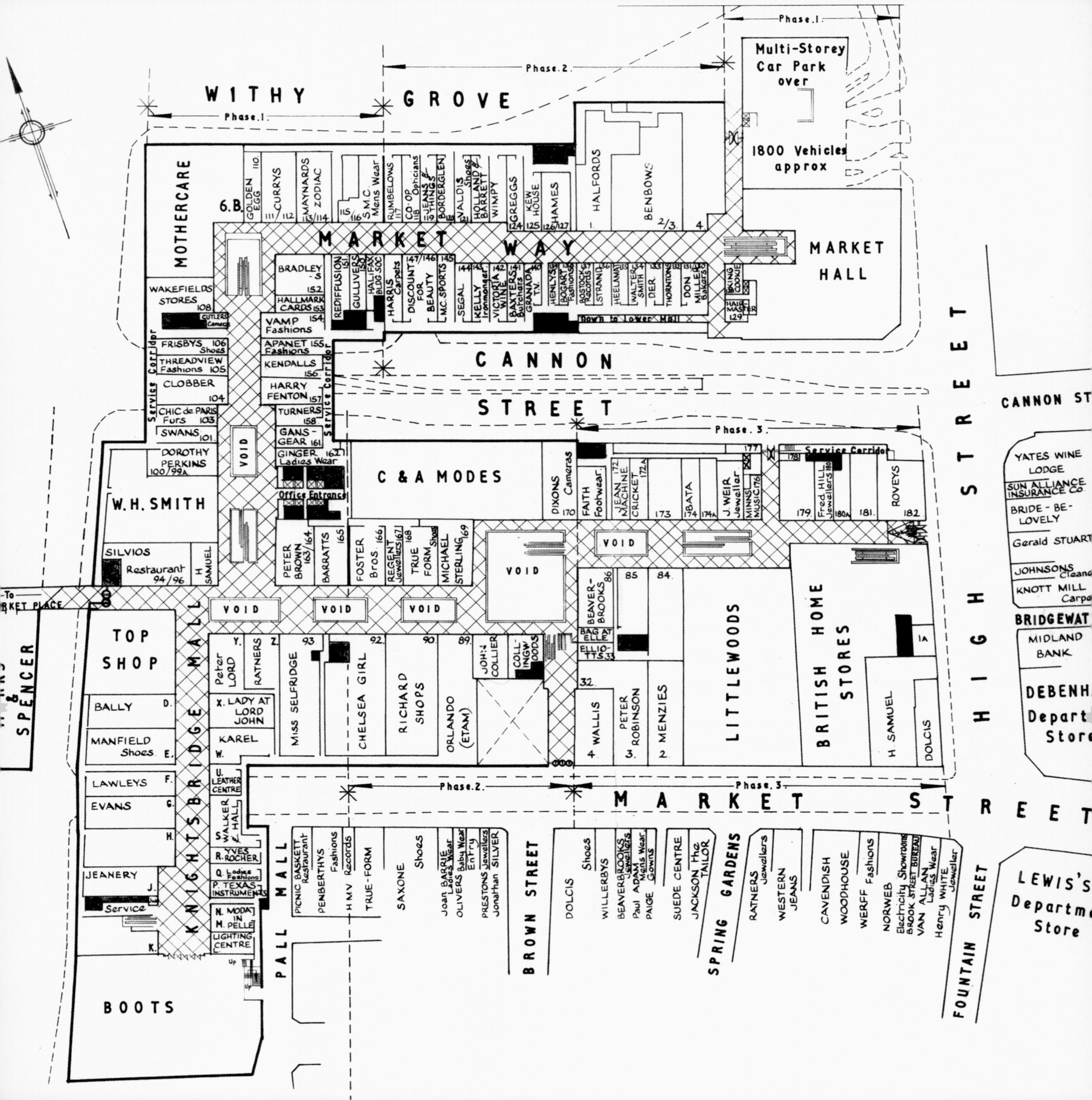

WITHY
GROVE
Phase.1.
Phase.2.
Multi-Storey
Car Park
over
1800 Vehicles
approx
MOTHERCARE
6.B.
GOLDEN EGG 110 111
CURRYS 112
MAYNARDS ZODIAC 113/114
115 116
S.M.C. Mens Wear
RUMBELOWS 117
CO-OP Opticians 118
JEANS & THINGS 119
BORDERGLEN 120
VALDIS Shoes 121
HOLLAND & BARRETT
WIMPY
GREGGS 124
KEW HOUSE 125
HAMES 126/127
HALFORDS 1.
BENBOWS 2/3.
4
MARKET WAY
MARKET HALL
BRADLEY-S 152
HALLMARK CARDS 153
REDIFFUSION 151
GULLIVERS 150
HALIFAX BLDG SOC
HARRIS Carpets
DISCOUNT FOR BEAUTY 147/146
M.C. SPORTS 145
SEGAL 144
KELLY Ironmonger 143
VICTORIA WINE 142
BAXTERS Butchers 141
GRANADA T.V. 140
HENLYS
BOGART Fashions
BOSTOCK Records
STRAND
HEELAMAT
WALTER SMITH
DER
THORNTONS
DON MILLER Bakers
BAKING COOKIE
HAIR MASTER 129
Down to Lower Mall
WAKEFIELDS STORES 108
CUTLERS Cameras
FRISBYS Shoes 106
THREADVIEW Fashions 105
CLOBBER 104
CHIC de PARIS Furs 103
SWANS 101
DOROTHY PERKINS 100/99A
Service Corridor
VAMP Fashions 154
APANET Fashions 155
KENDALLS 156
HARRY FENTON 157
TURNERS 158
GANS-GEAR 161
GINGER Ladies Wear 162
Office Entrance
CANNON
STREET
Phase. 3.
CANNON ST
C & A MODES
DIXONS Cameras 170
FAITH Footwear
JEAN MACHINE 172
CRICKET 172A
173
BATA 174
174A
J. WEIR Jeweller
MINNS MUSIC 176
177
178
Service Corridor
179
Fred HILL Jewellers 180
180A
181
ROVEYS 182
VOID
W.H. SMITH
SILVIOS Restaurant 94/96
H SAMUEL
PETER BROWN 163/164
BARRATTS 165
FOSTER Bros. 166
REGENT Jewellers 167
TRUE FORM Shoes 168
MICHAEL STERLING 169
BEAVERBROOKS 86
85
84.
BAG AT ELLE
ELLIOTTS 33
LITTLEWOODS
BRITISH HOME STORES
1A
YATES WINE LODGE
SUN ALLIANCE INSURANCE Co.
BRIDE-BE-LOVELY
Gerald STUART
JOHNSONS Cleaners
KNOTT MILL Carpets
BRIDGEWAT
MIDLAND BANK
DEBENH
Depart
Store
HIGH STREET
To MARKET PLACE
SPENCER
TOP SHOP
BALLY D.
MANFIELD Shoes E.
LAWLEYS F.
EVANS G.
H.
JEANERY J.
Service
K.
BOOTS
KNIGHTSBRIDGE MALL
Peter LORD Y.
RATNERS Z.
X. LADY AT LORD JOHN
KAREL W.
U. LEATHER CENTRE
S. WALKER & HALL
R. YVES ROCHER
Q. Ladies Fashions
P. TEXAS INSTRUMENTS
N. MODA IN M. PELLE
LIGHTING CENTRE
Up
Miss SELFRIDGE 93
CHELSEA GIRL 92.
RICHARD SHOPS 90.
ORLANDO (ETAM) 89.
JOHN COLLIER
COLLINGWOODS
32.
WALLIS 4.
PETER ROBINSON 3.
MENZIES 2.
H. SAMUEL
DOLCIS
MARKET STREET
Phase.2.
Phase.3.
PALL MALL
PICNIC BASKETT Restaurant
PENBERTHYS Fashions
HMV Records
TRUE-FORM
SAXONE Shoes
Joan BARRIE Ladies Wear
OLIVERS Baby Wear Entry
PRESTONS Jewellers
Jonathan SILVER
BROWN STREET
DOLCIS Shoes
WILLERBYS
BEAVERBROOKS Jewellers
Paul ADAM Mens Wear
PAIGE Gowns
SUEDE CENTRE
JACKSON the TAILOR
SPRING GARDENS
RATNERS Jewellers
WESTERN JEANS
CAVENDISH
WOODHOUSE
WERFF Fashions
NORWEB Electricity Showrooms
BROOK STREET BUREAU
VAN ALLAN Ladies Wear
Henry WHITE Jeweller
FOUNTAIN STREET
LEWIS'S
Departm
Store

1978

The Arndale: shopping behemoth

Historically, the hub of Manchester's shopping has been Market Street. However, it was long recognised as overly congested and an inadequate thoroughfare for a regional metropolis. The surrounding narrow streets were in need of redevelopment to provide space for larger retail premises and easier pedestrian movement. From the early nineteenth century onwards, many schemes were advanced to widen and straighten Market Street, and some changes were indeed made to the medieval street, but it had to wait until the 1950s before serious plans were drawn up to completely transform this part of the city centre. By the mid 1960s, a Comprehensive Redevelopment Area had been approved for the area bounded by Corporation Street and High Street, and from Withy Grove to Market Street. A massive new retail centre, straddling Cannon Street, was to be built, enclosing over 1,000,000 square feet of space and providing modern premises for upwards of 200 shops at cost of about £15 million.

This would become the Arndale Centre, and when it opened it was proclaimed to be the largest indoor shopping centre in Europe. The ambitious project was undertaken by a partnership of the City Corporation and specialist shopping centre developers, Town & City Properties, with financial support from the Prudential Assurance Company. The detailed design for the scheme was undertaken by Hugh Wilson and Lewis Womersley, the architects responsible for several major redevelopment projects in Manchester.

The leading lights in Town & City Properties were the Bradford-based team of entrepreneurs, Arnold Hagenbach and Sam Chippindale – who amalgamated their first and last names to create the *'Arn'-'dale'* moniker that would become synonymous with their big covered shopping centres. Starting with their first shopping centre in Jarrow in 1958, they became successful in city-centre retail development and by the late 1980s had built 24 Arndales up and down Britain. The one in

OPPOSITE. Healey & Baker, Manchester, *Arndale Shopping Centre upper shopping mall tenancy plan* (1978) [AUTH]

Manchester was very much their flagship development.

Construction started in the early 1970s, and it was opened in several phases. It brought significant disruption to the heart of Manchester for many years until its completion in 1978. The main mall was spread over two levels and included a walkway over Market Street to a large new Boots store that was built on the site of the Guardian Newspaper offices. Also included was a market hall that was managed by the Council, an integrated bus station on Cannon Street and a 1600-vehicle multistorey car park. Rising above the centre was a substantial 21-storey office block. Beneath the public mall was a service level with a roadway for delivery trucks. The original plan had included a link to an underground station on the planned Picc-Vic rail route, a major 1970s' transport scheme for Manchester city centre but one that ultimately would not be built.

When it opened in 1978, most retail units had already been filled. Stores included many of the major retail names of the 1970s such as C&A, Littlewoods, British Homes Stores and Mothercare. The occupancy plan shows that directly adjacent to the Arndale were three big department stores – Debenhams and Lewis's on one side and Marks & Spencer on the other.

The Manchester Arndale was a major presence in the centre of the city, impossible to ignore and transformative in its impact on circulation and commerce. Nearby, on Oldham Street many shops lost trade and were forced to close. But on its own terms, the Arndale Centre was undoubtedly commercially successful. As a retail destination, it regularly attracted more than 100,000 shoppers per day, offering a wide range of shops in an enclosed and clean environment, with easy transport links and plentiful car parking.

However, the Arndale was also a rather unloved hulk. It had many critics who had harsh views on its uninspiring exterior and alienating effects on the surroundings streets. Particularly problematic were the expansive blank façades made from pre-cast concrete panels that were clad in tiles of a distinctly unattractive yellowish hue. The resulting edifices had something of the appearance of a municipal lavatory, and for evermore the Arndale would be ridiculed for looking like a toilet.

The scale of the Arndale gave the public of Manchester and the wider region the closest thing to an American-style retail mall that they would see for 20 years. However, the interior shopping areas were functional at best and the lack of natural light meant they felt claustrophobic to some. The navigability of the centre was criticised for its monotone design and numerous dead-end corridors. There was no obvious circular route that took in all the shops. The retail offer was large but constricted in scope, favouring the national chains able to pay high rents. The Arndale also lacked a sense of place and symbolic connection to Manchester's past. It was a typical soulless shopping space with its rows of mass market identikit stores, with little individuality or spontaneity. But through the 1980s many teenagers' Saturdays were spent happily 'hanging out' in the Arndale, and for many families and older shoppers in particular, the safe indoor environment with seating and toilets was a pleasant place to spend a few hours browsing – or sleeping.

There have been various substantial refurbishments of the Arndale to update its decor and meet changing fashions in retailing. The most dramatic was the reconstruction following the IRA bomb of 1996. The southern half of the Arndale was extensively refashioned with the reviled yellow tiles replaced by sandstone and glass cladding. A new glass bridge over Corporation Street linked to Marks & Spencer. The internal bus station on Cannon Street was replaced by a new transport interchange on Shudehill.

The Arndale continues to thrive despite much competition, not least from large retail parks, out-of-town supermarkets and in particular the Trafford Centre, a massive £600 million American-style shopping mall that opened in 1998 and is strategically positioned next to the M60 motorway at Dumplington. The Manchester Arndale remains full, with most of the big-name retail brands represented, although very few of those shown on the 1978 plan have survived. Boots is the only major store to have kept its original position over the years; Mothercare's store of 1978 has become a Next and TK Maxx now occupies the Littlewoods space fronting onto Market Street.

OPPOSITE. Manchester Arndale, *Store locator map* (*c.*2016) [Manchester Arndale]

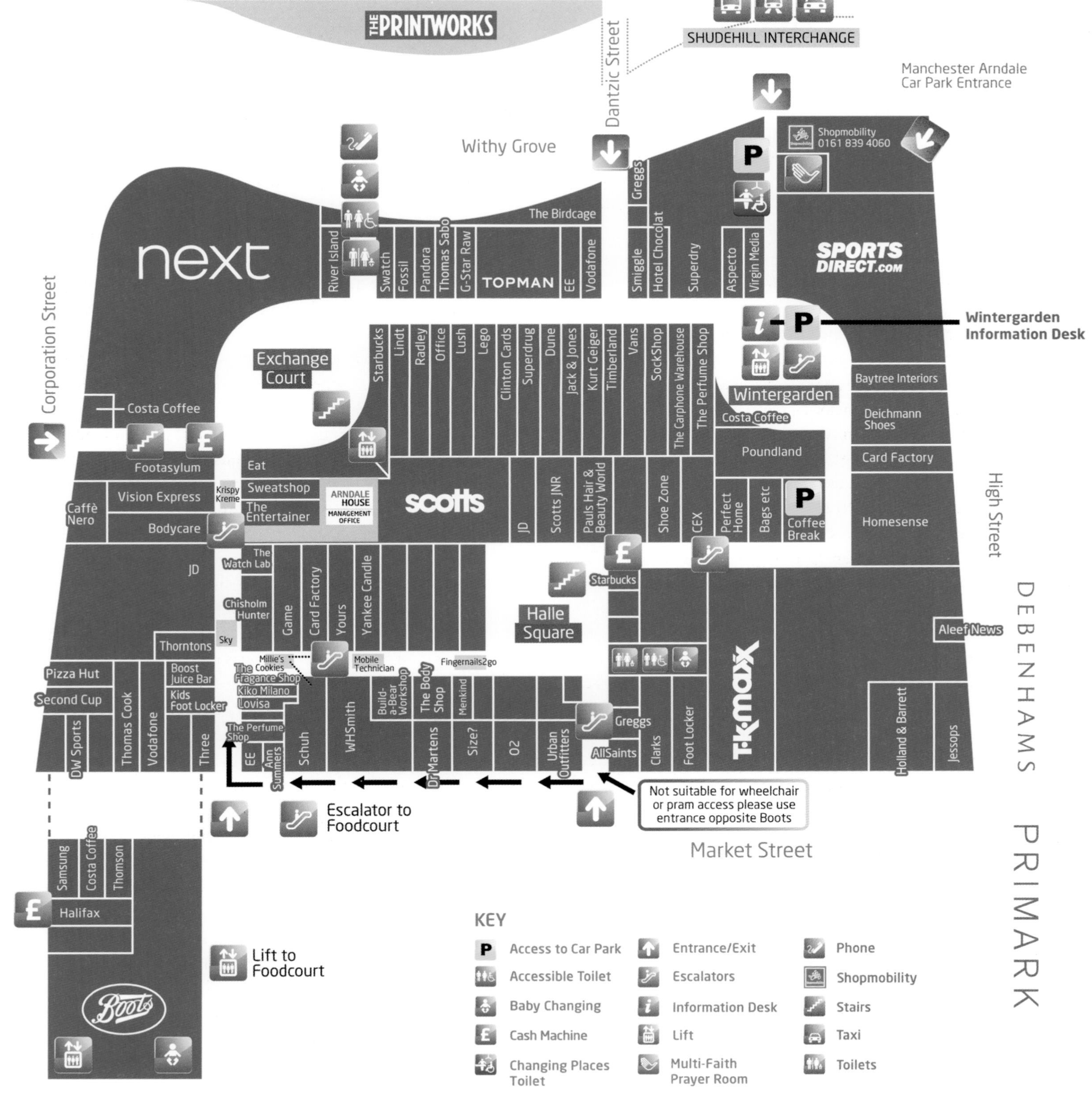

THE PRINTWORKS
Shudehill Interchange
Dantzic Street
Manchester Arndale Car Park Entrance
Withy Grove
Shopmobility 0161 839 4060
next
The Birdcage
River Island
Swatch
Fossil
Pandora
Thomas Sabo
G-Star Raw
TOPMAN
EE
Vodafone
Greggs
Smiggle
Hotel Chocolat
Superdry
Aspecto
Virgin Media
SPORTS DIRECT.com
Corporation Street
Exchange Court
Starbucks
Lindt
Radley
Office
Lush
Lego
Clinton Cards
Superdrug
Dune
Jack & Jones
Kurt Geiger
Timberland
Vans
SockShop
The Carphone Warehouse
The Perfume Shop
Wintergarden Information Desk
Wintergarden
Costa Coffee
Baytree Interiors
Deichmann Shoes
Card Factory
Poundland
Homesense
High Street
SELFRIDGES&CO
Costa Coffee
Footasylum
Eat
Sweatshop
Krispy Kreme
Vision Express
Caffè Nero
Bodycare
The Entertainer
ARNDALE HOUSE MANAGEMENT OFFICE
scotts
JD
Scotts JNR
Pauls Hair & Beauty World
Shoe Zone
CEX
Perfect Home
Bags etc
Coffee Break
JD
The Watch Lab
Chisholm Hunter
Sky
Game
Card Factory
Yours
Yankee Candle
Halle Square
Starbucks
Thorntons
Pizza Hut
Second Cup
Boost Juice Bar
Kids Foot Locker
Thomas Cook
Vodafone
DW Sports
Three
Millie's Cookies
The Fragance Shop
Kiko Milano
Lovisa
The Perfume Shop
EE
Ann Summers
Schuh
Mobile Technician
WHSmith
Build-a-Bear Workshop
The Body Shop
Menkind
Fingernails2go
Dr Martens
Size?
O2
Urban Outfitters
Greggs
AllSaints
Clarks
Foot Locker
TK maxx
Aleef News
Holland & Barrett
Jessops
DEBENHAMS
PRIMARK
M&S EST. 1884
Escalator to Foodcourt
Not suitable for wheelchair or pram access please use entrance opposite Boots
Market Street
Samsung
Costa Coffee
Thomson
Halifax
Lift to Foodcourt
Boots
KEY
Access to Car Park
Accessible Toilet
Baby Changing
Cash Machine
Changing Places Toilet
Entrance/Exit
Escalators
Information Desk
Lift
Multi-Faith Prayer Room
Phone
Shopmobility
Stairs
Taxi
Toilets

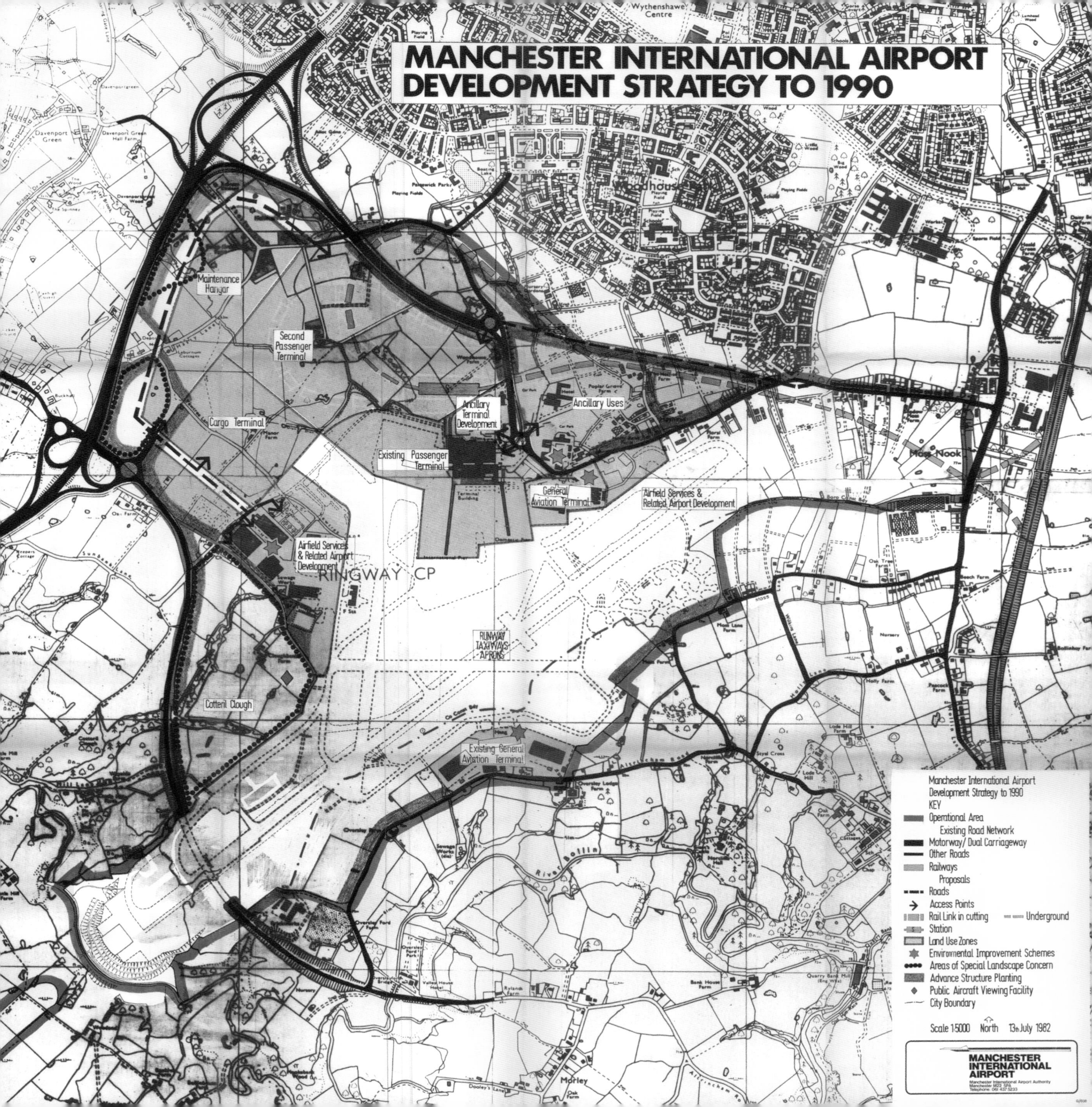
MANCHESTER INTERNATIONAL AIRPORT
DEVELOPMENT STRATEGY TO 1990
Wythenshawe Centre
Maintenance Hangar
Second Passenger Terminal
Cargo Terminal
Ancillary Terminal Development
Ancillary Uses
Existing Passenger Terminal
General Aviation Terminal
Airfield Services & Related Airport Development
Airfield Services & Related Airport Development
RINGWAY CP
RUNWAY TAXIWAYS APRONS
Cotteril Clough
Existing General Aviation Terminal
Moss Nook
River Bollin
Morley
Quarry Bank Mill
Manchester International Airport
Development Strategy to 1990
KEY
Operational Area
Existing Road Network
Motorway/ Dual Carriageway
Other Roads
Railways
Proposals
Roads
Access Points
Rail Link in cutting
Underground
Station
Land Use Zones
Environmental Improvement Schemes
Areas of Special Landscape Concern
Advance Structure Planting
Public Aircraft Viewing Facility
City Boundary
Scale 1:5000
North
13th July 1982
MANCHESTER INTERNATIONAL AIRPORT
Manchester International Airport Authority
Manchester M22 5PA
Telephone: 061 437 5233

1982a

Taking to the skies: Manchester Airport

Flying has become routine for so many people that it is easy to forget how technically sophisticated airports are, able to effect the efficient transition from ground to air for millions of people. As demand for air travel has grown, so airport terminals have been greatly expanded in complexity and scale. They became true icons of technological progress and global mobility by the late twentieth century, in much the same way as the vast railway stations were symbols of the Victorian age.

Manchester's airport goes back to the beginnings of passenger aviation before the First World War. Between 1911 and 1915, flights operated from a landing field in Trafford Park. During the interwar years, the Council made use of a site on the Hough End estate in Chorlton, then a site in Wythenshawe, and at Barton in the early 1930s. These were later all deemed inadequate for various reasons, and the Council decided to proceed with a new aerodrome 10 miles south of the city centre, on open farmland in the Cheshire plain. Manchester's new airport, Ringway, opened in 1938 and had, for its time, a substantial and well-designed modernist-style terminal with attached hangar. Initially the landing strip was grass, but a concrete runway was soon laid down.

However, the airport only operated for a short time before the outbreak of war meant it was requisitioned for military use and used as the training school for the newly formed Parachute Regiment. Civilian flights restarted at Ringway in 1946, and the Council successfully resisted nationalisation of the airport and maintained its local municipal control. The City Council has long regarded investment in airport facilities as vital to the commercial future of the city, and there has been a consistent emphasis on the airport as an engine for growth in the region. Ringway was distinctive as it continued to be local-authority owned and managed.

As traffic increased, plans were developed for a huge expansion in the 1950s, sweeping away the 1930s' terminal

---

OPPOSITE. MIAA, *Manchester International Airport development strategy* . . . (1982) [Lloyd Robinson]

to create an airport capable of attracting international flights and handling several million passengers a year. Construction started in 1958 on a large multilevel terminal with two long piers for domestic and international departure gates, and a nine-storey administrative block capped by a control tower. To handle more and larger planes, an extensive concrete apron area was also built. The £2.8 million development opened with considerable fanfare in 1962; it propelled the newly named Manchester International Airport into the top rank of airports, where it has remained ever since.

When the new airport terminal was opened in 1962, it was still the era of mid-sized propeller planes, but growing demand and technological advances soon led to much larger jetliners carrying several hundred passengers. To cope with this, the airport has undergone many phases of new building and redevelopment. The arrival of jumbo jets in the early 1970s necessitated the addition of a third pier able to accommodate these massive aircraft. A large multistorey car park was built behind the terminal building. In 1972, the airport was linked directly with the newly opened M56 motorway, providing fast road access to the city centre as well as the M6. At the same time, the aviation industry developed foreign package holidays. This was a time when many families got their first experience of flying.

Through subsequent decades, the airport cemented its position as the most significant airport outside the London area, and its Category 1 runway means it can handle the largest aircraft. As it grew, it also continued to be owned by Manchester Council and profits were returned to the city.

The map featured here was produced for strategic planning, and it provides an outline description of all the buildings and significant permanent structures and their purpose, along with access routes and locations of key future developments. The map also reveals how Manchester International, like other airports, is quite distinctive in its spatial layout when seen from above – a dense concentration of structures around the terminal contrasting with extensive open space alongside the runway to provide a safe operating zone for fast-moving aircraft.

It is evident from the map how the airport was still largely surrounded by agricultural land in the early 1980s, but in decades to come many of these fields would be swallowed up. Some legacy buildings from wartime usage and post-war expansion, plus those associated with aircraft manufacturers are also evident – airports tend to be an accretion of structures and with a tendency to sprawl outwards, as stands are added, more surface car parking is needed, and runways extended. The map is an anticipatory snapshot view of the airport that was continuously developing.

In the 1990s, new phases of large-scale expansion were pushed through, including a completely new terminal and a highly controversial second runway. Manchester Airport and aviation more generally, was beginning to face increasingly tough scrutiny over its environmental impacts. A direct railway connection was eventually opened in 1993, and it has proved to be a successful way to link the airport to many cities across the north of the country. The affordability and wider availability of air travel has required airports such as Manchester to continue to expand, to the point where they have almost become small cities in their own right.

In 2001, the business grew substantially through the acquisition of a rival, East Midlands Airport, and it was restructured and managed by a holding company Manchester Airports Group (MAG). This operates commercially but remains one-third owned by Manchester City Council while the other nine local authorities in Greater Manchester own a further 29 per cent. In 2013 MAG, expanded its portfolio further with the purchase of Stansted Airport. The dividends of its successful commercial airport operations continue to support the residents of Greater Manchester.

In 1939, the original Ringway airport handled about 4,000 travellers flying to less than half-a-dozen destinations, whereas in 2017 well over 27 million passengers passed through the three airport terminals to scores of different cities across Europe and around the world. Moreover, the airport continues to be essential to the economic potential of the Manchester region and a strategic asset for the whole of the north of England.

OPPOSITE. MIAA, *Finding your way through Manchester International Airport* (*c*.1985) [AUTH]

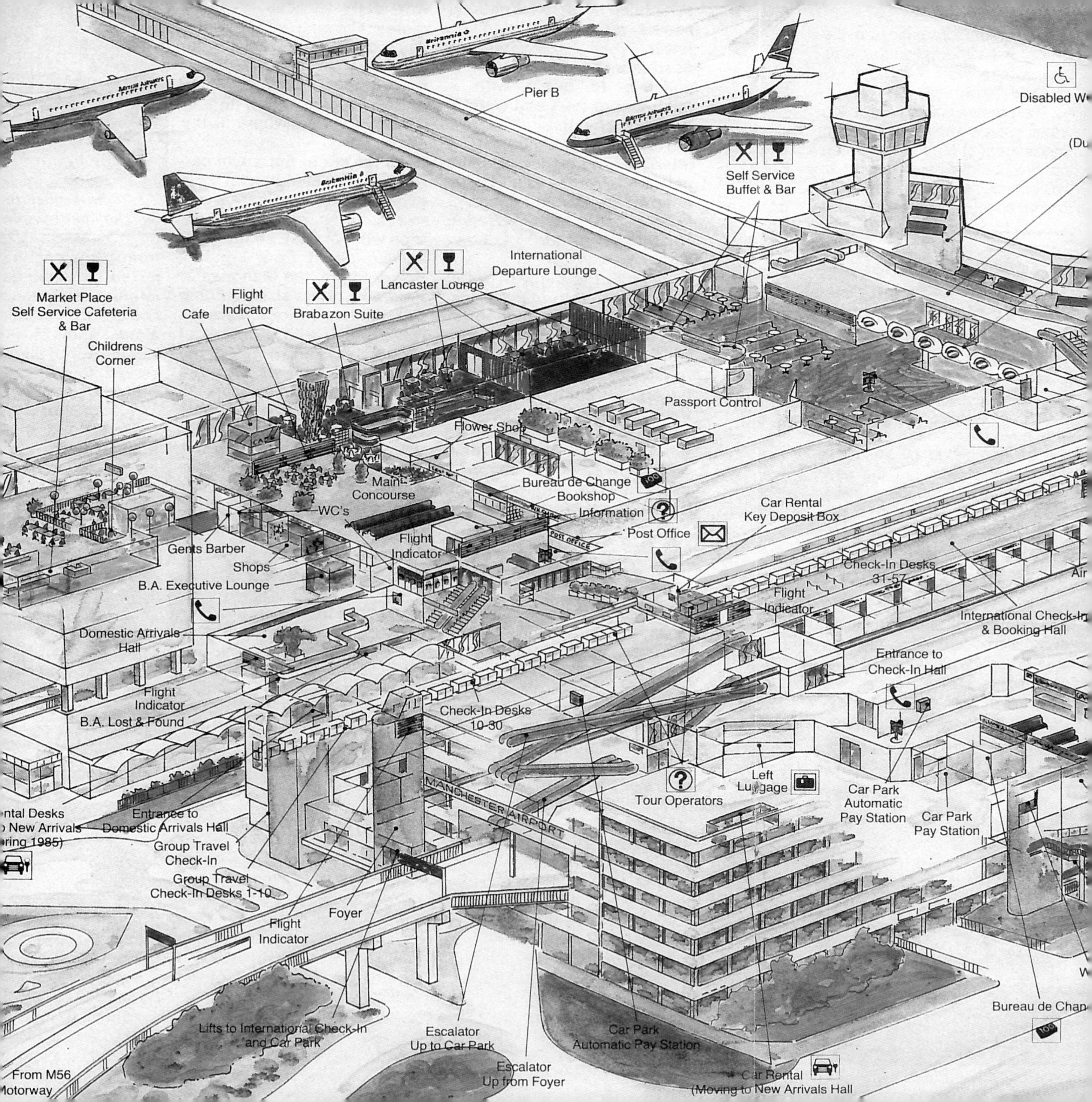
Pier B
Self Service
Buffet & Bar
International
Departure Lounge
Lancaster Lounge
Brabazon Suite
Market Place
Self Service Cafeteria
& Bar
Flight
Indicator
Cafe
Childrens
Corner
Passport Control
Flower Shop
Main
Concourse
Bureau de Change
Bookshop
Information
WC's
Car Rental
Key Deposit Box
Post Office
POST OFFICE
Gents Barber
Shops
Flight
Indicator
Check-In Desks
31-57
B.A. Executive Lounge
Flight
Indicator
International Check-In
& Booking Hall
Domestic Arrivals
Hall
Entrance to
Check-In Hall
Flight
Indicator
B.A. Lost & Found
Check-In Desks
10-30
MANCHESTER AIRPORT
Left
Luggage
Tour Operators
Car Park
Automatic
Pay Station
Car Park
Pay Station
Entrance to
Domestic Arrivals Hall
Group Travel
Check-In
Group Travel
Check-In Desks 1-10
Foyer
Flight
Indicator
Lifts to International Check-In
and Car Park
Escalator
Up to Car Park
Escalator
Up from Foyer
Car Park
Automatic Pay Station
Car Rental
(Moving to New Arrivals Hall
From M56
Motorway

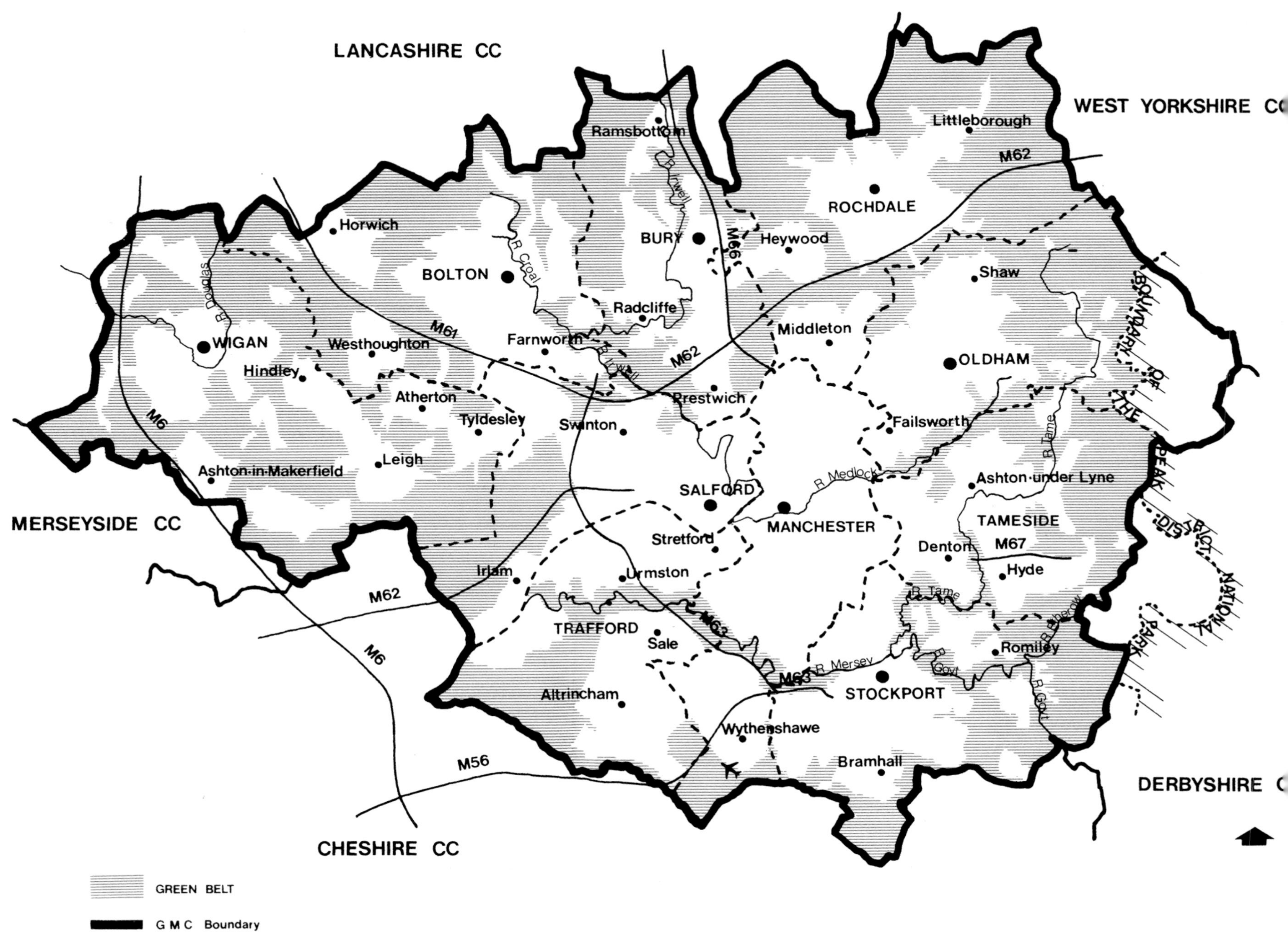

GREEN BELT

G M C Boundary

District Boundaries

Adjoining County Boundaries

M6 Motorways

R Tame Rivers

Manchester International Airport

This map can only show the suggested Green Belt areas in a very broad way. The large scale plans (on display at the exhibition) should be refered to for the details.

Scale 0 1 2 3 4 miles
0 1 2 3 4 5 kilometres
(Approx)

1982b

Defining the green belt

One of the most influential elements of post-war planning has been the designation of green belts. The power to restrain the spread of housing estates and commercial development has helped to preserve open land within easy access for city dwellers. Decisions on where to draw the boundary of green belts and the degree to which the rules have been enforced have shaped the growth of urban Britain for better or worse.

Concern about the extent to which cities were spreading outwards came to the fore in the 1920s and 1930s when speculative building by private developers aimed to attract middle-class families out from cities. In the interwar years, around 700,000 new private homes were built on the edge of towns and in the countryside. These often used a roster of 'off-the-peg' architectural designs, which has given the interwar suburbs across Britain a recognisable visual character. Before the tightening of planning law in 1947, there was little that local councils could do to prevent new housing estates springing up if landowners were willing to sell to the developers.

The pace and scale of housing sprawl around London in particular led to calls for control. Initial efforts were made by London County Council in the mid 1930s to buy plots of open land to protect them from residential developers. The objective was 'to provide a reserve supply of public open spaces and of recreational areas and to establish a green belt or girdle of open space lands, not necessarily continuous, but as readily accessible from the completely urbanised area of London as practicable'. This strategy was formalised by the 1938 Green Belt Act.

In Manchester's case, by the 1930s there was virtually continuous urban development for many miles outwards in all directions from the Town Hall. Only the moss land west of Salford, and the Pennine hills to the north and east worked as natural barriers to relentless continuous urban growth. Consequently, there was much pressure from developers to build to the south of Manchester, and soon housing estates stretched

---

OPPOSITE. Greater Manchester Council, *Green Belt sketch map* (c.1982) [David Kaiserman]

far into the Cheshire plain along arterial roads and around commuter railway stations. This residential sprawl was exacerbated by the City Council itself through their plans to construct tens of thousands of social homes at Wythenshawe in the 1930s and the post-war years.

There was growing concern about the damage caused by urban sprawl, with unplanned speculative house building encroaching on scenic sites and swamping small historic settlements in Cheshire. Local planners argued for a strategic approach to the planning of the region and sought to develop a formal green belt – in a roughly 10-mile radius around Manchester Town Hall, covering about 20,000 acres. This, it was argued, would protect 'all high lands, wooded areas, and other places of natural beauty'.

After the Second World War, there was a radical change in how urban development was to be directed with significant state control enacted in the 1947 Town and Country Planning Act, which created a legally enforceable system of planning permission. In 1955, Whitehall codified the principles of green belts and gave powers to local authorities to create them. However, across the Manchester region there was uncertainty about how best to define its green belt. Difficulties in decision-making were compounded by pressures to site 'new towns' in the 1950s and the pressing need for overspill municipal housing estates in the 1960s to cater for the numbers of households displaced by slum clearance.

When Greater Manchester Council (GMC) was created in 1974, it inherited an incomplete 'patchwork-quilt' of green belt designations from its subsidiary local authorities. There was a need to rationalise and bring consistency to the approach for the whole Greater Manchester region. Defining the green belt boundaries became part of the larger comprehensive structure planning process in the late 1970s, but this took many years of data gathering, deliberation, consultation and draft publications. The County Structure Plan was eventually formally adopted in 1981, and the *Greater Manchester Green Belt Local Plan* came into effect in 1984. It argued that the purposes of the green belt were: to check further growth of the large built-up areas of Greater Manchester; prevent neighbouring towns and urban areas from merging into each other; and preserve the special character and identity of towns and villages in Greater Manchester.

The green belt comprised extensive areas of open land but also included narrow fingers and wedges of land between settlements, particularly running along the river valleys and canals. There was an extensive green belt around Wigan and north and east of Rochdale and Oldham. Manchester itself had very little green belt land within its boundaries, except for Heaton Park and the river corridor of the Mersey. It was somewhat ironic that Manchester Airport was surrounded on many sides by green belt land, a large chunk of which would be lost with the controversial building of the second runway in the 1990s and subsequent expansion of airport-related activities. Following the abolition of GMC in 1986, the defence of green belt boundaries was carried forward by the ten separate district councils and only relatively minor amendments were introduced subsequently through Local Plans and Unitary Development Plans.

By 2015, the green belt covered about 47 per cent of Greater Manchester, amounting to over 59,000 hectares of land. Inevitably, there are wide variations in the amounts across the ten districts, in large part reflecting their distance from the core of the conurbation and their relationship to the Pennines: 63 per cent of Rochdale is green belt; Bolton, Wigan and Bury each have over 50 per cent; whereas Salford and Trafford have less than 40 per cent, and Manchester has a mere 15 per cent. However, in recent years, there has been increasing pressure to build on green belt land, and this has been facilitated by governmental changes to the planning system primarily aimed at stimulating new housing. The Greater Manchester Spatial Framework proposed an 8 per cent reduction in the conurbation's green belt. There have been widespread and well-orchestrated protests in response to this with local groups trying to defend the existing green belt lands. This can be regarded partly as 'nimbyism' on the part of local people, yet the scale of new development in some areas is overwhelming. Protest groups argue there are alternative brownfield sites and scope for creative sustainable housing solutions within cities rather than turning open land into bland cul-de-sacs and car-dependent suburbia.

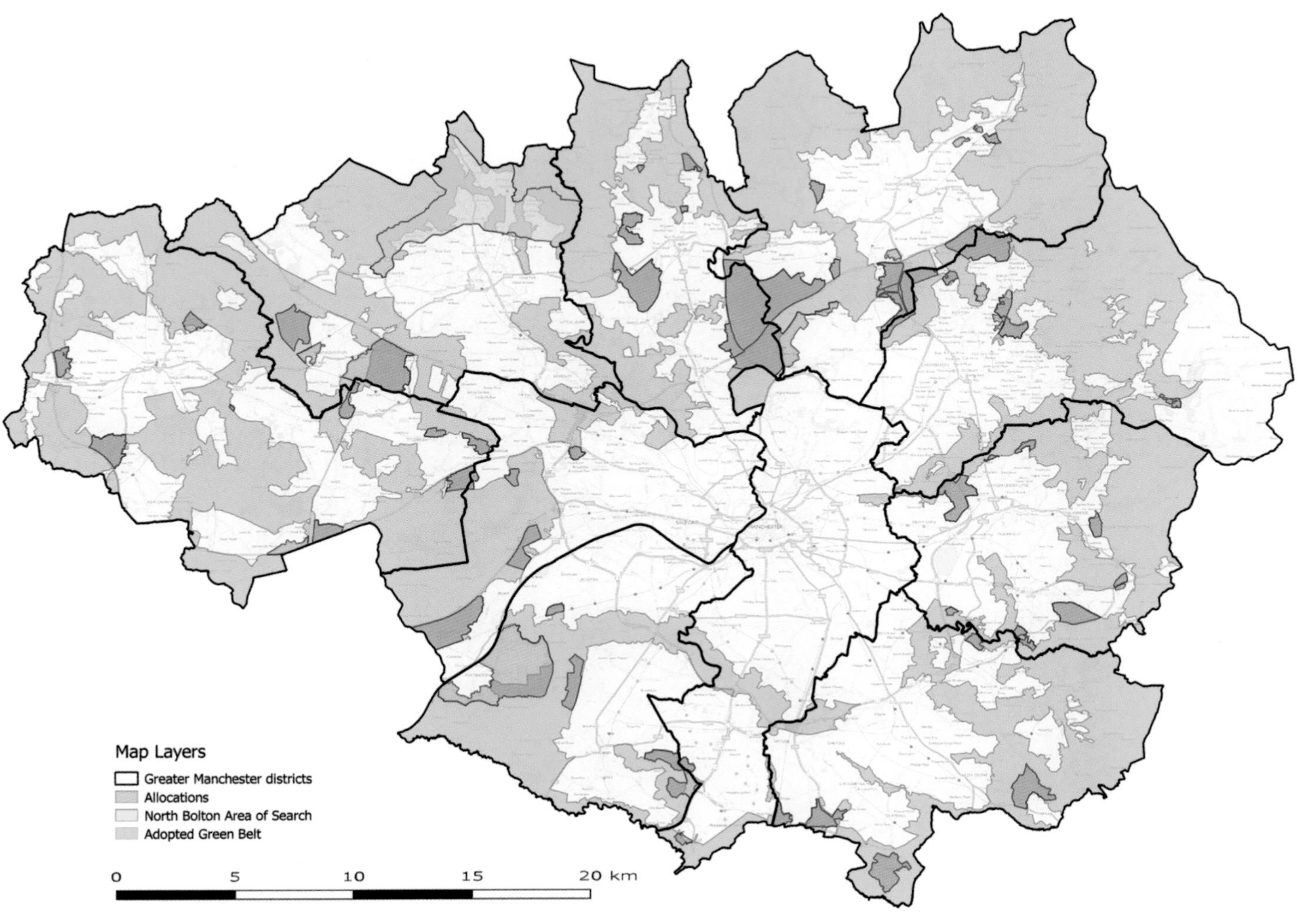

Greater Manchester Combined Authority, *Proposed allocations and green belt* (2016) [Greater Manchester Combined Authority]

Leaving aside problematic issues around constricted housing supply and social equity, green belt policies have been useful in encouraging greater sustainability of cities through maintaining 'green lungs' for recreation, in places helping to preserve wildlife habitats and ecologically significant corridors, and also encouraging more compact urbanism and the reuse of brownfield sites.

9A
9B
8
7a
7b
7c
6

1985

Inventing Salford Quays and MediaCity

The long saga of transforming derelict docks into Salford Quays has proved a triumph of regeneration. There were three distinctive elements to the redevelopment: Salford Quays itself, Trafford Park on the opposite side of the Ship Canal and, most recently, MediaCityUK, which effectively joined the former two together. The development focused on the Port of Manchester at the head of the Ship Canal. In the 1950s, the port was the third busiest in the UK, but its trade declined rapidly in the 1970s as local industries contracted, trade links with Continental Europe grew at the expense of North Atlantic trade and the upper reaches of the canal proved no longer navigable for the ever-larger container ships. The port closed in 1982 and the docks were left, in the words of Felicity Goodey, as a 'dirty, stinking wasteland in the midst of an already impoverished conurbation'. On the opposite side of the canal, the collapse of the once thriving industries of Trafford Park was part of that dereliction.

The initial attempt to tackle the dereliction was the declaration of Salford Quays and Trafford Park as an Enterprise Zone in 1981, but this attracted little new industrial investment and failed to reverse the area's fortunes. In 1983, Salford City Council bought most of the dock area from the Ship Canal Company and, despite the bleak and shabby nature of the area, Peter Hunter of the London-based firm of Shepheard, Epstein & Hunter saw scope for creating waterside venues and inspired the redevelopment of the derelict docks. The area was rebranded as 'Salford Quays', and its redevelopment was begun by Urban Waterside. Over a two-year period, the severe pollution of the water was tackled by a major programme of aeration, which eventually allowed the introduction of fish and held out the promise that the docks could be developed for water sports. The four docks (numbers 6–9) were reconfigured with dams or locks linking the piers between them; dock 7, for example, was divided into a series of small basins.

Shepheard, Epstein & Hunter, *Salford Quays development plan . . .* (1985) [AUTH]

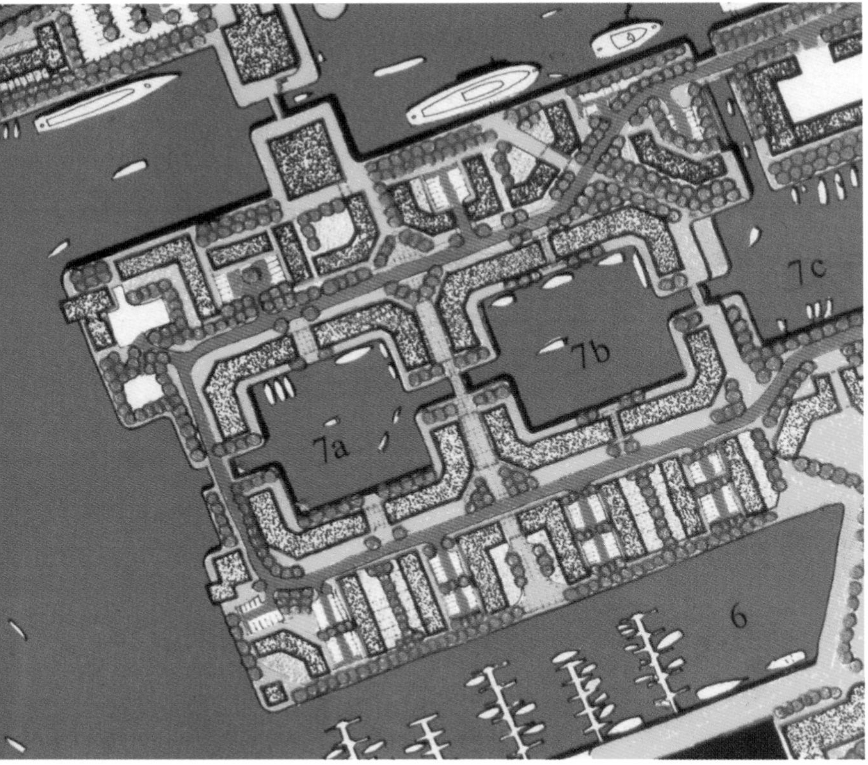

Dock 7 divided into separate small basins.

Many observers thought it implausible that this bleak area, with cold winds whipping off the water, could ever provide an attractive working and residential environment. They were to be proved wrong. The dilapidated warehouses and sheds were replaced by residential blocks, and the docks gradually became a successful office and recreational area. Merchants Quay was the first of the residential areas, built in the late 1980s and comprising town houses, mews and low-rise apartments. The first high-rise block, Imperial Point, was built on Pier 8 in 2001. The first hotel, the Copthorne, was subsequently joined by a Holiday Inn, a Premier Inn, a Travelodge and a Marriott Hotel.

The initial master plan reserved Pier 8 for an arts-based development, and an international competition was held from which The Lowry, the jewel in the crown, emerged as the winner. Designed by Michael Wilford, it opened in 2000 and was built with funding from the Arts Council, Heritage Lottery and other sources. It won the Royal Fine Arts Commission Building of the Year in 2001. It comprises two theatres, studio spaces, gallery space, cafes, bars and a restaurant; and it houses a large permanent display of the work of L.S. Lowry. The building itself sits at the end of Pier 8, largely surrounded by water and with the appearance of a large ship breasting the waters. Opposite it is a large retail centre, the Lowry Outlet Mall, which opened in 2001.

Earlier, on the opposite side of the canal, a major regeneration of Trafford Park took place. This was the work of the Trafford Park Development Corporation, set up by the government in 1987 and including not only the industrial park, but parts of Stretford, Salford Quays and the former steelworks at Irlam, now called Northbank. Over a period of 11 years, the Corporation succeeded in reconfiguring the area, attracting some 1,000 new companies and creating almost 30,000 jobs so that the Park once again became a major employment centre. Its flagship developments include the Quay West office development on Wharfside and the Northbank Industrial Park. The area also saw another iconic building, the Imperial War Museum North, which opened in 2002, sitting directly opposite The Lowry across the canal. It was designed by Daniel Libeskind as a metaphor for a world shattered by war with three shards – air, earth and water – housing exhibition space and restaurants. The two dramatic buildings have helped to transform the townscape of the Quays and have become a significant attraction for visitors.

The third major development, on Pier 9, was MediaCityUK, which resulted from the BBC's decision in 2004 to relocate a number of its programme teams to the Manchester area. The shortlist of four potential sites included one close to the existing Granada Studios site in Manchester and the last remaining large undeveloped site in the Quays on Pier 9. The decision to choose the Salford site caused Manchester some anguish (and for a long time left a large undeveloped hole where the BBC studios had sat on Oxford Street), but it has proved an inspired decision. Construction started in 2007 and the BBC, which began its move in 2011, now occupies three buildings, each housing teams responsible for TV and radio programme such

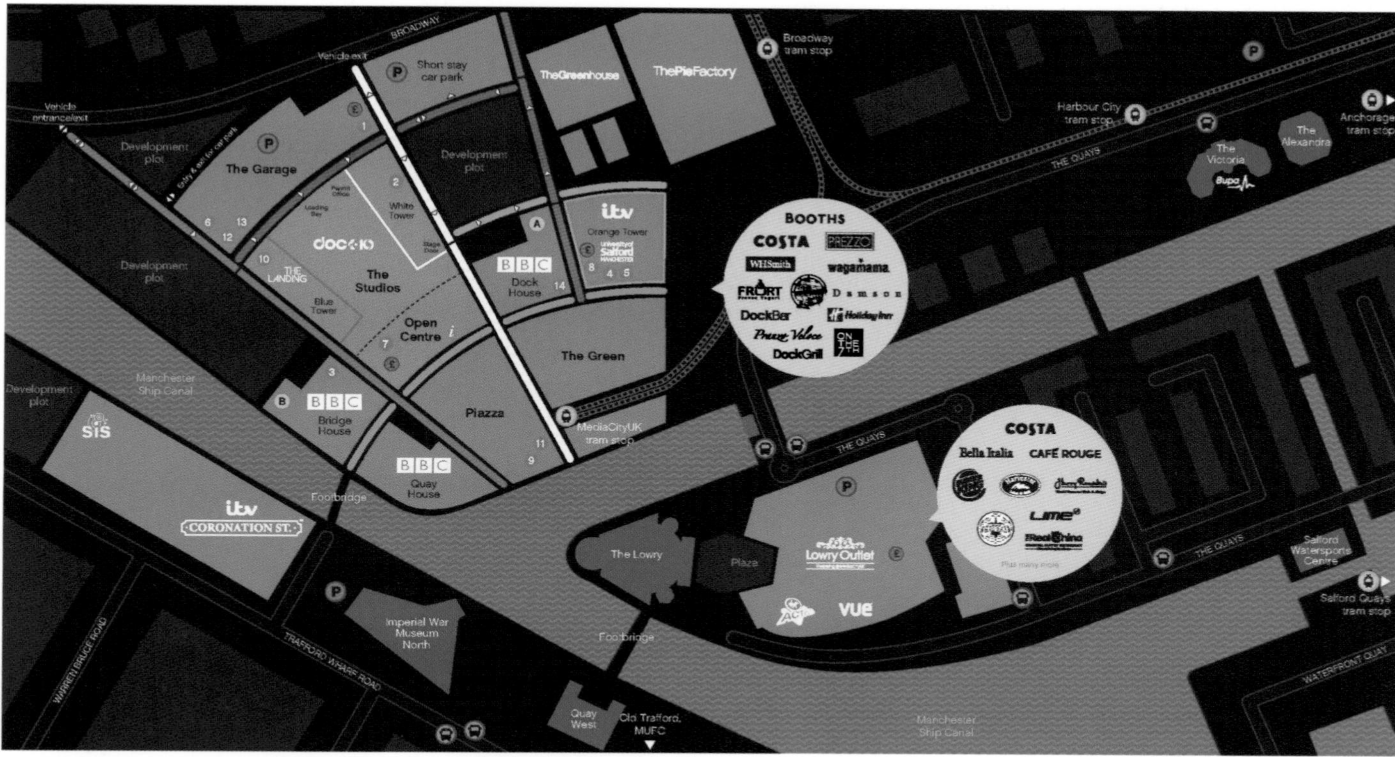

MediaCityUK, *MediaCityUK wayfinder* (2017) [MediaCityUK]

as *BBC Breakfast*, *Match of the Day*, Radio 5 Live, *North West Tonight*, *Mastermind*, *Blue Peter* and *Dragons' Den*. Another building houses Sports Information Services, a news-gathering company. The Studios comprise seven high-definition studios, the biggest of which claims to be the largest in Europe. Salford University, with its well-established expertise in television production and media studies, moved its media-related teaching and research activities to the site in 2011. Two tower blocks provide over 350 apartments adjacent to the studio facilities. ITV Granada had originally decided not to move to the Quays, but changed its mind and began a move to a new production facility on Trafford Wharf in 2013.

A key element in the Salford Quays development was the opening in 1999 of a Metrolink line, which connected the area by tram to the centre of Manchester. Later, in 2010, a station was opened on Pier 9 to serve MediaCity directly. The ultimate piece of the jigsaw is the striking Lowry footbridge, which crosses the canal between The Lowry and the Quay West office on Trafford Wharf. It is the final link in a circular walk that connects The Lowry, Salford Quays and MediaCity on the north bank of the canal, with Trafford Park and the Imperial War Museum North on the south bank.

The whole complex has become a key part of the Manchester conurbation, attracting significant visitor and tourist attraction, generating substantial employment, flaunting some dramatic architecture, contributing to the reversal of outward migration from the inner city and cementing Manchester's role as the largest cultural, creative and digital venue outside London. There has also been real benefit from the publicity given to Manchester by MediaCity, since many BBC news and current affairs reports now draw on local examples and local participants.

CASTLEFIELD

Britain's first Urban Heritage Park

For further information about Castlefield contact:
Castlefield Visitors' Centre
330 Deansgate, Castlefield,
Manchester. 061-832 4244

For further information about City Centre Manchester contact:
The Tourist Information Centre
Town Hall Extension
St. Peter's Square.
Manchester 061 234 3157/8

Granada TV

Castlefield's most famous residents. The Victoria and Albert warehouses contain the set for the Albion Market Series. The Coronation Street and Baker Street sets are visible from the Greater Manchester Museum of Science and Industry. These sets are not open at present to the public.

Air and Space Gallery

Part of the Greater Manchester Museum of Science and Industry. History of flight from earliest times to the present day. Large collection of space material on loan from NASA. Souvenir shop. Open seven days a week, small admission charge.

Urban Studies Centre

Situated above the Visitors' Centre and operated by Manchester Education Committee. Exhibitions and displays about the urban environment many with particular reference to Castlefield.

G-MEX

Manchester's new internationa[cut off] exhibition centre built in the t[cut off] hall of the former Central Railw[cut off] station. Provides more than 100,000 sq.ft. of uninterrupted exhibition space together with bars restaurants etc. G-MEX i[cut off] the first phase of the redevelopment of the whole Central Station site.

QUAY ST.
LOWER BYROM ST.
DEANSGATE
WATER ST.
LIVERPOOL RD.

The Mark Addy

Named after a famous nineteenth century boatman famed for his lifesaving on the River Irwell, the Mark Addy provides an opportunity to eat and drink and look out on Manchester's river. Landing stage and terrace.

Castlefield Gallery

Manchester's contemporary [cut off] gallery. Open Tuesday to Saturday 10.30 am to 5.00 pm Sunday 12.00 to 4.30 pm. Talk[cut off] videos and workshops accompany every exhibition.

Greater Manchester Museum of Science and Industry

The fastest growing museum in Europe and already one of the largest. Currently open are the 1830 station buildings of the world's first passenger railway station, the Power Hall telling story of industrial power from earliest times to the present day, the Lower Byrom Street Warehouse with displays of textiles, printing and papermaking, The National Electricity Gallery, a museum shop and "Chuffers" restaurant in a railway carriage. Still to come are the Greater Manchester story and the Manchester Underground exhibitions (1987) and a Science Centre (1988). Open seven days a week, admission free.

The Roman Fort

The North Gateway of the Roman Fort has been imaginatively reconstructed on its original site. Within the fort the guardroom has been furnished as it might have been in Roman times. Information about visits to the guardroom may be obtained from the Castlefield Visitors' Centre. Excavations of Manchester's Roman past are still continuing and tours can also be arranged through the Visitors' Centre. Adjoining the Fort is a mural illustrating the history of Castlefield from Roman times to the present day.

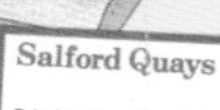

Salford Quays

Linked to Castlefield by the river Irwell and the Manchester Ship Canal, Salford docks and the Ship Canal were opened in 1894. In recent years traffic has declined and work is in hand to convert the area to other uses including a marina, an international hotel and a nine-studio cinema.

Castlefield Visitors' Centre

The Centre provides an introduction to the whole of Castlefield. It includes displays and exhibitions about the area and also temporary exhibitions about the City and its development. Open Monday to Saturday 10.00 am to 5.00 pm and on Sundays from April to September.

The Canals

The Bridgewater canal was built between 1759 and 1761 to bring coal from the mines at Worsley to the growing town of Manchester. It was the first to be cut across open country rather than following the course of an existing river. This was the beginning of the network of canals that spread throughout England. Today commercial traffic has ceased and the canal is used for leisure traffic. In the next few years the canal basin where the Bridgewater meets the Rochdale Canal and the River Medlock will be made more attractive as a leisure and recreation area.

 Disabled access available.

1986

Castlefield: urban heritage

Castlefield can claim to be Manchester's most historic site. It encapsulates the site of the original Roman fort of Mamucium, the canal basin where the Bridgewater Canal and Rochdale Canal meet in an elaborate series of canal arms, Liverpool Road Station, the world's first rail passenger terminus and railway warehouse, and four massive viaducts of later railways that march majestically across the area. Few sites anywhere in the world can boast a complex of iconic buildings spanning such pioneering examples of the early canal and railway ages.

However, in the 1960s, the area was largely derelict and its historic importance overlooked. St Matthew's Church, designed by Charles Barry and built on Liverpool Road, had been demolished in 1951, although its Sunday school was saved (and renovated as an office). The canals were effectively disused; most were silted up and some canal arms had been filled in. The closure of Central Station in 1969 meant that most of the rail viaducts were abandoned. The canal basin, hidden behind buildings on Liverpool Road, was a dilapidated and neglected part of the city.

However, from 1972, the then Greater Manchester Council (GMC) sponsored a number of archaeological digs, which led to the recognition of the significance not only of the Roman fort, but especially of Liverpool Road Station and its 1830s' warehouse, both of which now have listed status. In 1979, the area was declared a conservation area, within which the large number of listed buildings include the dramatic sweeps of the mostly disused railway viaducts. In 1982, Castlefield was designated an urban heritage park. Following the excavations, the north gate and part of the wall of the Roman fort were reconstructed. Most significantly, the Liverpool Road Station complex was converted into the Museum of Science and Industry and was soon followed by the conversion of the Lower Campfield Market building into the museum's Air and Space Gallery. Together, the two museum buildings have housed

---

OPPOSITE. Castlefield Urban Heritage Park, *Castlefield* (*c.*1986) [AUTH]

railway engines and rolling stock, aircraft, and steam, water, electric and gas engines. Among the highlights are historic stationary steam engines, an Avro Shackleton aircraft and other Avro aircraft built locally at Chadderton and Woodford. The revitalisation of the area was driven not only by GMC but also by the Civic Trust, the Georgian Group, the Victorian Society and the Manchester Region Industrial Archaeology Society.

The overall aim of the makeover of Castlefield was to create a new tourist and visitor attraction, capitalising on this wealth of historic sites and buildings. The fact that Granada Television's Studios' Tour was adjacent to the area – with its *Coronation Street* set and House of Commons reconstruction among a variety of other attractions – added hugely to the tourist and visitor potential of Castlefield. Central Station – which had become increasingly dilapidated, damaged by fire and used as a car park – was acquired by GMC and opened in 1986 as the Greater Manchester Exhibition and Conference Centre or G-Mex (later renamed 'Manchester Central' to reflect its railway history).

GMC was abolished in 1986, but in 1988 the government set up a 'mini-Urban Development Corporation' – Central Manchester Development Corporation (CMDC) – covering the crescent of land wrapped around the southern part of the city centre. This included Castlefield, so CMDC was able to build on the earlier work of GMC in the area. CMDC was controversial since it was essentially a creature of central government not the local authority; it could act as its own planning authority and reported directly to the Department of Environment, thereby bypassing the Council's decision-making processes. Government and the private sector dominated the Board. It was chaired by James Grigor, seconded from Ciba-Geigy, and its Chief Executive was John Glester, a senior civil servant from the Government Office for the North West. This was all a far cry from the later government regeneration strategies, which started with City Challenge in Hulme in 1992, in which the local authority was centrally involved as the key player in the process. Nevertheless, by a process of *legerdemain*, the City Council played a significant role in Castlefield. The MP David Trippier, who had been the Conservative leader of Rochdale Council, became Minister for Inner Cities after the 1987 election reshuffle and had the task of establishing CMDC. He and the then Labour leader of Manchester, Graham Stringer, formed an alliance in which both were committed to Manchester's betterment and were conscious that nothing would happen without giving the private sector confidence to invest. Despite the hostility of the City Council, Stringer sat on the UDC Board and achieved a *modus vivendi* in which the city's Planning Department played a powerful 'consultancy' role, in effect working collaboratively with the UDC in steering the course of developments. The key private-sector members of the Board subsequently played key roles in the regeneration of the city – for example, John Whittaker, Chief Executive of Peel Holdings, which owned the Manchester Ship Canal; Alan Cockshaw of AMEC; and Bob Scott, the theatre impresario who was influential in persuading the Board to support the city's 1996 Olympic bid.

CMDC's strategy built directly on GMC's earlier aim of strengthening the tourism and leisure base of Castlefield. Its initial work focused on improving the environment of the area – excavating filled-in canal arms and dredging and clearing debris and rubbish from others. By 1993, the Inland Waterways Association was able to hold a boat festival on the newly restored canals, which attracted over 300,000 spectators. Major developments included the budget Castlefield Hotel with a spa, health centre and indoor running track; the Outdoor Events arena with covered seating for 450 spectators; the Victoria and Albert Hotel, converted from an 1844 warehouse and opened by Granada Television. The Corporation gave particular emphasis to the quality of its refurbishment, using cobbles and York stone rather than concrete and tarmac, high-quality street furniture and cast-iron bridge work to complement the neighbouring Victorian buildings. The Castlefield Management Company was created in 1992 to provide services, run events and maintain the environmental quality of the area; and an Urban Ranger service was set up to help visitors and oversee the whole area.

At its outset, CMDC had been expected to promote industrial development and generate income from land sales, but its timing coincided with the recession of the early 1990s and the collapse of land values. In response, it gave much greater emphasis to housing. Across the whole CMDC area, 2,583 housing units were built (as against an initially expected target of only 471). This was effectively the start of repopulating the city centre, which has seen a huge number of apartments spring up in central sites across the city. Castlefield played an important role in this since, under CMDC, many of the handsome historic warehouses were converted into upmarket apartments, and new housing was built aimed both at retirees and young households, capitalising on the allure of waterside sites. Other warehouses were converted into design studios and offices. One of the most successful of the local entrepreneurs was Jim Ramsbottom whose Castlefield Estates company developed several significant projects: workspaces in Eastgate, a converted rag mop factory; the 1825 Merchants Warehouse, which had suffered fire damage but was converted for creative media and technology companies; and Dukes 92, a popular pub whose name was taken from its location adjacent to the final lock on the Rochdale Canal.

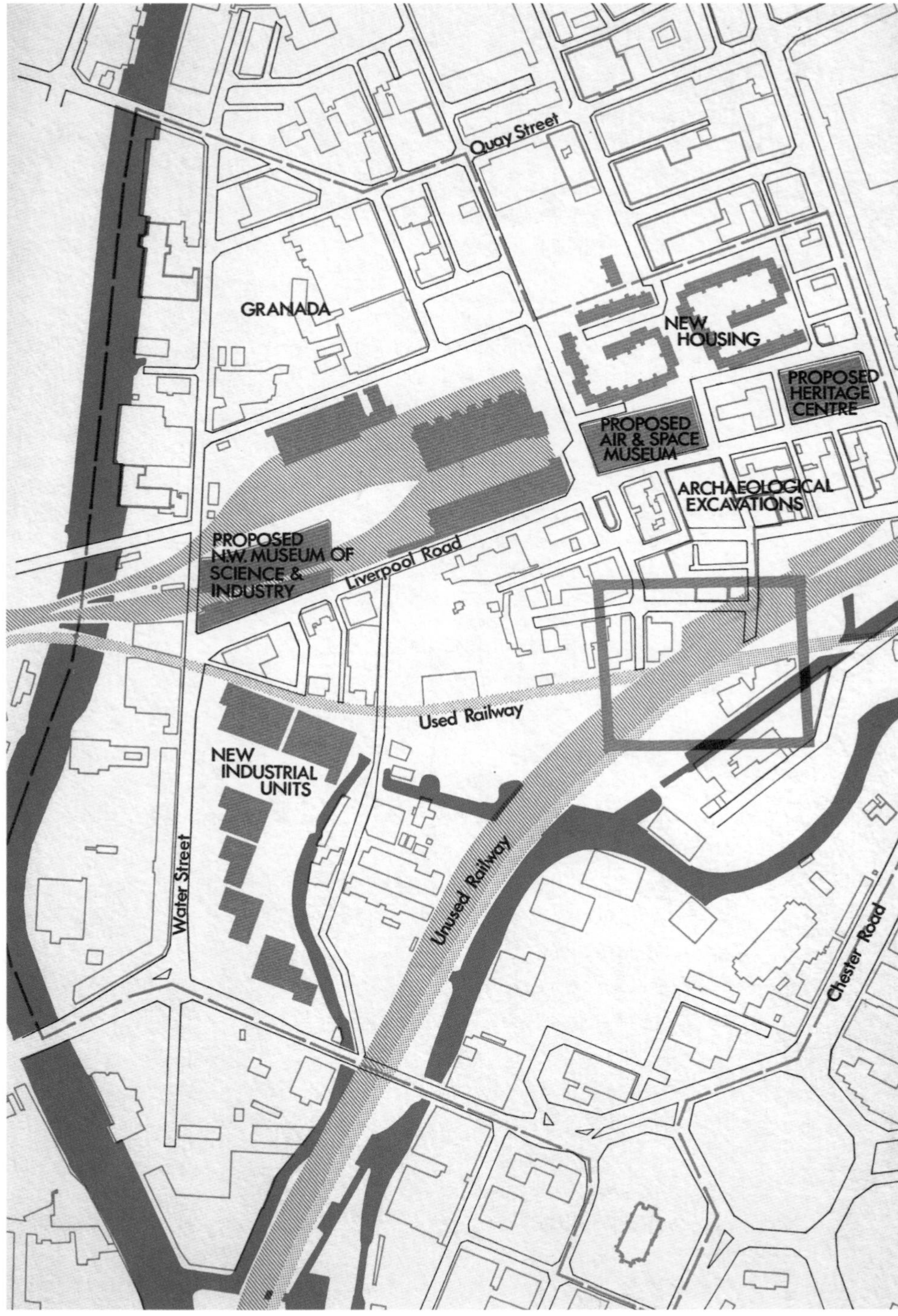

Manchester City Council, *Castlefield* (c.1980) [AUTH]. The dashed red line outlines the Conservation Area. The empty red square is the site of the Roman fort. The Irwell is on the left-hand side.

1996a

Reconfiguring the city centre after the IRA bomb

The bomb which the IRA exploded at the junction of Market Street and Corporation Street on Saturday 15 June 1996 could not have come at a worse time for Manchester since the city was faced with competition from the imminent opening of the huge new Trafford Centre shopping complex at Dumplington. On the other hand, by the early 1990s Manchester's centre had become down at heel and increasingly unappealing both to business and shoppers so, despite the trauma and the extensive physical and commercial damage caused by the bomb, it offered a potential spur to the revitalisation of the city centre.

The 1,500-kilogram bomb was the largest to have been detonated in Britain since the war. It left a huge trail of destruction: several buildings in the area had to be demolished; some 700 businesses were affected, many of which never reopened; few windows survived within a half-mile radius; and some 300 people were injured. Amazingly, there were no deaths, in part thanks to the quick reaction of the police and emergency services who responded to the IRA telephone warning by evacuating over 70,000 people from the area. The worst damage was to three buildings close to the site of the explosion: Michael House, home to Marks & Spencer; Longridge House, the offices of Royal and Sun Alliance; and substantial parts of the large, commercially successful but unloved Arndale Centre owned by P&O.

The city clearly faced a massive challenge. It met it with a determined and imaginative response. Howard Bernstein, then Deputy Chief Executive, and Richard Leese, the new Leader of the Council, approached Michael Heseltine, the Deputy Prime Minister and long-time supporter of urban regeneration. Heseltine threw his support behind the city: within 11 days of the bomb, he had announced an international competition for the redevelopment of the bomb-damaged area and backed this

---

OPPOSITE. Greater Manchester Fire and Rescue Service (1996) [Greater Manchester Fire Service Museum] Corporation Street bomb damage. Marks & Spencer is on the right.

with the promise of government finance. Five bids were shortlisted, and the winning design came from a consortium put together by Ian Simpson Architects, which included the international design company EDAW. The ideas suggested by the rising local architects Ian Simpson and Rachel Haugh proved critical to the consortium's success, especially the proposal to drive a new street – which became New Cathedral Street – to link the long-neglected medieval core with the largely Georgian centre focused on St Ann's Square and King Street. The two areas had been separated by the line of buildings at the end of St Ann's Square, which cut off the Cathedral and the medieval core from the main circulation flows in the city centre. Driving a new street through to join the two was a stroke of genius and promised to transform the geometry of the central area.

The master plan was essentially a framework rather than a fixed plan. By winning, the two key members of the consortium were invited to work with the city and other partners to guide the reconstruction of the bombed area. EDAW led the implementation of the proposals, and Simpson and Haugh worked with the new task force, Manchester Millennium Ltd, which was set up by the Council to oversee the work. This was the second major example of the city using its arms-length public–private partnership model. The task force was chaired by Alan Cockshaw, chairman of the construction and engineering company AMEC, and included a mix of senior figures from the public and private sectors: Marianne Neville-Rolfe, Director of the Government Regional Office for the North West; David Trippier, High Sheriff of Lancashire and ex-MP; Richard Leese, Leader of the Council; Kath Robinson, Deputy Leader; and Tony Strachan, Agent of the Bank of England. The small project team was run by Howard Bernstein as Company Secretary and included secondees from key private-sector companies, notably Alison Nimmo who was seconded from KPMG to act as Project Director.

The speed and imaginativeness of the rebuilding was impressive. Much of the rebuilding was completed by 2000, although work continued until 2005.The most significant trigger in the process was Marks & Spencer's decision to reinvest in Manchester. It bought and demolished the neighbouring Longridge House and created new premises by 1999, the largest national branch of the store, thereby attracting investment from other high-quality chains. The store was later shared with Selfridges. The Corn Exchange, whose dome had been displaced by the blast, reopened in 2000 as The Triangle, a high-quality retail complex. The owners of the building summarily ejected the previous tenants who had included a number of small market businesses, many of whom moved to the Northern Quarter and boosted the evolution of that area as a slightly bohemian part of the city. Exchange Square was designed with a cascading water feature and some rather incongruous windmills. The Printworks, which had housed some of the major national newspapers, was converted into an entertainment centre with a multiplex cinema. The Royal Exchange, whose glass dome had been displaced by the explosion, was reconfigured over a period of more than two years with National Lottery funding. The Arndale Centre began its rebuilding by removing and remodelling its frontage, thereby helping to change its external image from having been 'the world's biggest public toilet' – dominated by its sickly ceramic tiles – into a less despised building and a profitable key part of the retail core.

Subsequently, a second competition was held for the design of the Urbis Building, which was again won by Simpson and Haugh and sited so as to create a large open space between the Corn Exchange and Chetham's School. This became Cathedral Gardens, designed by the architectural firm BDP to seat the Cathedral and Chetham's Library and Music School in a green landscaped context with a flowing water feature. The EDAW master plan of 1999 shows both the original bomb-site redevelopment and Cathedral Gardens.

As well as New Cathedral Street, which attracted a Harvey Nichols store and other major retailers, other of Simpson and Haugh's ideas also came to fruition. One key was the physical relocation of two historic pubs – the Old Wellington and Sinclair's Oyster Bar – to create a new Shambles Square sitting at the end of New Cathedral Street. Another was the construction of the dramatic No. 1 Deansgate, high-quality apartments which attracted some of the well-heeled football players of the city and showed again the potential of city-centre living which

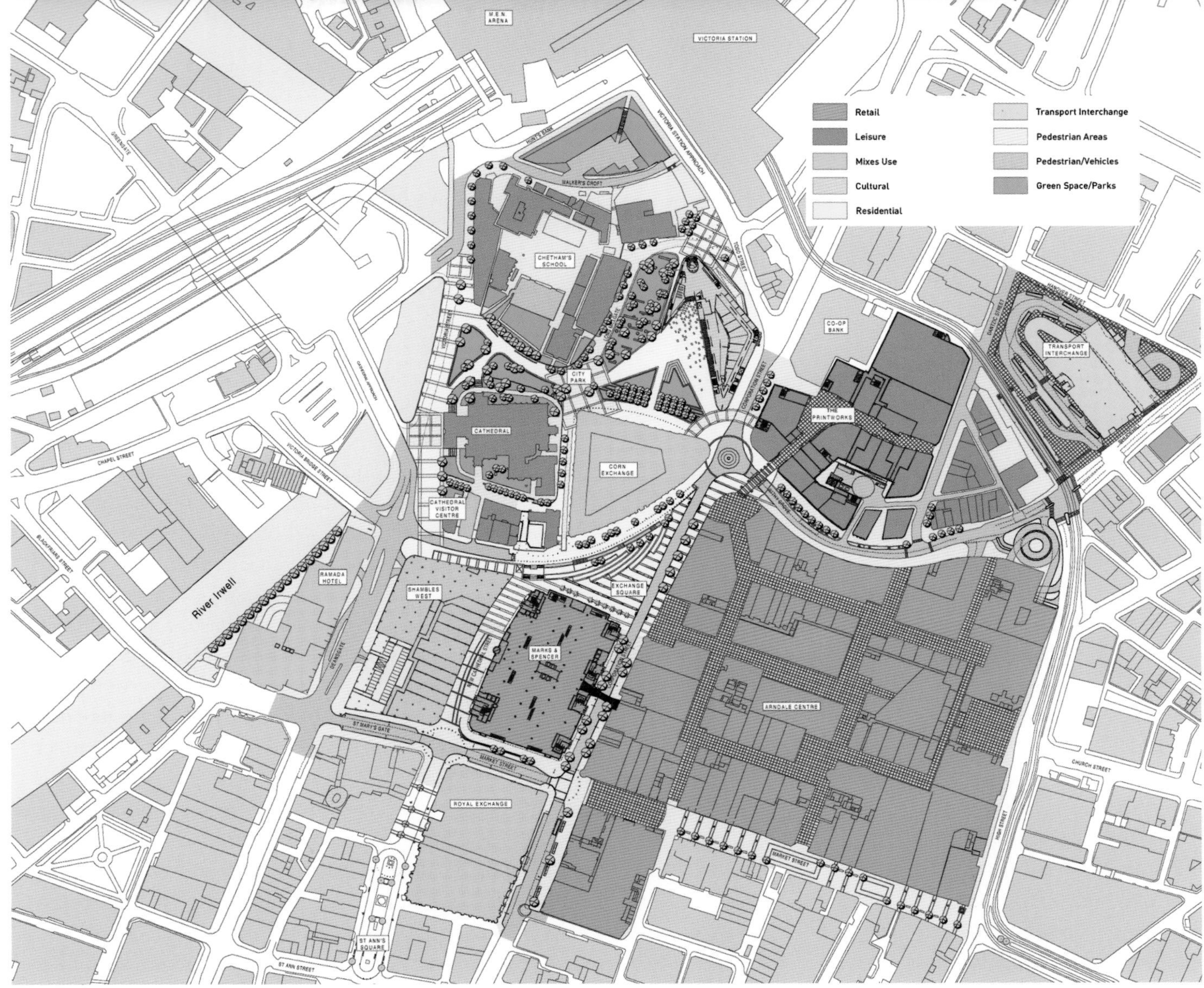

Manchester Millennium (1999) *Manchester city centre masterplan* [EDAW] Covers both the original bomb-damaged site with New Cathedral Street and Exchange Square, and the later development of Urbis and Cathedral Gardens.

had started with the work of the Central Manchester Development Corporation a decade earlier. A third was the striking Urbis Building, which started life in 2002 as a rather uncertain celebration of urbanism with a somewhat incoherent mix of displays – although it was later converted to house the National Museum of Football, which has a rather clearer purpose and greater visitor appeal.

The whole reconstruction project was a major success. Not only did it ensure the continuation of some of the key retail outlets of the centre, not least the Arndale Centre, but it created a long-overdue revitalisation of the city centre, which included some of the new iconic buildings that have helped to transform the image of the city. It dramatically connected the medieval and Georgian nodes of the city, and it introduced new leisure and entertainment attractions which helped to boost the already growing flow of urban tourists into Manchester. Perhaps above all, it demonstrated once again how successful the city was in using its network of partners and that funders could rely on Manchester to deliver big projects on time and to budget.

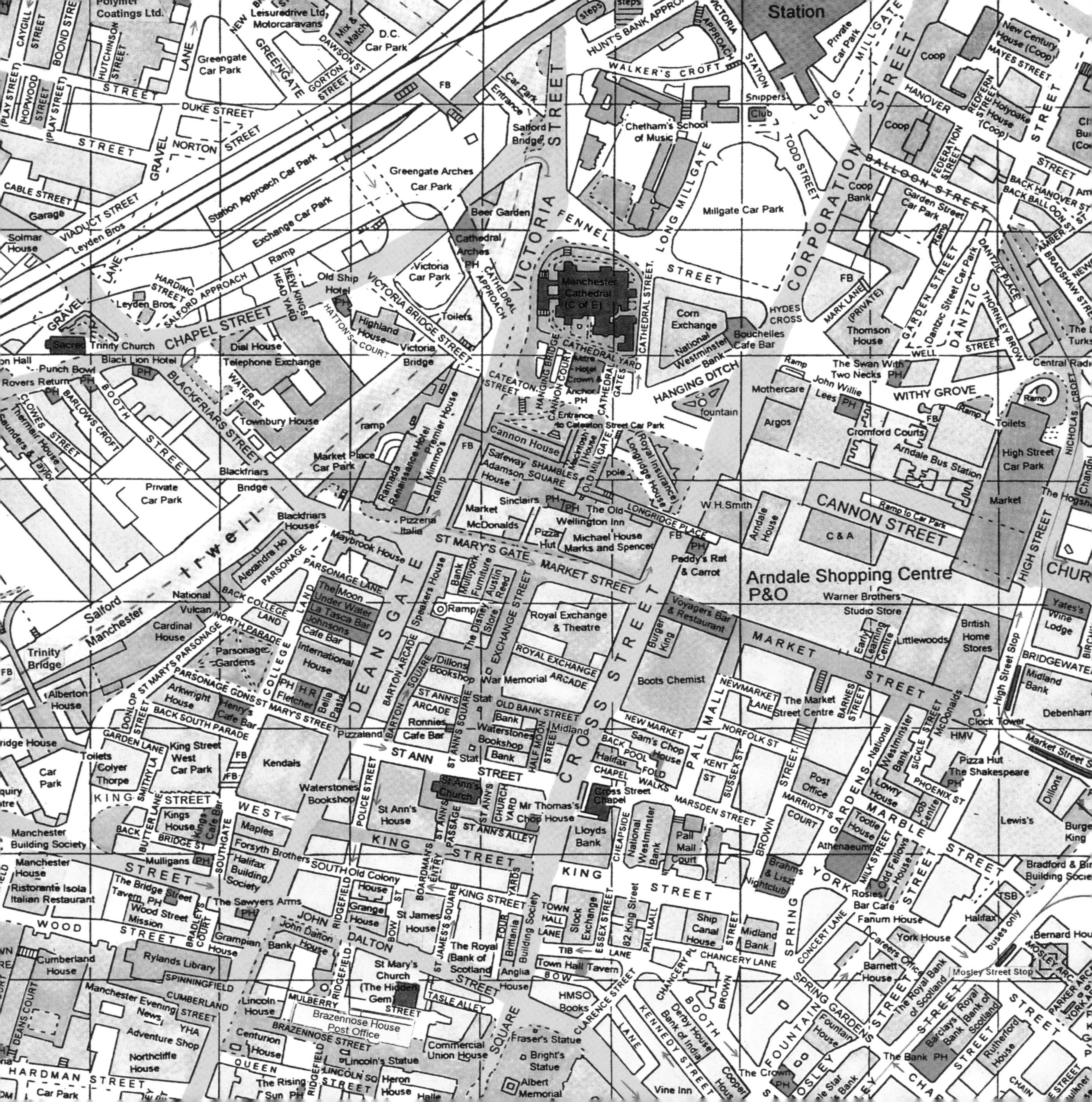

Station
Polymer Coatings Ltd.
Leisuredrive Ltd. Motorcaravans
CAYGILL STREET
BOOND STREET
HUTCHINSON STREET
NEW BRIDGE
Mix & Match
D.C. Car Park
Greengate Car Park
GREENGATE
DAWSON ST
GORTON STREET
HUNT'S BANK APPROACH
VICTORIA STATION APPROACH
WALKER'S CROFT
Private Car Park
MILLGATE
New Century House (Coop)
MAYES STREET
Coop
HANOVER
REDFERN STREET
Holyoake House (Coop)
FB
Car Park
Entrance
Salford Bridge
Snippers
Club
Chetham's School of Music
TODD STREET
LONG STREET
(PLAY STREET)
HOPWOOD
DUKE STREET
NORTON STREET
GRAVEL LANE
Greengate Arches Car Park
Station Approach Car Park
CABLE STREET
Garage
VIADUCT STREET
Leyden Bros
Exchange Car Park
Solmar House
Ramp
Beer Garden
Cathedral Arches PH
VICTORIA STREET
FENNEL STREET
LONG MILLGATE
Millgate Car Park
CORPORATION STREET
BALLOON STREET
FEDERATION STREET
Coop Bank
Garden Street Car Park
BACK HANOVER ST
BACK BALLOON ST
AMBER
BRADSHAW STREET
Victoria Car Park
CATHEDRAL APPROACH
HARDING STREET
SALFORD APPROACH
NEW KINGS HEAD YARD
Old Ship Hotel PH
HATTON'S COURT
VICTORIA BRIDGE STREET
Manchester Cathedral (C of E)
CATHEDRAL STREET
Corn Exchange
HYDES CROSS
FB
MARK LANE (PRIVATE)
GARDEN STREET
Dantzic Street Car Park
DANTZIC STREET
THORNLEY BROW
DANTZIC PLACE
Leyden Bros
CHAPEL STREET
GRAVEL LANE
Sacred Trinity Church
Highland House
Toilets
Victoria Bridge
National Westminster Bank
Bouchelles Cafe Bar
Thomson House
WELL
Dial House
Telephone Exchange
CATHEDRAL YARD
Mitre Hotel
Crown & Anchor PH
HANGING BRIDGE
CANNON COURT
CATHEDRAL GATES
HANGING DITCH
fountain
Ramp
The Swan With Two Necks PH
Central Rail
Punch Bowl
Black Lion Hotel
Rovers Return PH
BOOTH STREET
BLACKFRIARS STREET
WATER ST
CATEATON STREET
Entrance to Cateaton Street Car Park
Mothercare
John Willie Lees PH
WITHY GROVE
Ramp
Toilets
NICHOLAS CROFT
CLOWES STREET
Thermal House
Saunders & Taylor
BARLOWS CROFT
Townbury House
ramp
Premier House
Renaissance Hotel
Ramada
Mimmo's
Cannon House
Mcintosh House
OLD MILLGATE
Royal Insurance
Longridge House
Argos
Cromford Courts
Arndale Bus Station
High Street Car Park
Market Place Car Park
Blackfriars Bridge
FB
Safeway
Adamson House
SHAMBLES SQUARE
pole
Private Car Park
Ramp
Market
McDonalds
Sinclairs PH
The Old Wellington Inn
LONGRIDGE PLACE
W.H.Smith
Arndale House
CANNON STREET
Ramp to Car Park
C & A
Market
The Hogshead
HIGH STREET
CHURCH
trwell
Blackfriars House
Pizzeria Italia
Pizza Hut
Michael House
Marks and Spencer
FB
PH
Paddy's Rat & Carrot
Maybrook House
ST MARY'S GATE
MARKET STREET
Arndale Shopping Centre
P&O
Alexandra Ho
PARSONAGE
PARSONAGE LANE
Speakers House
Bank
Mulberry
Furniture
Austin Reed
EXCHANGE STREET
Warner Brothers Studio Store
Yates's Wine Lodge
Salford
Manchester
National Vulcan
BACK COLLEGE LAND
The Moon Under Water
La Tasca Bar
Johnsons Cafe Bar
DEANSGATE
Ramp
The Disney Store
Royal Exchange & Theatre
Voyagers Bar & Restaurant
Burger King
CROSS STREET
Early Learning Centre
Littlewoods
British Home Stores
Trinity Bridge
Cardinal House
NORTH PARADE
Parsonage Gardens
COLLEGE LAND
International House
BARTON ARCADE
BARTON SQUARE
Dillons Bookshop
War Memorial
ROYAL EXCHANGE ARCADE
Boots Chemist
MARKET STREET
High Street Stop
BRIDGEWATER
Midland Bank
Debenhams
FB
Alberton House
ST MARY'S PARSONAGE
Arkwright House
PARSONAGE GDNS
Henry's Cafe Bar
PH
H.R. Fletcher
Bella Pasta
ST ANN'S ARCADE
Stat
OLD BANK STREET
NEWMARKET LANE
The Market Street Centre
BARNES STREET
NEW MARKET
PALL MALL
NORFOLK ST
McDonalds
SICKLE STREET
Clock Tower
Market Street Stop
DONLOP STREET
BACK SOUTH PARADE
ST MARY'S STREET
Pizzaland
Ronnies Cafe Bar
Bank
Waterstones Bookshop
Midland
HALF MOON STREET
Sam's Chop House
BACK POOL FOLD
Westminster Bank
National
HMV
Bridge House
GARDEN LANE
Toilets
Colyer Thorpe
King Street West Car Park
FB
Kendals
ST ANN STREET
ST ANN'S SQUARE
Stat
Bank
Halifax
CHAPEL WALKS
KENT ST
SUSSEX ST
Post Office
Lowry House
Pizza Hut
The Shakespeare PH
PHOENIX ST
Dillons
Car Park
SMITHY LA
KING STREET
BUTLER LANE
WEST
Waterstones Bookshop
POLICE STREET
St Ann's Church
St Ann's House
CHURCH YARD
Mr Thomas's Chop House
Cross Street Chapel
MARSDEN STREET
MARRIOTTS COURT
Tootle House
Job Centre
Lewis's
Burger King
Enquiry Centre
Manchester Building Society
BACK BRIDGE ST
Kings House
Kings Cafe Bar
SOUTHGATE
Maples
KING STREET
ST ANN'S PASSAGE
ST ANN'S ALLEY
Lloyds Bank
CHEAPSIDE
National Westminster Bank
Pall Mall Court
BROWN STREET
Athenaeum
GARDENS
MILK STREET
Odd Fellows House
MARBLE STREET
Mulligans PH
Forsyth Brothers
Halifax Building Society
SOUTH
Old Colony House
BOARDMAN'S ENTRY
KING STREET
Brahms & Liszt Nightclub
YORK STREET
Rosies Bar Cafe
Bradford & Bingley Building Society
Manchester House
Ristorante Isola Italian Restaurant
The Bridge Street Tavern PH
Wood Street Mission
The Sawyers Arms PH
BRADLEY'S COURT
Grange House
BOW ST
St James House
KING STREET
TOWN HALL LANE
Stock Exchange
ESSEX STREET
82 King Street
PALL MALL
Ship Canal House
Midland Bank
SPRING GARDENS
CONCERT LANE
Fanum House
Halifax
TSB
WOOD STREET
Grampian House
JOHN DALTON STREET
John Dalton House
Bank
RIDGEFIELD
ST JAMES'S SQUARE
The Royal Bank of Scotland
FOUR YARDS
Brittania
Building Society
Anglia House
TIB LANE
CHANCERY PL
CHANCERY LANE
York House
Careers Office
buses only
Bernard House
MOSLEY ARC.
Cumberland House
Rylands Library
SPINNINGFIELD
CUMBERLAND STREET
St Mary's Church (The Hidden Gem)
Town Hall Tavern
BOW LANE
Barnett House
The Royal Bank of Scotland
Mosley Street Stop
PARKER ARC
YORK ST
Manchester Evening News
YHA Adventure Shop
Northcliffe House
DEANS COURT
Lincoln House
MULBERRY STREET
TASLE ALLEY
Brazennose House Post Office
BRAZENNOSE STREET
HMSO Books
CLARENCE STREET
KENNEDY STREET
Derby House
Bank of India
BOOTH STREET
FOUNTAIN STREET
Fountain House
SPRING GARDENS
Barclays Bank
Royal Bank of Scotland
Rutherford House
Centurion House
Commercial Union House
SQUARE
Fraser's Statue
Bright's Statue
Lincoln's Statue
LINCOLN SQ
The Bank PH
The Crown PH
MOSLEY
Cooper House
CHAIN STREET
HARDMAN STREET
QUEEN STREET
The Rising Sun PH
Heron House
Albert Memorial
Vine Inn
Car Park

1996b

One man maps the city centre

Untrained in cartography but with a long-term interest in maps, Andrew Taylor single-handedly created one of the most useful recent street directory maps of Manchester's city centre. He had very limited funds but a good deal of ingenuity and lots of enthusiasm for the hours pounding the pavements on fieldwork. In some respects, his freelance mapping is analogous to the work of Phyllis Pearsall in the mid 1930s when she created her original A–Z street atlas for London.

Taylor had to give up geography in his second year at secondary school due to timetable limitations, but he continued to pursue an interest in cartography. As he said:

> . . . lunchtime at school gave me the opportunity to lose myself in the five-volume *Times Atlas of the World Mid-Century Edition* in the library. I used to spend school holidays hand-drawing extracts from this and other atlases for my own enjoyment. In 1967, *The Times* brought out a one-volume edition of the *World Atlas* and, by the following year, I had saved enough pocket money to buy a copy.

He moved to Manchester in 1986 and was surprised that there seemed to be no detailed city-centre map. The popular A–Z street atlases covered the whole of the city, but this meant that the map of the central district was typically cramped on a single page. In response, he decided to create his own map of central Manchester.

Starting with the basic street layout and building footprints from large-scale Ordnance Survey plans, Taylor supplemented this raw data with detailed contextual information gathered from other sources, including planning applications and architects' drawings. He also walked the streets himself, noting road restrictions, the names of office buildings, shops and pubs, and recent changes in land use. All of this data was mapped by

---

OPPOSITE. A. Taylor, *Manchester City Centre* (1996) [Andrew Taylor]

hand, and he experimented with various methods of labelling buildings and streets using Letraset and lettering stencils. Eventually, he typed every text label onto white paper in suitable font sizes and pasted them onto the base map using glue and forceps. He also tried different methods of shading building types and land uses, including watercolour, coloured pencils, fibre-tip pens and wax crayons. He eventually settled on using Royal Sovereign 'Magic Markers'.

After many months of laborious effort, his first map of the city centre was published in July 1996. It showed the obviously useful detail one would expect on a street directory map, but it also '[n]ames almost 200 office blocks. Marks pedestrian precincts. Marks tower blocks, footpaths and canal walkways.' The comprehensiveness, convenient size and cost made Taylor's map a step above what was otherwise publicly available and comparable in many respects to the best street directory maps produced for Manchester in the early nineteenth century by Pigot and Slater. The accuracy and legibility in naming all the side streets and small alleyways and local landmark details like statues, fountains and monuments, along with the office block names, made it particularly valuable for visitors to navigate through an unfamiliar city. It was perhaps rather less useful for public transport users (for example, the tram stops are not very obvious), and while it provided much detail on road restriction for drivers, car park locations were not prominently mapped.

Frustratingly, Taylor's effort to produce the 1996 map were rendered partly obsolete, almost before they had reached the shops, by the massive IRA bomb that was detonated in June of that year. The subsequent re-planning of the area introduced considerable changes to street layouts and land uses. In response, Taylor quickly revised his map and a second edition was on sale to the public in late spring 1998.

Taylor's cartographic technique evolved through trial and error, and through the multiple print editions his maps developed an increasingly professional look. The underlying high quality of the survey remained unchanged. Later updates also covered a slightly larger area. The eight different editions of his map found a market with tourists and professions alike, filling a niche not well served in the late 1900s and early 2000s by the Ordnance Survey or commercial rivals. Taylor also branched out to produce similarly detailed and useful street directory maps for Liverpool, Preston and York city centres.

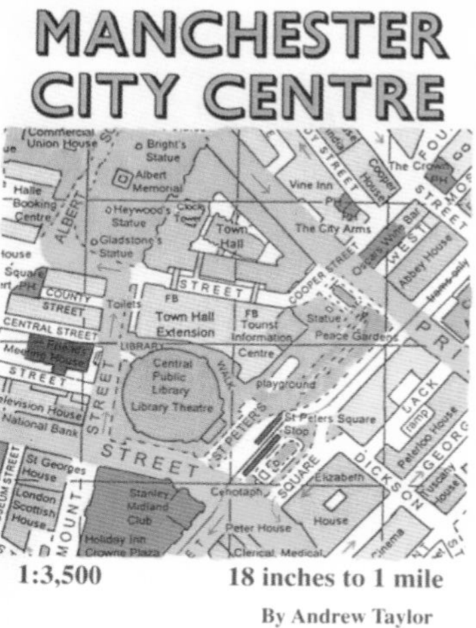

ABOVE. Taylor, Map cover.

OPPOSITE.
Marketing Manchester, *Manchester* (2006) [Marketing Manchester]

Taylor's last map of Manchester was published in 2013, but by this time the market for print street maps was shrinking significantly along with people's willingness to pay for high-quality cartography. Given the plethora of free maps, including those produced by Manchester's tourism promotion company, it was increasing difficult for Taylor to get volume sales. He made a start on compiling a ninth edition of his map, aiming to extend the coverage east and west by several hundred metres to encompass recently development at New Islington and around Salford Cathedral. However, this was never completed because so many people had shifted to using free Google street maps, which display buildings in great detail and pinpoint directions from their smartphones.

The first edition map from 1996 has numerous aspects of good cartographic design as well as a distinctive aesthetic, although Taylor himself is quite self-critical: 'I always feel it was the scruffiest edition.' Yet despite this, Taylor's map was a singular achievement, not least since he was working full time as a biomedical scientist in a hospital laboratory in Manchester. He demonstrated that the motivated amateur still had a place in map-making. Moreover, the eight maps of Manchester city centre that he created at a time of considerable change to the fabric of the city is a sequence that is now a useful historical source. And the first edition Manchester City Centre map from 1996 has become something of a collector's item.

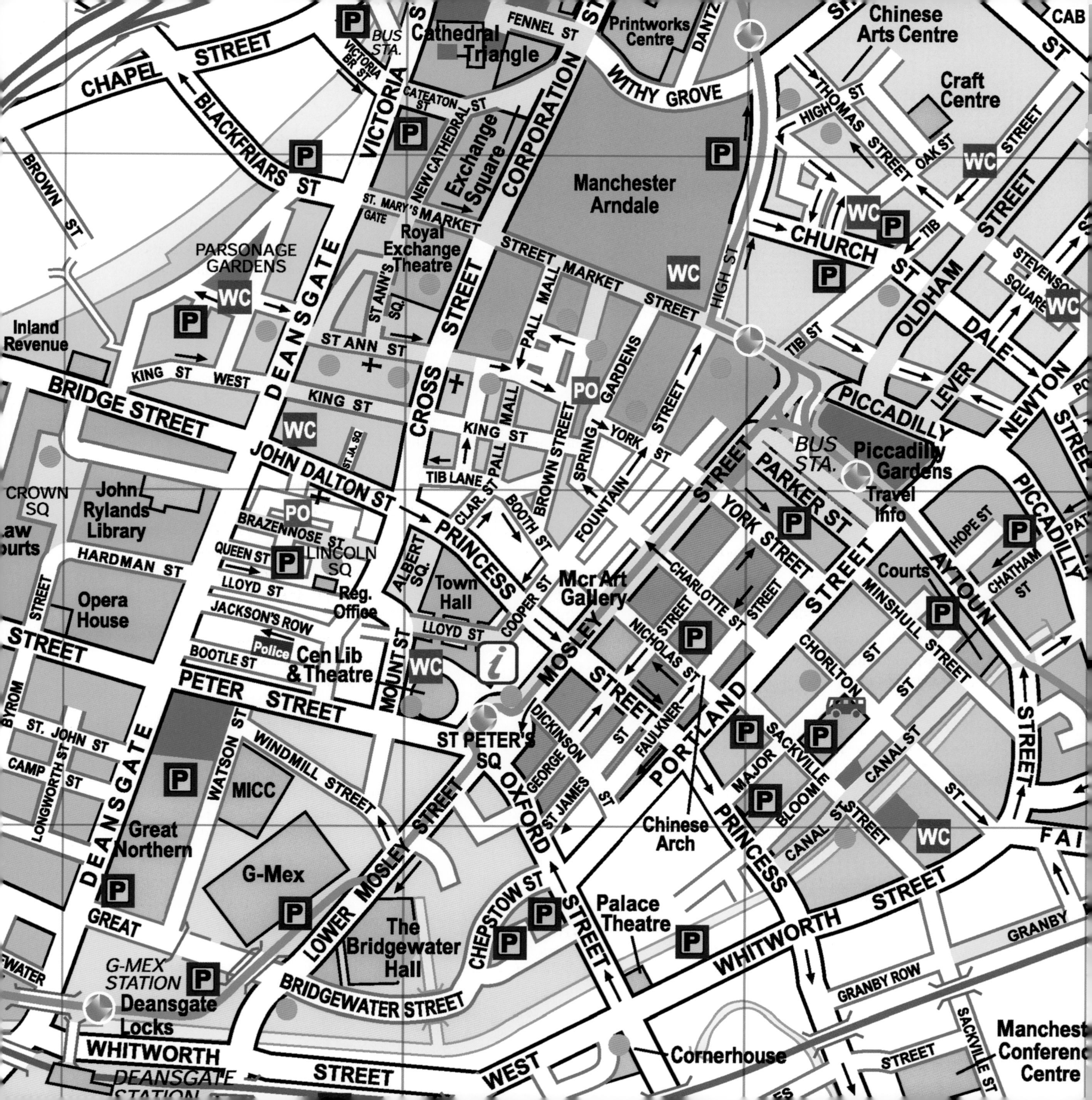

Chinese Arts Centre
Printworks Centre
Cathedral Triangle
FENNEL ST
BUS STA.
VICTORIA BR ST
CHAPEL STREET
BLACKFRIARS ST
BROWN ST
VICTORIA
CATEATON ST
NEW CATHEDRAL ST
Exchange Square
CORPORATION
WITHY GROVE
Craft Centre
THOMAS ST
HIGH ST
MASON STREET
OAK ST
Manchester Arndale
ST. MARY'S GATE
MARKET ST
Royal Exchange Theatre
PARSONAGE GARDENS
DEANSGATE
ST ANN'S SQ.
ST ANN ST
CROSS STREET
MARKET STREET
CHURCH ST
TIB ST
OLDHAM ST
STEVENSON SQUARE
DALE ST
NEWTON ST
Inland Revenue
KING ST WEST
BRIDGE STREET
KING ST
PALL MALL
PO
GARDENS
SPRING GARDENS
YORK ST
PICCADILLY
LEVER ST
BUS STA.
Piccadilly Gardens
Travel Info
CROWN SQ
John Rylands Library
Law Courts
JOHN DALTON ST
ST JA. SQ
TIB LANE
CLAR. ST
BOOTH ST
BROWN STREET
FOUNTAIN ST
MOSLEY STREET
YORK STREET
PARKER ST
HOPE ST
PICCADILLY
BRAZENNOSE ST
QUEEN ST
LINCOLN SQ
HARDMAN ST
LLOYD ST
PRINCESS ST
ALBERT SQ.
Town Hall
Mcr Art Gallery
CHARLOTTE ST
COOPER ST
Courts
AYTOUN ST
CHATHAM ST
Opera House
JACKSON'S ROW
Reg. Office
LLOYD ST
NICHOLAS ST
MINSHULL STREET
CHORLTON ST
STREET
Police
Cen Lib & Theatre
BOOTLE ST
MOUNT ST
PETER STREET
DICKINSON ST
FAULKNER ST
PORTLAND ST
SACKVILLE ST
CANAL ST
BYROM ST
ST. JOHN ST
CAMP ST
LONGWORTH ST
WATSON ST
WINDMILL STREET
MICC
ST PETER'S SQ
OXFORD
GEORGE ST
ST JAMES ST
MAJOR ST
BLOOM ST
Chinese Arch
Great Northern
DEANSGATE
G-Mex
LOWER MOSLEY STREET
The Bridgewater Hall
CHEPSTOW ST
OXFORD STREET
Palace Theatre
PRINCESS ST
WHITWORTH STREET
GRANBY ROW
GRANBY ROW
GREAT
G-MEX STATION
Deansgate Locks
BRIDGEWATER STREET
WHITWORTH STREET
DEANSGATE STATION
WEST
Cornerhouse
STREET
SACKVILLE ST
Manchester Conference Centre

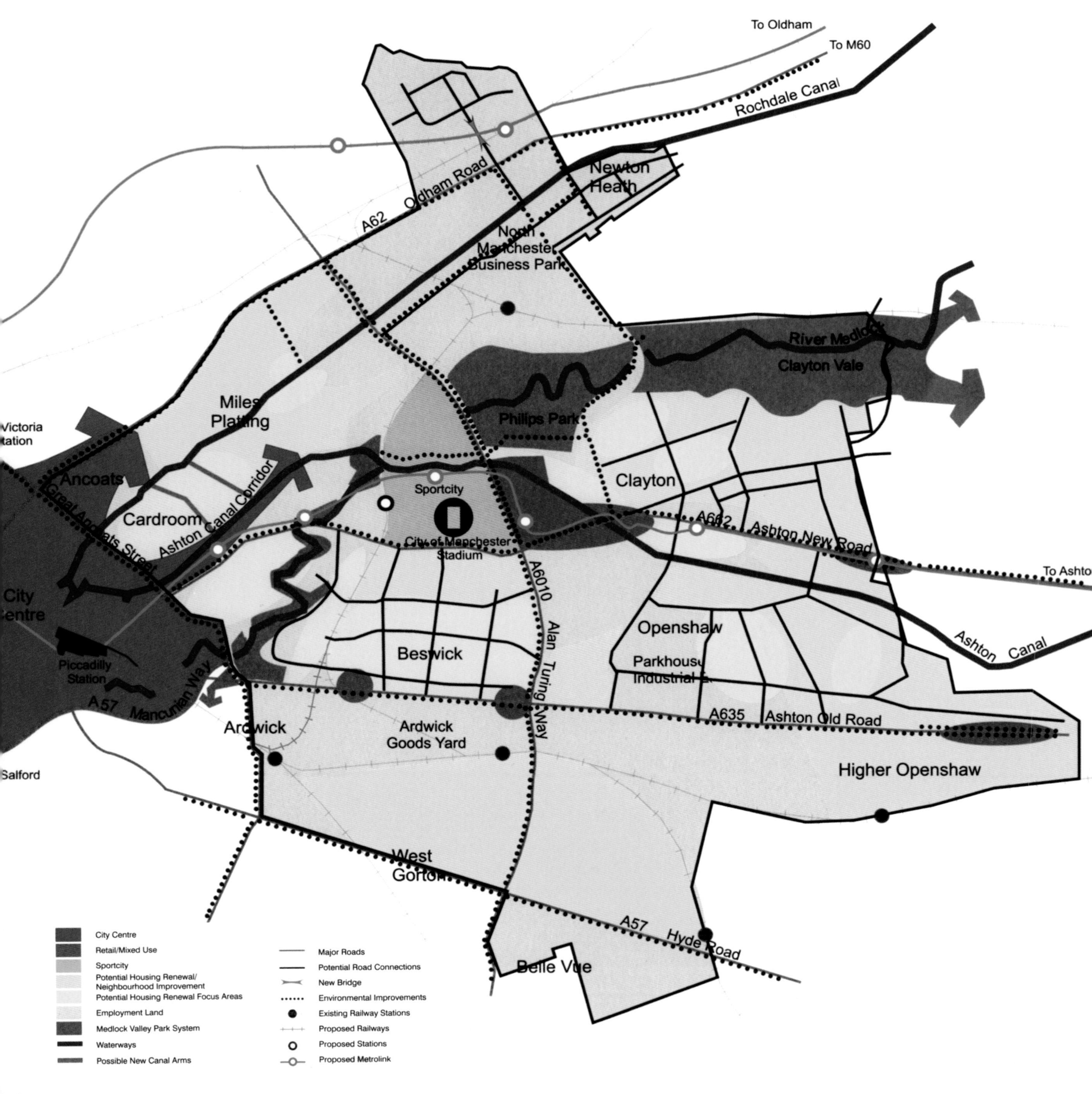
To Oldham
To M60
Rochdale Canal
Oldham Road
A62
Newton Heath
North Manchester Business Park
River Medlock
Clayton Vale
Philips Park
Miles Platting
Victoria Station
Ancoats
Great Ancoats Street
Cardroom
Ashton Canal Corridor
Sportcity
City of Manchester Stadium
Clayton
A662
Ashton New Road
To Ashton
A6010
Alan Turing Way
City Centre
Openshaw
Ashton Canal
Beswick
Parkhouse Industrial E
Piccadilly Station
A57
Mancunian Way
Ardwick
Ardwick Goods Yard
A635
Ashton Old Road
Higher Openshaw
Salford
West Gorton
A57
Hyde Road
Belle Vue
City Centre
Retail/Mixed Use
Sportcity
Potential Housing Renewal/ Neighbourhood Improvement
Potential Housing Renewal Focus Areas
Employment Land
Medlock Valley Park System
Waterways
Possible New Canal Arms
Major Roads
Potential Road Connections
New Bridge
Environmental Improvements
Existing Railway Stations
Proposed Railways
Proposed Stations
Proposed Metrolink

2001

Regenerating east Manchester

The regeneration of east Manchester is a powerful example of the effectiveness of the 'Manchester model' of public–private collaboration in which the Council plays an ostensibly hands-off role in order to generate developer confidence. It is also an example of the city's strategy of focusing regeneration on large areas, with big projects that can create momentum and high profile.

East Manchester has been called 'the epicentre of the world's first industrial revolution' – with some justification since it includes the legacy of the textile mills of Ancoats and Miles Platting, and the sites of heavy engineering and chemical industries across its wider area. In the decade from 1975, the area lost 60 per cent of its employment. Population fell from 164,000 in 1951 to 62,000 by 2001. Large swathes were left derelict and the housing market virtually collapsed, with falling values and vacancy rates as high as 20 per cent. The Council progressively targeted more and more government regeneration funding streams at the area – Round 4 of the Single Regeneration Budget (SRB) and various zonal programmes (Sure Start, Health Action, Sport Action, Education Action) – but its boldest and most significant move was to bid simultaneously in 1998 for two programmes, New Deal for Communities (NDC) and Round 5 of SRB, both of which were targeted at Beswick and Openshaw. It won both and handled the funds from the two as a single pot of resources totalling £76 million, which was used for 'social' projects with a high degree of involvement from the local community.

An Urban Regeneration Company, New East Manchester (NEM), was then established with overall responsibility for strategy across the huge area of well over 1,000 hectares. It was a partnership between the City Council, the Northwest Regional Development Agency (NWDA) and the Homes and Communities Agency. NEM was the third example of the 'Manchester model': Alan Cockshaw of AMEC was its initial

---

OPPOSITE. New East Manchester, *Masterplan* (2001) [AUTH]

Chair; Marianne Neville-Rolfe, formerly Director of the Government Office for the North West, was its first Chief Executive; Richard Leese was Deputy Chair; and Howard Bernstein was Secretary.

The scale and ambition of the regeneration helped to create a momentum and political imperative that attracted private investment and a series of further public resources. For example, new housing areas in Beswick and Openshaw were accelerated by the Housing Market Renewal programme, which helped to bring new population into the area and re-created a functioning housing market.

Most of the regeneration began to take shape before the 2007–8 financial crisis caused an inevitable hiccup. Severe reductions in public-sector funding and the closure of the NWDA brought the closure of NEM, with responsibility for redevelopment passing to the City Council.

Four examples illustrate the transformation of the area. First, was North Manchester Business Park (later called Central Business Park), which attracted investment from hi-tech firms such as Fujitsu and involvement in IT from the local universities and FE college. Second, the core of the initial regeneration focused on 'Sportcity' creating a new centre at the heart of the area. The cycling velodrome had existed since 1994 as the highly successful home of British cycling. To it was added the stadium, built for the 2002 Commonwealth Games, which subsequently became the Etihad, home of Manchester City FC whose owners have contributed significantly to the regeneration of the wider area. The surrounding site was developed with the largest concentration of sporting venues in Europe, including the National Squash Centre, the Manchester Regional Arena, the English Institute of Sport, the regional Tennis Centre and a state-of-the-art gymnasium. Opposite the site is the Asda Eastlands Super Centre and a complex of new apartment buildings, which houses largely young professional households.

The third and fourth examples are even more dramatic. Ancoats Urban Village comprises the core of the city's historic textile manufacturing area, which was left with numerous decaying and abandoned sites. Its transformation started with a Townscape Heritage Initiative, which attracted private and public investment. The most dramatic of its historic buildings is the series of huge mills fronting the Rochdale Canal, which English Heritage calls 'The finest complex of Georgian mills in Europe'. Murrays' Mill was originally the largest factory in the world. Others had been built alongside it, including McConnel & Kennedy's Mill. Lying between Jersey Street and Redhill Street, there are now eight listed mills. Urban Splash, the innovative local firm founded by Tom Bloxham and Jonathan Falkingham, refurbished some of the mill spaces for apartments and renovated much of the historic complex. Four further listed buildings shown on the plan have also been renovated. The Iceplant, built in 1850 on Blossom Street, was the first commercial ice-making plant in northern England. Its ice was used both by stallholders in the Smithfield Market and by numerous local Italian ice cream makers who located in the building. Nearby is St Peter's Church, built in Italianate style in 1859–60 with a striking interior with slender cast-iron pillars and huge windows which flood the interior with light. Now refurbished, it is used as a rehearsal venue by the Hallé Orchestra. To the north of the village is Victoria Square built in 1889 as the city's first municipal housing and adjacent to it is the refurbished and delicately renamed Anita Street (originally called Sanitary Street). Finally, the Daily Express Building, opened in 1939 with its striking futurist art-deco design with curved corners and cladding of white glazed windows and black vitriolite panels, is now converted into offices.

Finally, south of the Rochdale Canal is the area which can claim to be the world's first industrial suburb. It was originally a dense mix of small terraced houses, small workshops, industrial buildings and pubs. In the 1970s, a clearance programme led to their replacement by local-authority houses on the Cardroom Estate, which rapidly became a notorious 'sink estate' with high rates of crime and antisocial behaviour. It experienced a familiar self-reinforcing downward trajectory as people left the area, and shops, pubs and the local primary school closed as the population fell. By 2000, only half of the

OPPOSITE. Sportcity, showing both the stadium and the velodrome.

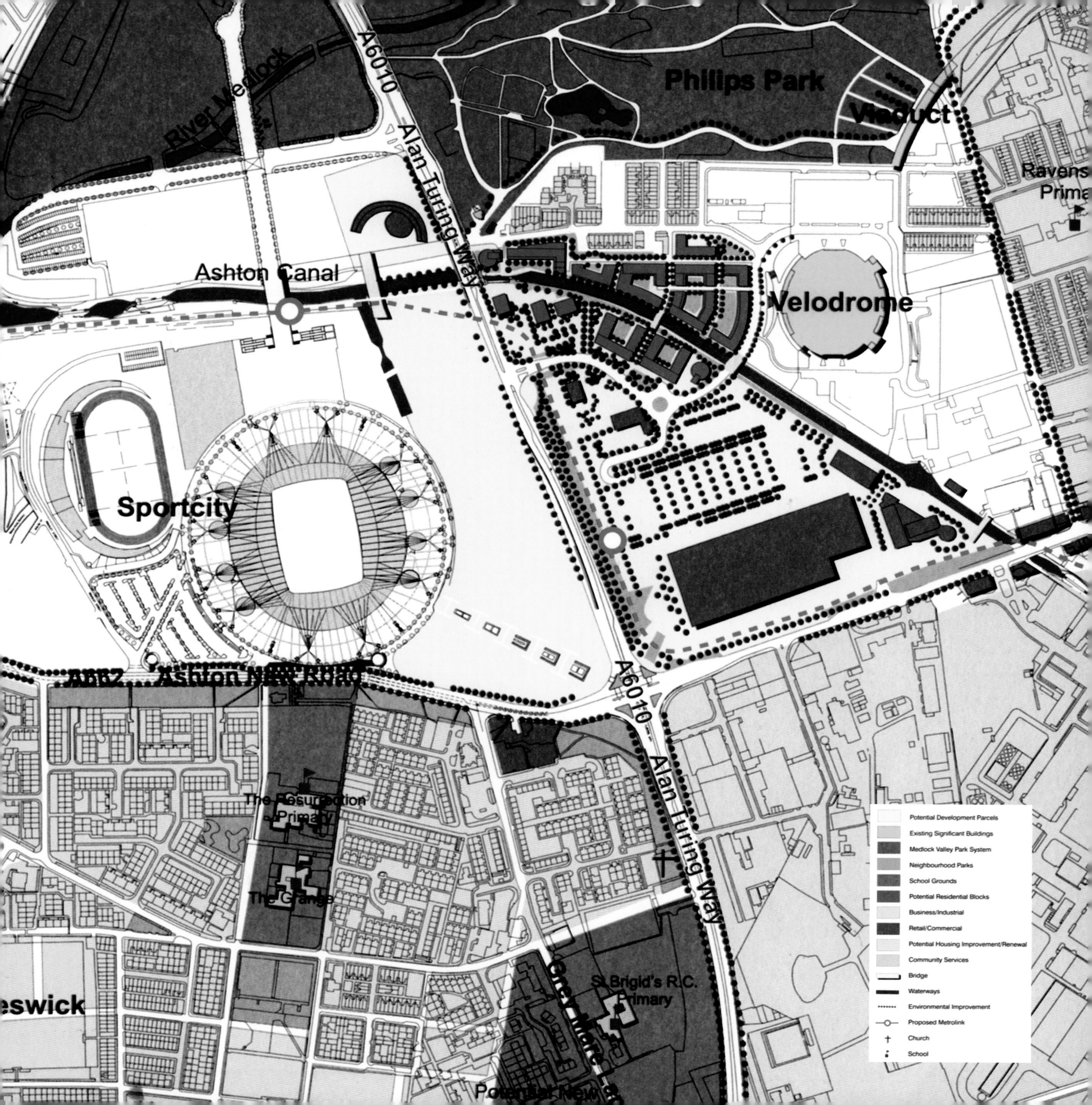

Philips Park
Viaduct
River Medlock
A6010
Alan Turing Way
Ashton Canal
Velodrome
Sportcity
Ashton New Road
A6010
Alan Turing Way
The Resurrection Primary
The Grange
St Brigid's R.C. Primary
Potential Development Parcels
Existing Significant Buildings
Medlock Valley Park System
Neighbourhood Parks
School Grounds
Potential Residential Blocks
Business/Industrial
Retail/Commercial
Potential Housing Improvement/Renewal
Community Services
Bridge
Waterways
Environmental Improvement
Proposed Metrolink
Church
School

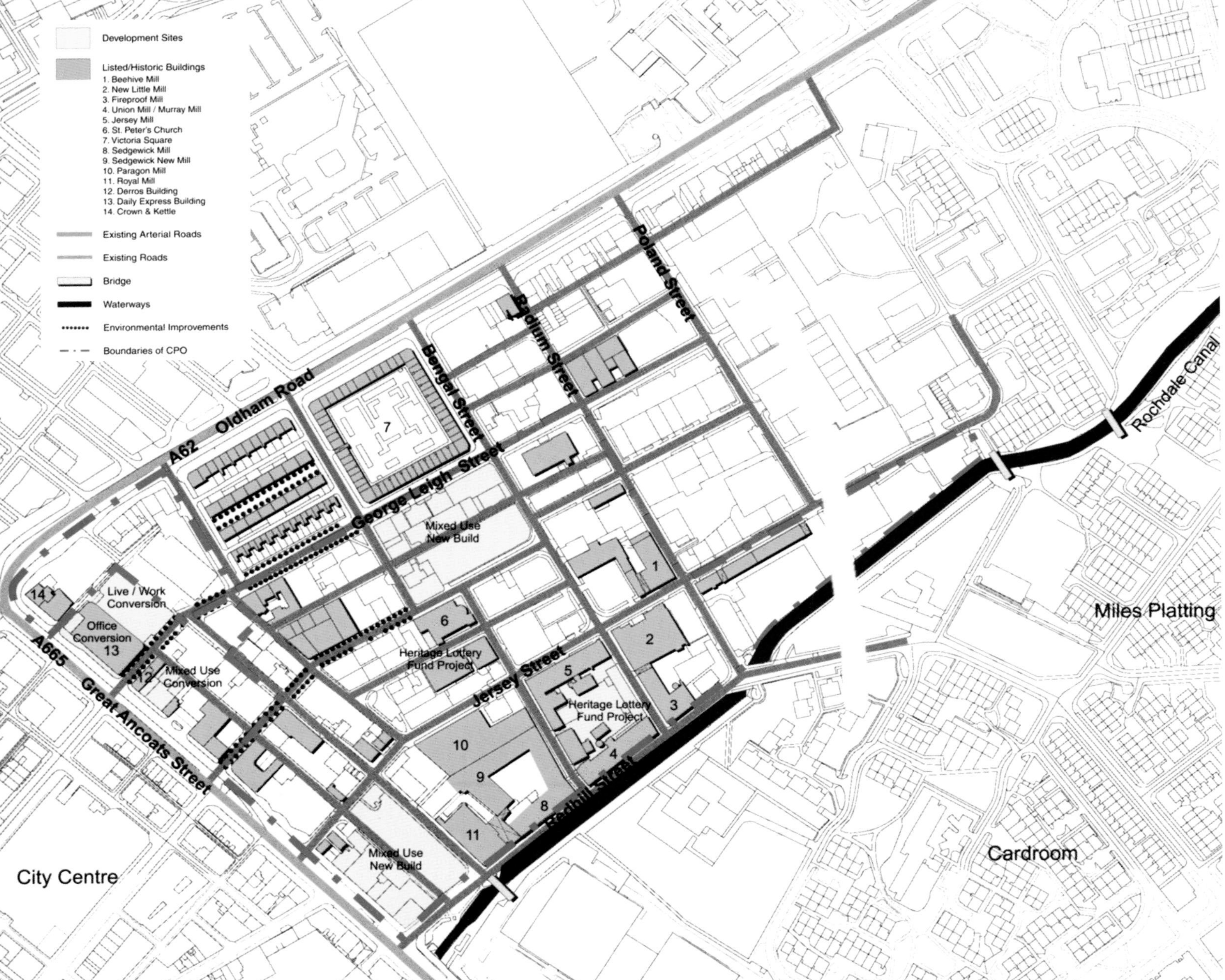

No fewer than nine of the listed buildings in Ancoats Urban Village are mills. St Peter's Church (6) sits in the centre. Victoria Buildings (7) was the city's first major municipal housing development.

original 200 houses were still occupied. The redevelopment of the district was led by Urban Splash. Renamed New Islington, drawing on the neighbourhood's original name, it is one of the country's Millennium Community Programme areas. The plan for the area capitalised on the water frontages of the Rochdale and Ashton canals. Work started in 2003 and much of the initial effort focused on decontaminating land that had been severely polluted by industrial waste. Among the completed developments is a new waterway joining the two canals and providing a marina and an eco-park. The most celebrated (or at least most photographed) building is the Chips building, designed by Will Alsop and resembling a set of three fat chips

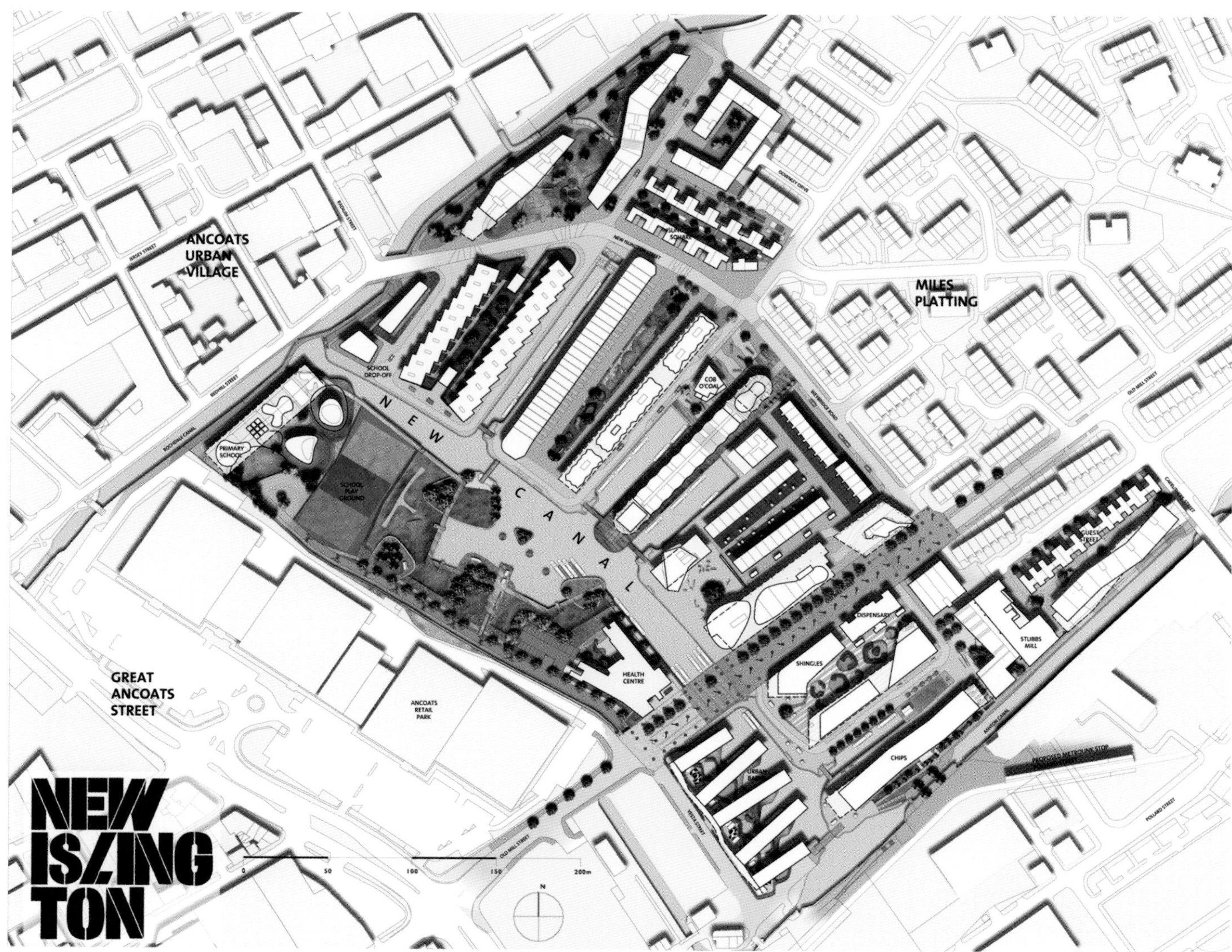

Urban Splash, *New Islington* (*c.*2005) [Urban Splash] The new waterway link creating a marina between the Rochdale and Ashton canals is prominent.

resting on a plate. Its apartments have attracted young professionals who work in the nearby city centre. Redevelopment also brought new facilities, a primary care centre and a free school supported by Manchester Grammar School. The Grade II listed Ancoats Dispensary, built in the 1870s and the last remaining building of the historic Ancoats Hospital that was originally founded in 1828, was saved with Heritage Lottery funds and a major private donation, in the hope of converting the crumbling shell into a community facility with a focus on mental health. James Kay, one of the hospital's founders, wrote his 1832 *Moral and Physical Condition of the Working Classes* based in part on his experiences in working there.

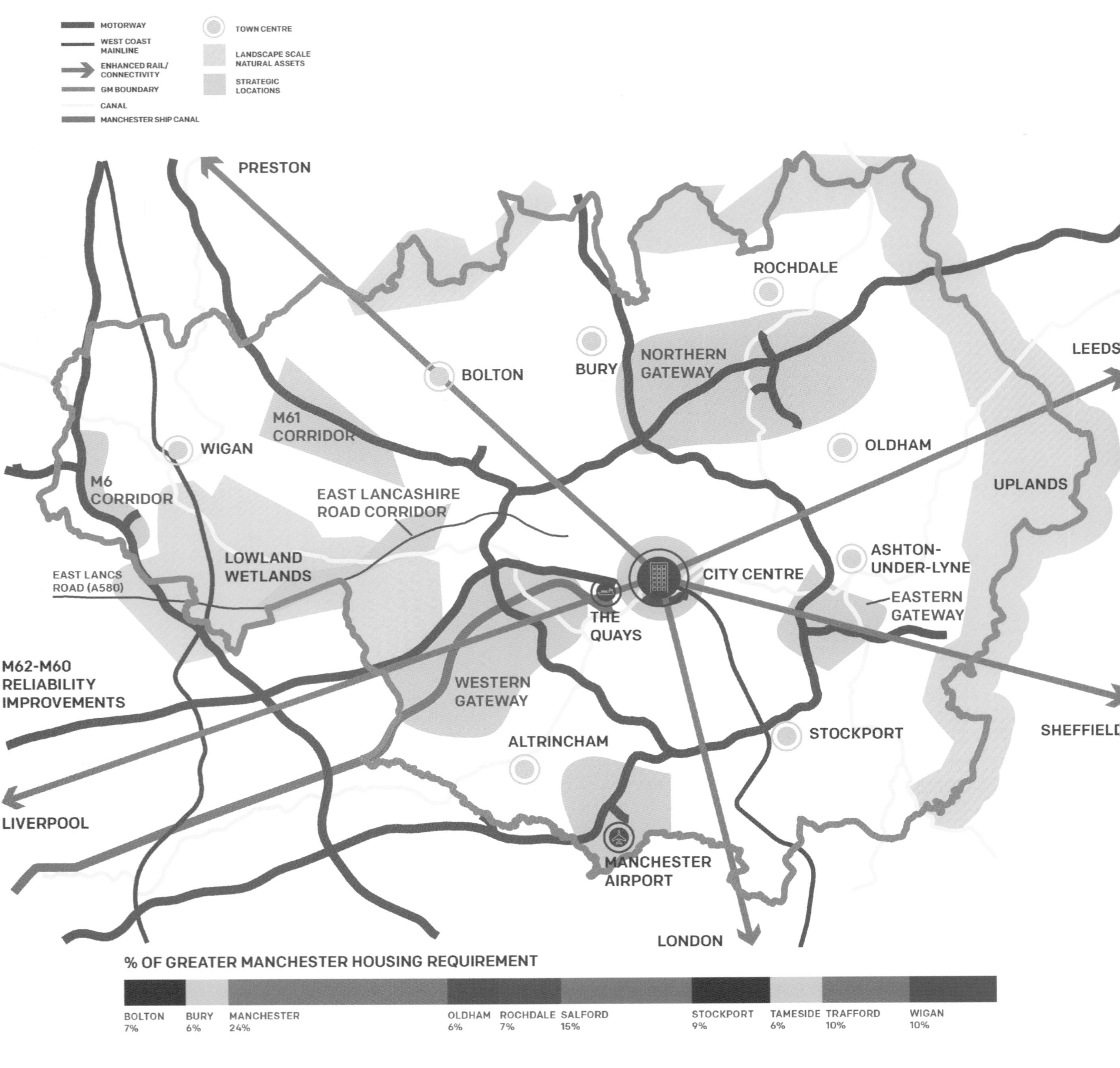
MOTORWAY
WEST COAST MAINLINE
ENHANCED RAIL/ CONNECTIVITY
GM BOUNDARY
CANAL
MANCHESTER SHIP CANAL
TOWN CENTRE
LANDSCAPE SCALE NATURAL ASSETS
STRATEGIC LOCATIONS
PRESTON
ROCHDALE
LEEDS
BOLTON
BURY
NORTHERN GATEWAY
OLDHAM
M61 CORRIDOR
WIGAN
M6 CORRIDOR
EAST LANCASHIRE ROAD CORRIDOR
UPLANDS
LOWLAND WETLANDS
EAST LANCS ROAD (A580)
CITY CENTRE
ASHTON-UNDER-LYNE
EASTERN GATEWAY
THE QUAYS
M62-M60 RELIABILITY IMPROVEMENTS
WESTERN GATEWAY
ALTRINCHAM
STOCKPORT
SHEFFIELD
LIVERPOOL
MANCHESTER AIRPORT
LONDON
% OF GREATER MANCHESTER HOUSING REQUIREMENT
BOLTON 7%
BURY 6%
MANCHESTER 24%
OLDHAM 6%
ROCHDALE 7%
SALFORD 15%
STOCKPORT 9%
TAMESIDE 6%
TRAFFORD 10%
WIGAN 10%

2016

Recognising the wider city region

The Spatial Framework for the conurbation was produced by the Greater Manchester Combined Authority, which is now the formal administrative body covering Manchester's city region. City regions have long been talked about, but have only recently come into being in Britain. Manchester led the way. The administrative relevance of functional city regions was recognised in the 1945 City of Manchester Plan, which ended, rather stirringly, with a plea for governance at a conurbation-wide scale: 'The physical development of this conurbation is technically a single problem: ideally it should be planned by a single overall authority.' These may well be the words of the planning expert Derek Senior, who did much of the writing of the Plan and was a passionate advocate of the concept of city regions. As a member of Redcliffe-Maud's 1960s' Royal Commission on Local Government in England, Senior produced a Memorandum of Dissent in which he argued for a two-tier system with 35 city regions and 148 other districts. This would certainly have made sense for Manchester where, within a radius of 10 miles of the city, there were then 85 local authorities. The subsequent history of transport planning for SELNEC (South East Lancashire and North East Cheshire), the significance of AGMA (Association of Greater Manchester Authorities) and the recent creation of the current Combined Authority, each reinforce the argument about the logic of city regions.

Indeed, the significance of the city's wider catchment area was evident even in the eighteenth century. It was underlain by the textile industry, which saw Manchester increasingly playing the role of coordinator of the business of textile firms in the outlying towns. The city's warehouses, the Port of Manchester and above all its Royal Exchange played key roles for all the scattered manufacturers in the wider region – for the import of raw cotton, the storage and display of the finished textile products, and the export of textiles.

Among British provincial cities, Manchester has proved the

OPPOSITE. Greater Manchester Combined Authority, *Key diagram* (2016) [Greater Manchester Combined Authority] most alert in realising the potential of its wider functional region. In part, this is a product of the fact that the conurbation is essentially monocentric, focused on Manchester itself, but also – ironically – it is a function of the high degree of 'balkanisation' of the area. As the central node of the conurbation, Manchester is far less dominant in terms of its population than any other of England's core cities. Greater Manchester is divided into no fewer than ten metropolitan districts. In 2016, the city of Manchester had only 19 per cent of the total population of its wider conurbation, whereas Sheffield had 42 per cent of its conurbation, Birmingham had 39 per cent, Liverpool and Leeds each had 34 per cent, and Newcastle had 26 per cent. The administrative fragmentation of the Manchester conurbation was, ironically, a spur to greater collaboration between the ten districts.

There have been numerous examples of clashes between some of the constituent districts of Greater Manchester: a major dispute between Trafford and Manchester over the development of the Trafford Shopping Centre; Manchester clearly felt its nose put out of joint when Salford was selected as the site of the BBC's move to the North West; Wigan and Manchester both expressed interest in an Olympic-sized swimming pool, and a senior Manchester politician dismissed Wigan as 'that muddy little hole'; and the tight administrative boundaries of Manchester have long been a thorn in the city's side, as was clear from its battles with Cheshire over the building of Wythenshawe.

Yet, set against this, there have been numerous examples of effective collaboration between the districts: the airport was long owned jointly by all the districts, bringing significant financial benefit to all ten (and now, with the creation of the Manchester Airports Group, Manchester has a 35 per cent stake and the other nine share 29 per cent); and Manchester opted to include Salford and Trafford – and later Tameside – in its response to the government's invitation to develop its City Pride document.

However, the most significant collaboration came in 1986 when the metropolitan regions, except for Greater London, were abolished. In response, the Greater Manchester districts chose to establish AGMA, which has played a key role in cementing the idea of the conurbation as a single collaborative entity. The Labour government declared Leeds and Manchester as the first two formal city regions, and subsequently, in response to the Coalition government's invitation to local authorities to create Local Economic Partnerships, the ten districts opted for their LEP boundary to be coterminous with Greater Manchester and hence to reinforce the coherence of the conurbation.

Manchester persuaded the government to recognise the ten as the first formal Combined Authority, which was created in 2011, and Andy Burnham was elected as Mayor in 2017. One of the early products of the Combined Authority was its Spatial Framework, which outlined a strategy for growth across the whole of the conurbation. The key diagram provides a visual summary of the strategy. It tries to strike a delicate balance between further strengthening the city centre – which incorporates Salford as well as Manchester and which it sees as the principal business location in the country outside London – with supporting the other main town centres to enhance their roles as local economic drivers. It sees Salford Quays/MediaCity and the airport making disproportionate contributions to economic growth: the Quays continuing to develop its potential as a unique economic, tourist and residential location; and the airport growing as a key driver of growth with more connections to the emerging markets of the Far East.

Investment in transport is seen as one of the keys to the conurbation's growth, not least in the context of the ambitions of the Northern Powerhouse, but also recognising the need to improve connectivity between outlying centres and the employment prospects of the city centre. It identifies three major areas of potential growth in 'gateways', which could deliver new space both for employment and for housing alongside major improvements in transport. It also controversially proposes the release of sizeable areas of green belt to help in building 227,000 additional houses. One key feature of the plan is its stress on providing a 'green infrastructure network' across the conurbation. This focuses on increasing tree cover, improving water quality in river valleys and canals, and enhancing parks,

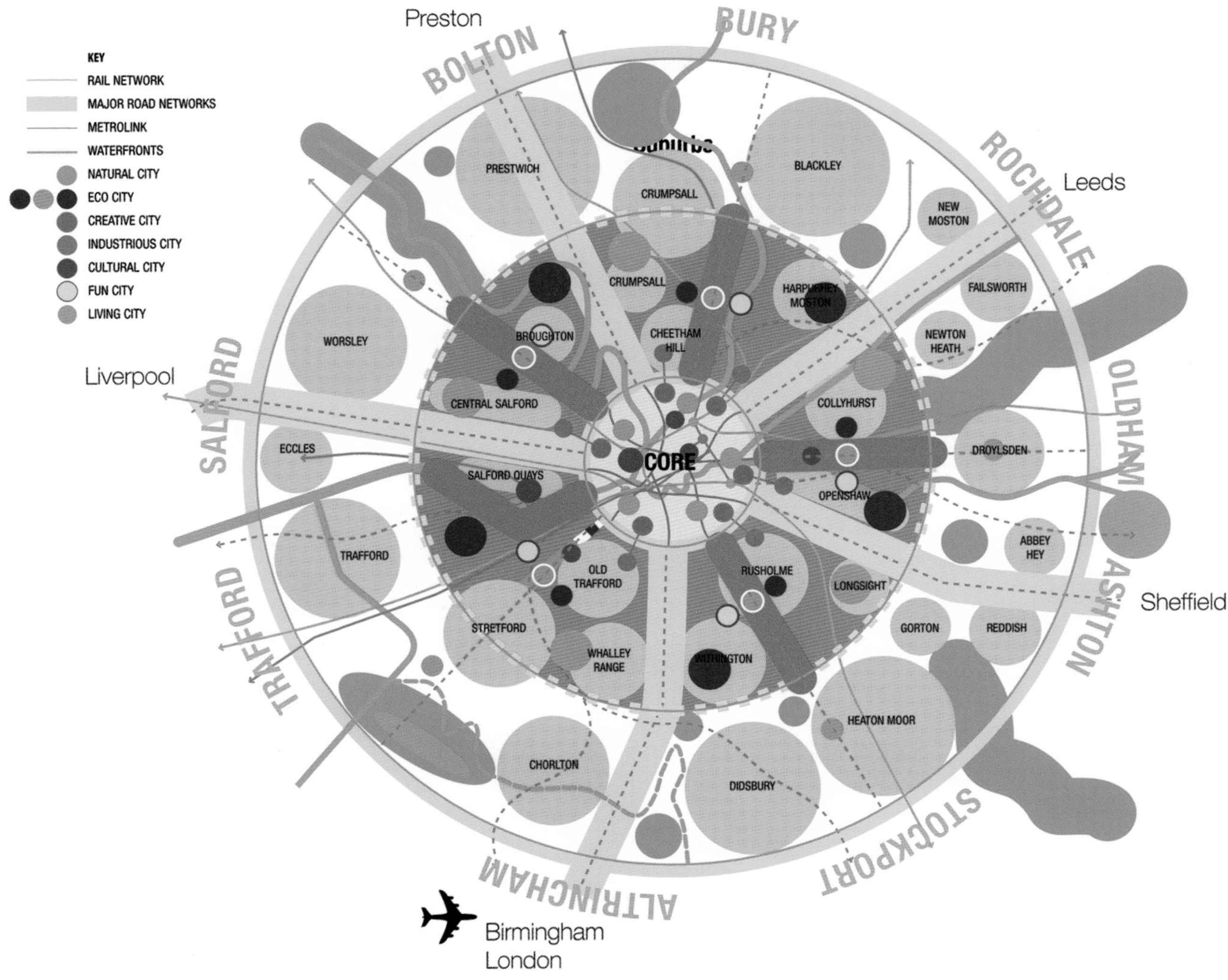

Building Design Partnership, *Schematic plan for Greater Manchester* (2011) [BDP]

open spaces, the uplands and the lowland wetlands that are scattered throughout the conurbation.

Strategic thinking for the whole of the functional city region is one of the keys to Manchester's future vitality. The functional footprint of the city stretches well beyond Greater Manchester itself, as the flows of journeys-to-work demonstrate. The difference since the nineteenth century is that now the conurbation is virtually a single continuous built-up area rather than a series of semi-separate settlements. BDP's schematic plan of the overall conurbation suggests the interdependence between its parts. The expansion of Metrolink – from having been a single line between Altrincham and Bury into a genuine network – has helped to transform the conurbation into a more cohesive whole and reinforces the view that, set within east–west links across the Pennines, Manchester is now the logical complement to the otherwise overweening dominance of London.

Further reading

Bergin, T., Pearce, D.N. and Shaw, S., *Salford: A City and Its Past* (Salford: The City of Salford, 1974).

Briggs, Asa, *Victorian Cities* (London: Odhams Press, 1963), chapter 3.

Brooks, R. and Dodge, M., *Making Post-War Manchester: Visions of an Unmade City* (Manchester: Modernist Society, 2016).

Brumhead, D. and Wyke, T. (eds), *Moving Manchester: Aspects of the History of Transport in the City and Region since 1700* (Manchester: Lancashire and Cheshire Antiquarian Society, 2004).

Busteed, M. and Hodgson, R., 'Angel Meadow: A Study of the Geography of Irish Settlement in Mid-Nineteenth Century Manchester', *The Manchester Geographer*, 14, no. 1 (1993), 3–26. Available online at https://www.mangeogsoc.org.uk/pdfs/manchestergeographer/MG_14_1_busteed.pdf.

Chadwick, G. 'The Face of the Industrial City: Two Looks at Manchester', in H.J. Dyos and M. Woolf (eds), *The Victorian City: Images and Realities,* Vol. 1 (London: Routledge & Kegan Paul, 1976), 247–56.

Chalklin, C.W., *The Provincial Towns of Georgian England: A Study of the Building Process 1740–1820* (London: Edward Arnold, 1974).

Conway, H., *People's Park: The Design and Development of Victorian Public Parks* (London: Cambridge University Press, 1991).

Davies, J. and Kent, A., *The Red Atlas: How the Soviet Union Secretly Mapped the World*, (Chicago: University of Chicago Press, 2017).

Deakin, D., *Wythenshawe: The Story of a Garden City* (Bognor Regis: Phillimore, 1989).

Dodge, M., 'Mapping the Geographies of Manchester's Housing Problems and the Twentieth Century Solutions', in W. Theakstone (ed.), *Manchester Geographies* (Manchester: Manchester Geographical Society, 2017), 17–34.

Farnie, D.A., *Manchester Ship Canal and the Rise of the Port of Manchester 1894–1975* (Manchester: Manchester University Press, 1980).

Goodey, F., 'The Sexiest Building in the North', *Manchester Memoirs*, 139 (2002), 22–7.

Harley, J.B., *A Map of the County of Lancashire, 1786, by William Yates* (Liverpool: Historic Society of Lancashire and Cheshire, 1968).

Harrison, M., 'Housing and Town Planning in Manchester Before 1914', in A. Sutcliffe (ed.) *British Town Planning: The Formative Years* (Leicester, Leicester University Press, 1981), 106–53.

Hartwell, C., *Manchester* (London: Yale University Press, 2002).

Hyde, R., 'Cartographers Versus the Demon Drink', *Map Collector*, 3 (1978), 22–7.

Hyde, R., *A Prospect of Britain: Town Panoramas of Samuel and Nathaniel Buck* (London: Pavilion Books, 1994).

James, G., *Manchester: A Football History* (Halifax: James Ward, 2008).

Kellett, J.R., *The Impact of Railways on Victorian Cities* (London: Routledge & Kegan Paul, 1969).

Kidd, A., *Manchester: A History* (Lancaster: Carnegie Publishing, 2006).

Kidd, A. and Wyke, T. (eds), *The Challenge of Cholera: Proceedings of the Manchester Special Board of Health 1831–1833* (Record Society of Lancashire and Cheshire, 2010).

Kidd, A. and Wyke, T. (eds) *Manchester: Making the Modern City* (Liverpool: Liverpool University Press, 2016).

Lee, J., *Maps and Plans of Manchester and Salford 1650–1843: A Handlist* (Altrincham: J. Sherratt, 1957).

Mosley, S., *The Chimney of the World: A History of Smoke Pollution in Victorian and Edwardian Manchester* (London: Routledge, 2013).

Nevell, M. and Wyke, T. (eds), *Bridgewater 250: The Archaeology of the World's First Industrial Canal* (Salford: University of Salford, 2012).

Nicholas, R., *City of Manchester Plan* (Norwich: Jarrold & Sons, 1945).

Nicholas, R., *District Regional Planning Proposals* (Norwich: Jarrold & Sons, 1945).

Nicholls, R., *Trafford Park: The First Hundred Years* (Bognor Regis: Phillimore, 1996).

Pergam, E.A., *The Manchester Art Treasures Exhibition of 1857: Entrepreneurs, Connoisseurs and the Public* (Farnham: Ashgate, 2011).

Parkinson-Bailey, J.J., *Manchester: An Architectural History* (Manchester: Manchester University Press, 2000).

Poole, R. (ed.), *Return to Peterloo* (Lancaster: Carnegie Publishing, 2014).

Ramsden, C., *Farewell Manchester: History of Manchester Racecourse* (London: J.A. Allen, 1966).

Robson, B., 'Maps and Mathematics: Ranking the English Boroughs for the 1832 Reform Act', *Journal of Historical Geography*, 46 (2014), 66–79.

Robson, B., 'Cross-Checking: A Method to Test the Comprehensiveness of Pigot's Nineteenth-Century Plans of Manchester & Salford', *The Cartographic Journal*, 54 (2017), 115–25.

Rodgers, H.B., 'The Suburban Growth of Victorian Manchester', *Journal of the Manchester Geographical Society*, 58 (1962), 1–12.

Rose, M.E., Falconer, K. and Holder, J., *Ancoats: Cradle of Industrialisation* (Swindon: English Heritage, 2011).

Rowley, G., *British Fire Insurance Plans* (Old Hatfield: Chas. Goad, 1984).

Scholefield, R.A., *Manchester Airport* (Stroud: Sutton Publishing, 1998).

Scott, R.D.H., *The Biggest Room in the World: A Short History of the Manchester Royal Exchange* (Manchester: Royal Exchange Theatre Trust, 1976).

Simon, E.D. and Inman, J., *The Rebuilding of Manchester* (London: Longmans Green, 1938).

Spiers, M., *Victoria Park, Manchester: A Nineteenth-century Suburb in its Social and Administrative Context* (Manchester: Chetham Society, 1976).

Sutton, K., 'Vestiges of the Pre-Urban Landscape in the Suburban Geography of South Manchester', in *Manchester Geographies* (Manchester: Manchester Geographical Society, 2017), 37–46.

Taylor, A., 'Mapping Manchester: One Man's Contribution to City Centre Maps', *The Cartographic Journal*, 41 (2004), 59–67.

Whitaker, H., *A Descriptive List of the Printed Maps of Lancashire 1577–1900* (Manchester: Chetham Society Series, 101, 1938).

Williams, G., *The Enterprising City Centre: Manchester's Development Challenge* (London: E. & F.N. Spon, 2003).

Williams, M. and Farnie, D.A., *Cotton Mills in Greater Manchester* (Lancaster: Carnegie Publishing, 1992).

Willock, Z., *John Jennison – Belle Vue: Relating to the Making and Growth of the Famous Zoological Gardens, Belle Vue, Manchester* (Manchester: Chetham Library, 2013).

Wyke, T. and Robson, B., 'A Tale of Two Maps: The Town Plans of Manchester by William Green and Charles Laurent', *Transactions of the Historic Society of Lancashire and Cheshire*, 165 (2016), 19–38.

Wyke, T. and Rudyard, N., *Manchester Theatres*, Bibliography of North West England, Vol. 16 (Manchester: Manchester Central Library, 1994).

Index